RADIONUCLIDES
IN THE STUDY OF MARINE PROCESSES

This volume is the proceedings of an International Symposium to be held at the University of East Anglia, Norwich, UK, 10–13 September 1991.

Scientific Committee

M. S. Baxter	*International Atomic Energy Agency, Monaco*
P. Buat-Ménard	*Centre National de la Recherche Scientifique, France*
R. Cherry	*University of Cape Town, South Africa*
E. Duursma	*Monaco*
S. W. Fowler	*International Atomic Energy Agency, Monaco*
J.-C. Guary	*Institut National des Techniques de la Mer, France*
P. Guegueniat	*Commissariat à l'Energie Atomique, France*
E. Holm	*Lund University, Sweden*
T. D. Jickells	*University of East Anglia, UK*
M. Kelly	*Lancaster University, UK*
P. J. Kershaw	*Ministry of Agriculture, Fisheries and Food, UK*
D. Prandle	*Proudman Oceanographic Laboratory, UK*
P. Santschi	*Texas A & M University at Galveston, USA*
E. H. Schulte	*Commissariat à l'Energie Atomique, France*
J. Smith	*Bedford Institute of Oceanography, Canada*
J. Thompson	*Institute of Oceanographic Sciences, UK*
R. Wollast	*Université Libre de Bruxelles, Belgium*
D. S. Woodhead	*Ministry of Agriculture, Fisheries and Food, UK*

The Symposium was sponsored by

British Nuclear Fuels plc
Commissariat à l'Energie Atomique
Norwich City Council
Intergovernmental Oceanographic Commission

RADIONUCLIDES IN THE STUDY OF MARINE PROCESSES

Edited by

P. J. KERSHAW

and

D. S. WOODHEAD

*Ministry of Agriculture, Fisheries and Food,
Directorate of Fisheries Research, Lowestoft, UK*

ELSEVIER APPLIED SCIENCE
LONDON and NEW YORK

ELSEVIER SCIENCE PUBLISHERS LTD
Crown House, Linton Road, Barking, Essex IG11 8JU, England

Sole Distributor in the USA and Canada
ELSEVIER SCIENCE PUBLISHING CO., INC.
655 Avenue of the Americas, New York, NY 10010, USA

WITH 73 TABLES AND 117 ILLUSTRATIONS

© 1991 ELSEVIER SCIENCE PUBLISHERS LTD
© 1991 CROWN COPYRIGHT—pp. 373, 377, 378, 388

British Library Cataloguing in Publication Data

Radionuclides in the study of marine processes.
I. Kershaw, P. J.   II. Woodhead, D. S.
551.460028

ISBN 1-85166-707-5

Library of Congress CIP data applied for

No responsibility is assumed by the Publisher for any injury and/or damage to persons or property as a matter of products liability, negligence or otherwise, or from any use or operation of any methods, products, instructions or ideas contained in the material herein.

Special regulations for readers in the USA

This publication has been registered with the Copyright Clearance Center Inc. (CCC), Salem, Massachusetts. Information can be obtained from the CCC about conditions under which photocopies of parts of this publication may be made in the USA. All other copyright questions, including photocopying outside the USA, should be referred to the publisher.

All rights reserved. No part of this publication may be reproduced, stored in a retrieval system, or transmitted in any form or by any means, electronic, mechanical, photocopying, recording, or otherwise, without the prior written permission of the publisher.

Printed in Northern Ireland at The Universities Press (Belfast) Ltd.

PREFACE

The Norwich Symposium, 'Radionuclides in the Study of Marine Processes', is a sequel to the very successful conference held at Cherbourg, France in June 1987. The international character of the meeting has been maintained with thirty-eight contributions, from seventeen countries, being accepted for oral presentation.

For many years, the radioactive properties of the naturally occurring radionuclides have been used to determine their distributions in the marine environment and, more generally, to gain an understanding of the dynamic processes which control their behaviour in attaining these distributions. More recently the inputs from human activities of both natural and artificial (i.e. man-made) radionuclides have provided additional opportunities for the study of marine processes on local, regional and global scales. Because the sources of artificial radionuclides are often reasonably well defined in space and time, and because a wider range of elements is represented, new viewpoints for the study of processes have become available. Although it is outside the scope of this Symposium, it must be acknowledged that the radiological protection requirements for radioactive waste disposal practices to be based on a sound scientific understanding of radionuclide behaviour in the sea have also provided a very strong impetus for the studies.

The point has frequently been well made that radionuclides are rarely ideal tracers for marine processes—H-3 (as tritiated water) for water movement and C-14 (initially as carbon dioxide) for the marine carbon cycle are, perhaps, notable exceptions—because the particular process under study may change the physical and/or chemical state of the radionuclide leading to its involvement in other processes. Thus, the distribution and behaviour of any radionuclide is the integrated response to a variety of processes and great care is required in the interpretation of the observations for the purpose of understanding a specific aspect. Most importantly, additional complementary information from non-radionuclide sources is frequently necessary to develop a full understanding. The interdisciplinary nature of the approaches required is apparent from the studies reported in this book.

The primary objective of the Symposium is to provide a forum for an open discussion of the insights concerning processes in the marine environment which can be gained from studies of radionuclide behaviour. To this end, ample time has been allocated to permit discussion of the papers which have been grouped within the following principal themes:

- the uses of radionuclides as tracers of water transport;
- scavenging and particulate transport processes in the oceans as deduced from radionuclide behaviour;
- processes in the seabed; and
- radionuclides in biological systems.

The manuscripts of thirty-two papers have been subjected to peer review by the Scientific Committee and the amended versions are included in these conference proceedings. Abstracts of the remaining six papers, which have been offered for publication elsewhere, are also included, as are those of the poster presentations.

As a tribute to the memory of the late Dr B. Patel, and in recognition of the contributions he made to the study of radionuclide behaviour in the tropical marine environment, the paper he had prepared, in collaboration with his wife Dr S. Patel, for presentation at the Symposium has been included in this collection.

P. J. KERSHAW
D. S. WOODHEAD

CONTENTS

Preface v

The Uses of Radionuclides as Tracers of Water Transport

Radiocaesium and plutonium in Atlantic surface waters from 73°N to 72°S 3
   E. Holm, P. Roos, R. B. R. Persson, R. Bojanowski, A. Aarkrog, S. P. Nielsen and H. D. Livingston

Radioactive tracers in the Greenland Sea 12
   H. Dahlgaard, Q. J. Chen and S. P. Nielsen

Short- and long-range transport of caesium and plutonium in Spanish Mediterranean waters (*Abstract*) 23
   J. A. Sánchez-Cabeza, J. Molero, A. Morán, M. Solano, A. Vidal-Quadras, P. I. Mitchell and M. Blanco

Radionuclides in water and suspended particulate matter from the North Sea. 24
   H. Nies, H. Albrecht and J. Herrmann

Studies on the speciation of plutonium and americium in the western Irish Sea 37
   P. I. Mitchell, J. Vives Batlle, T. P. Ryan, W. R. Schell, J. A. Sánchez-Cabeza and A. Vidal-Quadras

Anthropogenic ^{14}C as a tracer in western UK coastal waters 52
   F. H. Begg, M. S. Baxter, G. T. Cook, E. M. Scott and M. McCartney

Radiocaesium in local and regional coastal water modelling exercises 61
   P. E. Bradley, E. M. Scott, M. S. Baxter and D. J. Ellett

Mathematical model of ^{125}Sb transport and dispersion in the Channel 74
   J. C. Salomon, P. Guegueniat and M. Breton

The dispersion of ^{137}Cs from Sellafield and Chernobyl in the NW European Shelf Seas 84
  D. Prandle and J. Beechey

Gas exchange at the air–sea interface: a technique for radon measurements in seawater 94
  G. Queirraza, M. Roveri, R. Delfanti and C. Papucci

Scavenging and Particulate Transport Processes as Deduced from Radionuclide Behaviour

Are thorium scavenging and particle fluxes in the ocean regulated by coagulation? 107
  P. H. Santschi and B. D. Honeyman

^{234}Th: an ambiguous tracer of biogenic particle export from northwestern Mediterranean surface waters 116
  C. E. Lambert, S. Fowler, J. C. Miquel, P. Buat-Ménard, F. Dulac, H. V. Nguyen, S. Schmidt, J. L. Reyss and J. La Rosa

The interaction between hydrography and the scavenging of ^{230}Th and ^{231}Pa around the polar front, Antarctica (*Abstract*) 129
  M. M. Rutgers van der Loeff

Particle/solution partitioning of thorium isotopes in Framvaren Fjord: insights into sorption kinetics in a super-anoxic environment . 130
  B. A. McKee, J. F. Todd and W. S. Moore

The use of ^{210}Po/^{210}Pb disequilibria in the study of the fate of marine particulate matter 142
  G. D. Ritchie and G. B. Shimmield

Study of the scavenging of trace metals in marine systems using radionuclides 154
  R. Wollast and M. Loijens

Lead-210 balance and implications for particle transport on the continental shelf, mid-Atlantic Bight, USA (*Abstract*) 164
  M. P. Bacon, R. A. Belastock and M. H. Bothner

Surface complexation modelling of plutonium adsorption on sediments of the Esk estuary, Cumbria 165
  D. R. Turner, S. Knox, F. Penedo, J. G. Titley, J. Hamilton-Taylor, M. Kelly and G. L. Williams

Processes in the Seabed

The determination of ^{240}Pu/^{239}Pu atomic ratios and ^{237}Np concentrations within marine sediments 177
   K. E. Sampson, R. D. Scott, M. S. Baxter and R. C. Hutton

A new tracer technique for *in situ* experimental study of bioturbation processes 187
   G. Gontier, M. Gérino, G. Stora and J.-P. Melquiond

Le ^{137}Cs: traceur de la dynamique sédimentaire sur le prodelta du Rhône (Méditerranée nord-occidentale) 197
   J. M. Fernandez, C. Badie, Z. Zhen et I. Arnal

^{10}Be and ^{9}Be in submarine hydrothermal systems (*Abstract*) 209
   D. L. Bourlés, G. M. Raisbeck, F. Yiou and J. M. Edmond

The role of radiogenic heavy minerals in coastal sedimentology ... 210
   R. J. de Meijer, H. M. E. Lesscher, M. B. Greenfield and L. W. Put

The use of ^{210}Pb as an indicator of biological processes affecting the flux and sediment geochemistry of organic carbon in the NE Atlantic   222
   T. Brand and G. Shimmield

The total alpha particle radioactivity for some components of marine ecosystems 234
   E. I. Hamilton, R. Williams and P. J. Kershaw

Transuranics contribution off Palomares coast: tracing history and routes to the marine environment 245
   L. Romero, A. M. Lobo, E. Holm and J. A. Sánchez

The use of radionuclides (unsupported ^{210}Pb, ^{7}Be and ^{137}Cs) in describing the mixing characteristics of estuarine sediments 255
   R. J. Clifton

Plutonium, americium and radiocaesium in sea water, sediments and coastal soils in Carlingford Lough 265
   P. I. Mitchell, J. Vives Batlle, T. P. Ryan, C. McEnri, S. Long, M. O'Colmain, J. D. Cunningham, J. J. Caulfield, R. A. Larmour and F. K. Ledgerwood

Radioecology of cobalt-60 under tropical environmental conditions . 276
   B. Patel and S. Patel

Radionuclides in Biological Systems

Thorium isotopes as tracers of particle dynamics and carbon export from the euphotic zone (*Abstract*) 285
J. K. Cochran, K. O. Buesseler, M. P. Bacon and H. D. Livingston

Interannual variation in transuranic flux at the VERTEX time-series station in the northeast Pacific and its relationship to biological activity 286
S. W. Fowler, L. F. Small, J. La Rosa, J.-J. Lopez and J. L. Teyssie

Enhanced deposition of Chernobyl radiocesium by plankton in the Norwegian Sea: evidence from sediment trap deployments 299
M. Baumann and G. Wefer

Polonium-210 and lead-210 in marine organisms: allometric relationships and their significance. 309
R. D. Cherry and M. Heyraud

Radiotracers in the study of marine food chains: the use of compartmental analysis and analog modelling in measuring utilization rates of particulate organic matter by benthic invertebrates 319
A. Grémare, J. M. Amouroux and F. Charles

Natural and artificial radioactivity in coastal regions of the UK ... 329
P. McDonald, G. T. Cook and M. S. Baxter

The ecological half-life of Cs-137 in Japanese coastal marine biota .. 340
Y. Tateda and J. Misonou

The use of Pb-210/Ra-226 and Th-228/Ra-228 disequilibria in the ageing of otoliths of marine fish. 350
J. N. Smith, R. Nelson and S. E. Campana

Some priorities in marine radioactivity research (*Abstract*) 360
M. S. Baxter

Posters

Characterizing the mixing of water flowing into the North Sea using artificial gamma emitters 363
P. Bailly du Bois, P. Guegueniat, R. Gandon and R. Léon

Transport d'éléments à l'état de traces dans les eaux côtières de la Manche: étude de la distribution spatiale d'un traceur radioactif (^{106}Ru–Rh) dans les moules et les *Fucus*. 364
P. Germain et J. C. Salomon

Studies on the transport of coastal water from the English Channel to the Baltic Sea using radioactive tracers (MAST project) 365
    H. Dahlgaard, H. Nies, A. W. van Weers, P. Guegueniat and P. Kershaw

Studies of Chernobyl ^{90}Sr and Cs isotopes in the northwest Black Sea 366
    K. O. Buesseler, H. D. Livingston, S. A. Casso, W. R. Curtis, J. A. Broadway, G. G. Polikarpov, L. G. Kulibakina and A. Karachintsev

The scope for studies of artificial radionuclides in the seas around the USSR. 367
    G. G. Polikarpov

Teneurs en $^{239+240}$Pu, ^{137}Cs, ^{90}Sr des eaux de mer au voisinage de l'atoll de Mururoa et en Polynésie Française 368
    Y. Bourlat et J. Rancher

Uranium-series nuclides in sediment, water and biota of Saqvaqjuac Inlet, a subarctic estuary, NW coast Hudson Bay 369
    G. J. Brunskill, R. H. Hesslein and H. Welch

Les mesures directes ininterrompues gamma-spectrométriques dans le milieu de mer: un des méthodes perspectives de l'étude et du contrôle de la radioactivité de l'eau maritime 370
    O. V. Rumiantcev

Iodine and its valency forms as the tracers of water movement . . . 371
    S. K. Novikova

An automated radiochemical purification procedure for the mass spectrometric analyses of thorium and plutonium in environmental samples. 372
    K. O. Buesseler

A preliminary study to assess the effect of some seawater components on the speciation of plutonium 373
    D. McCubbin and K. S. Leonard

Particle cycling rate constants from nutrient particle and thorium isotope data in the NW Atlantic Ocean 374
    R. J. Murnane and J. K. Cochran

Chernobyl radionuclides as tracers of sedimentation processes in the northern Adriatic Sea (Italy) 375
    R. Delfanti, V. Fiore, C. Papucci, L. Moretti, E. Tesini, S. Salvi, S. Bortoluzzi, M. Nocente and P. Spezzano

A study of Pb-210 method applicability to Guanabara Bay sedimentation rates 376
  I. Moreira, J. M. Godoy and L. B. Mendes

Artificial radionuclides in the surface sediment of the Irish Sea . . . 377
  M. McCartney, D. C. Denoon, P. J. Kershaw and D. S. Woodhead

Radiocaesium in north-east Irish Sea sediments: interstitial water chemistry 378
  S. J. Malcolm

Correlation between particle size and radioactivity in inter-tidal sediments in Northern Ireland 379
  F. K. Ledgerwood and R. A. Larmour

^{210}Pb chronologies of recent sediments in tidal lakes of the Chenier Marshes of Louisiana 380
  J. R. Meriwether, X. Xu, R. Lee, W. Sheu, M. Broussard, S. F. Burns, R. H. Thompson and J. N. Beck

The chemical associations and behaviour of long-lived actinide elements in coastal soils and sediments 381
  M. C. Graham, M. S. Baxter, F. R. Livens and R. D. Scott

Saltmarshes: an interface for exposure to radionuclides from marine discharges at Sellafield 382
  S. B. Bradley and S. R. Jones

Evaluation of metal dynamics in sediments of a tropical coastal lagoon by means of radiotracers and sequential extractions 383
  H. Fernandes, A. E. de Oliveira, S. Patchineelam, K. Cardoso and L. Holanda

^{210}Pb uptake by a tropical brown seaweed (Padina gymnospora) . . 384
  V. F. Magalhães, C. S. Karez, W. C. Pfeiffer and J. R. D. Guimarães

Strombus pugilis (Mollusc: gastropode) as a potential indicator of Co-60 in a marine ecosystem 385
  R. B. C. Moraes and L. M. Mayr

The uptake and distribution of α-emitting radionuclides in marine organisms 386
  P. McDonald, M. S. Baxter and S. W. Fowler

Abyssal Module Incubator 6000 meters (AMI 6000), un dispositif de marquage par l'eau d'organismes marins de grandes profondeurs . . 387
    G. Gontier, D. Calmet et S. Charmasson

Radionuclides in the study of marine processes: is there a role for monitoring? 388
    G. J. Hunt

Comparison of genotoxic responses of some benthic marine invertebrates to radiation. 389
    F. L. Harrison, R. E. Martinelli and J. P. Knezovich

A model of diadromous fish spawning 390
    N. C. Luckyanov

Index of Contributors 391

The Uses of Radionuclides as Tracers of Water Transport

RADIOCAESIUM AND PLUTONIUM IN ATLANTIC SURFACE WATERS FROM 73 °N TO 72 °S.

E. HOLM, P. ROOS, R.B.R. PERSSON
Department of Radiation Physics
Lund University
Lund, Sweden

and

R. BOJANOWSKI
Institute of Oceanology
Polish Academy of Sciences
Sopot, Poland

and

A. AARKROG, S.P. NIELSEN
Risø National Laboratory
Roskilde, Denmark

and

H. D. LIVINGSTON
Woods Hole Oceanographic Institution
Massachusetts, USA

ABSTRACT

During the recent Swedish Antarctic research Expedition (SWEDARP 89/89) samples of surface sea water were collected on board the Stena Arctica during steaming between Gothenburg, Sweden and the Antarctic Peninsula. Radio chemical separation was performed for radio caesium on 200 l samples and for plutonium on samples between 200 and 1500 l. The results are compared with those of the GEOSECS expedition in the North and South Atlantic in 1972-73 and the Polish expedition in 1977-78. They show that radio caesium has behaved rather conservatively and that the decrease in surface water concentrations during 16 years mainly is due to physical decay. On the other hand levels of $^{239+240}Pu$ have decreased by a factor of 4-5 giving a half life of 7-8 years in open Atlantic surface waters.

INTRODUCTION

Radiocaesium (^{134}Cs, ^{137}Cs) has been used as a tracer for the transport of water masses, especially from European reprocessing facilities into the Arctic (1, 2). It has been concluded that the major reason for the decrease of concentration in the surface waters was dilution while sedimentation and biological removal have played a minor role in the Arctic. In coastal areas as well as in a marginal sea such as the Baltic Sea, the mean residence time is relatively short and a significant fraction of the radio caesium is associated with the sediments (3).

The source term plays an important role and it was shown that even radiocaesium from the Chernobyl accident was, at least in the initial stage, rapidly removed from the upper mixed layer (4).

Plutonium is known to behave less conservatively in the water column than radiocaesium. Plutonium from nuclear test fallout occurs mainly in the +V or +VI valence states in sea water forming soluble carbonate complexes while plutonium from reprocessing plants mainly occurs in a lower oxidation state (+IV) and is deposited close to the source (5, 6). The mean residence time of plutonium in the upper mixed layer and in coastal waters is important for estimating the dose commitment to humans through the marine food chain from a radiological accident or controlled release of radioactivity.

In this paper the fate and distribution of radiocaesium and plutonium during 16 years in the surface waters of the North and South Atlantic are discussed using data from the GEOSECS expedition in 1972-73 (7), the Polish expedition in 1977-78 and the Swedish Antarctic Research Expedition in 1988-89.

SAMPLING AND MEASUREMENTS

Samples of surface sea water were collected on board the Stena Arctica during steaming between Gothenburg, Sweden and the Antarctic Peninsula in October and November, 1988, in the Weddell Sea and the west side of the Antarctic Peninsula down to 72 °S in February and March, 1989 and between the Antarctic Peninsula and Gothenburg during March and February, 1989.

The sampling from the the GEOSECS Atlantic expedition in 1972/73 has been described by Livingston et.al. (7) while data from the Polish expedition 1977/78 are still unpublished. Results from Atlantic and Pacific GEOSECS expeditions regarding composition, settling and dissolution and distribution of Pu, Th and Ra in the surface particulates have been described by Krishnaswami et.al.(8, 9). The sampling routes for the three expeditions are described in Fig. 1. We should here note that the Polish and the SWEDARP expeditions largely followed the same track while the GEOSECS expedition covered additional areas in the North West Atlantic and in the eastern part of the South Atlantic. The Swedish Antarctic Research Expedition went further south than either of the other two expeditions.

Caesium was sorbed on ammonium molybdophosphate from 200 l water samples at pH 2 using ^{134}Cs as a radio chemical yield determinant. At the same location radiocaesium was also collected, after filtration of

1000 to 5000 l samples by two sequential cotton wound cartridge filters impregnated by copper ferrocyanide. Comparison of the two methods will be made later but this allows also observations of the $^{134}Cs/^{137}Cs$ activity ratio. Filter samples were ashed and all samples were analyzed by HpGe gamma spectrometry for 2 - 5 days in a 60 ml calibrated geometry.

Actinides were preconcentrated on board from 200 or 1500 l samples by precipitation of hydroxides with ammonia after acidification and equilibration of the radio chemical yield determinants ^{242}Pu and ^{243}Am. Samples of 200 l from the circumpolar Antarctic waters were later pooled, in the laboratory to yield 600 or 800 l samples as the activity concentrations were expected to be very low in this region. Plutonium and americium were separated by radiochemical methods described previously (10) and analyzed for 7-10 days by solid state alpha spectrometry.

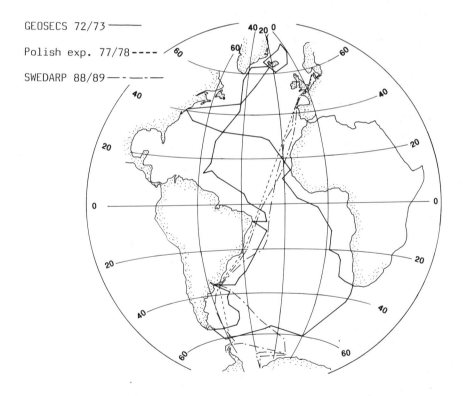

Figure 1. Cruise tracks in the North and South Atlantic; GEOSECS expedition in 1972/73; the Polish expedition in 1977/78 and the SWEDARP expedition in 1988/89

RESULTS AND DISCUSSION

Radiocaesium

<u>Caesium-137.</u> The results for ^{137}Cs from the three scientific expeditions are displayed in Fig. 2. In order to evaluate removal processes other than radioactive decay of ^{137}Cs, (dilution, sedimentation, biological removal), from the Atlantic surface waters during 1972 to 1989, all results were corrected for physical decay to February 1989.

The results from the three expeditions show, after correction for physical decay, an amazing agreement in the activity concentrations from 30°N to 70°S. The immediate conclusion is that radiocaesium from nuclear test fallout has behaved almost perfectly conservatively in open Atlantic surface water. We must, however, consider other input sources since 1973.

We can distinguish separate regions such as above 20 °N where the activity concentrations of ^{137}Cs are about 3 Bq m^{-3}; between 30 °S and 20 °N (about 2 Bq m^{-3}); between 65 °S and 45 °S (about 0.5 Bq m^{-3}); and south of 60 °S (about 0.3 Bq m^{-3}). Significant decreases of the concentrations occur when entering the South Sea current at 40 °S and the circumpolar Antarctic water at 55 °S. These rapid decreases are accompanied by decreases in water temperature.

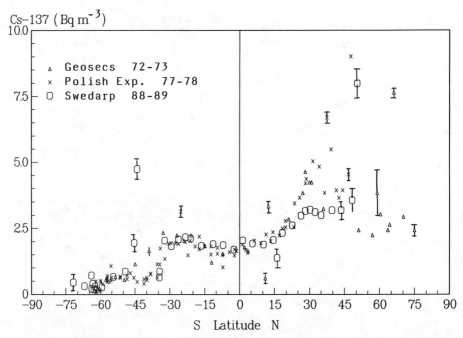

Figure 2. ^{137}Cs activity concentrations (Bq m^{-3}) in surface waters from the North and South Atlantic in 1972/73 (GEOSECS expedition), 1977/78 (Polish expedition) and 1988/89 (SWEDARP expedition). All data are corrected for physical decay to February 1989.

The GEOSECS expedition 1972/73 showed the characteristic peaks of the fallout distribution as a function of latitude, with the maxima at 40-50 °S and 35-45 °N for fallout material injected into the stratosphere. The data also reflect the relatively lower fallout in the southern hemisphere since most nuclear tests were carried out in the northern hemisphere. The Polish expedition and the SWEDARP expedition passed through the English Channel. In this area European reprocessing plants played the major role responsible for the activity concentrations of radiocaesium in the adjacent waters. Activity concentrations for ^{137}Cs of about 50-100 Bq m^{-3} (compared to 3-4 Bq m^{-3} in North Atlantic waters uninfluenced from these sources) were recorded and were not plotted in the figure.

The estimated fission yields of atmospheric nuclear tests since 1973, mainly Chinese and French tests, are about 6 Mt, compared to a total of 217 Mt since 1945 (11). The mean residence time of radioactive material injected into the stratosphere is about 11 months and remaining activity from older tests might give some contribution. The total deposited activity from 1973 to 1986 can be estimated as 6% of the total cumulative deposit in the Northern Hemisphere and 7% in the Southern Hemisphere. The cumulative deposit was measured over land and not sea and probably includes aeolian redistribution. Aeolian redistribution of radioactive material from land to sea is interesting and might be an important input to surface waters in certain areas, but few studies have been done.

Run off from rivers is less important, except for some marginal seas such as the Baltic Sea and the Black Sea in these cases due to the relatively large water input from rivers and drainage area contaminated by the Chernobyl accident.

If we take radioactive decay into account the deposition of ^{137}Cs since 1973 can be estimated to be about 10% or less of the total integrated delivery up to 1989. We can from this derive that the half life of ^{137}Cs in the surface waters of the North and South Atlantic is 100 years or longer, corrected for physical decay.

Caesium-134. Surprisingly, we find very little impact following the Chernobyl accident. The ^{134}Cs/^{137}Cs activity ratio in fallout from the Chernobyl accident is very charcteristic of this source term. It was about 0.47 in April/May 1986 and would be 0.17 in 1989. This ratio is significantly lower for European reprocessing plants effluent and essentially zero for nuclear test fallout today. ^{134}Cs was only found at a few locations north of 35°N, the most southerly latitude we would expect any impact of the Chernobyl accident, and in these cases is probably derived from European reprocessing plants rather than from the Chernobyl accident.

A rather rapid removal of radionuclides from the Chernobyl accident was observed in the Mediterranean and the Norwegian Sea (4, 12). In the coastal region of the Mediterranean the half life of radio caesium in macro algae was 40-100 days and only 15 days in the water (13). After one year the concentrations returned to pre-Chernobyl fallout levels. This is surprising since one would expect the equilibrium concentration to be significantly higher than before the Chernobyl accident based on nuclear test fallout relative to Chernobyl fallout. We can at this stage not exclude the possibility that fallout following the Chernobyl accident has a different behavior compared with nuclear test fallout.

Plutonium

Plutonium-239+240. The results for $^{239+240}$Pu from the GEOSECS and the SWEDARP expeditions are displayed in Fig.3. Results for plutonium from the Polish expedition are not available. The long physical half life of the plutonium isotopes makes it unnecessary to correct for physical decay between 1973 and 1989.

We can distinguish separate regions based on the plutonium results from the SWEDARP expedition. In the latitude band 50-25 °N the activity concentrations in the surface water are about 8 mBq m^{-3}; in the latitude band 25-5 °N about 3 mBq m^{-3}; in the latitude band 0 °S-60 °S about 1,5 mBq m^{-3} and slightly higher values were measured around the Antarctic Peninsula. As a comparison we can mention that based on the Swedish YMER-80 expedition the $^{239+240}$Pu activity concentrations in the Norwegian Sea and the Barents and Greenland Seas were in 1980 about 13+/- 2 mBq m^{-3} from 55 to 82 °N (1).

The general latitudinal distribution of plutonium is similar to that for radiocaesium reflecting the lower fallout in the southern hemisphere. For plutonium we do not observe the sharp decrease in concentration at 40 °S found for ^{137}Cs and in other respects the behavior of the radionuclides is quite different. The activity concentrations of plutonium decreased by a factor 4-5 during a 16 year period corresponding to a half life of 7-8 years in the surface waters of the North and South Atlantic. This is in

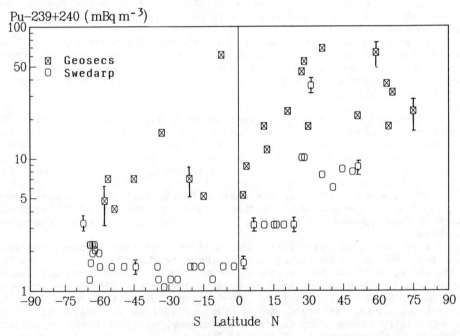

Figure 3. The distribution of $^{239+240}$Pu (mBq m^{-3}) in surface sea water in the North and South Atlantic in 1972/73 (GEOSECS expedition) and in 1988/89 (SWEDARP expedition).

fair agreement with estimate by Fowler et.al. (14) and Thein et.al. (15) who calculated a flux derived residence time of 2 to 13 years in the Northeast Pacific and 12.3 years in the Mediterranean, based on delivery rate data and mixed-layer transuranic inventories respectively.

The activity ratio Pu/Cs is highly variabile but has a value generally lower than that found in fresh nuclear test fallout ($1.2 \cdot 10^{-2}$). The activity ratio shows an interesting latitudinal distribution with a minimum of about $0.4 \cdot 10^{-3}$ at 30-40 °S and increases south and north of this region to higher values at 30-40 °N and at 65 °S ($2-3 \times 10^{-3}$).

It is well known that plutonium is associated with particulate matter to a much larger extent than is radiocaesium (10-20 % and ≤ 1 % respectively). Krishnaswami et.al.(8) found from the GEOSECS expedition, that 20% at least of the plutonium was associated with particulate matter in the polar region with decreasing values, down to 2 % found towards the equator. However no significant difference in the mean residence time of plutonium is observed in the surface waters from 50 °N to 65 °S in the present data set.

Pu-238: Plutonium-238 was detected in a few water samples from the southern hemisphere with values ranging from 14-22 % of those for $^{239+240}$Pu. In macro algae from the Antarctic peninsula we found an activity ratio of 0.27+/- 0.03 (n = 6, 1.S.E.) and in terrestrial samples (carpets of mosses and lichens) from the Antarctic peninsula area an activity ratio of 0.24+/- 0.02 (n =24, 1.S.E.).

In the southern hemisphere ^{238}Pu originates mainly from a satellite failure in 1964 (SNAP-9A) when 1 kg of ^{238}Pu metal re entered the atmosphere and burned up at high altitude over the Mozambique Channel. This event significantly increased both the environmental levels of ^{238}Pu in addition to the activity ratio to $^{239+240}$Pu,(which is typically 0.025 in fallout from nuclear tests), especially in the southern hemisphere. How much of the ^{238}Pu remains in the upper mixed layer and how the behavior of plutonium from this source is different from that of nuclear test fallout plutonium are difficult questions to answer. Our preliminary data indicate that the behavior of plutonium from the two source terms in the marine environment does not differ greatly in the marine environment.

CONCLUSIONS

Radiocaesium from nuclear detonation tests carried out in the late 1950's and the early 1960's shows a long residence time in surface waters of the North and South Atlantic Oceans. The halflife, corrected for physical decay, is estimated to be 100 years or longer. In open sea water radiocaesium can be regarded as conservative for evaluating oceanographic processes. From a radiological assessment point of view only physical decay has to be taken into account in dose commitment estimates.

Plutonium from the same source shows a half life of 7-8 years resulting in a $^{239+240}$Pu/^{137}Cs activity ratio in Atlantic surface waters in 1989 of ($3.3+/- 0.7$) 10^{-3} (n =37, 1 S.E.) compared to $1.2 \cdot 10^{-2}$ in fresh fallout. The activity ratio shows a high variability with latitude with a minimum at 30-40 °S. In the southern hemisphere ^{238}Pu can be detected as a result of the satellite failure in 1964. Isotopic ratios of plutonium constitute an important tool for identifying the source terms in this region.

ACKNOWLEDGEMENTS

This investigation was made possible by the Swedish Antarctic Research Expedition (SWEDARP). Assistance from the organizers, supporters, and leaders of the expedition is thereby gratefully acknowledged as is the financial support from the foundations Magn. Bergvalls Stiftelse and Carl Jönssons Understödsfond.

The following companies gave their support to this investigation in form of equipment: AB Tetra Pak, Alfa Laval AB, SAB-NIFE AB, Nordic Camera AB, Esselte Office AB, Nordiska Balzer AB, Arla Ekonomiska Förening, Christian Berner AB, Scanpump AB, Kiviks Musteri AB, AB Ventilationsutveckling, Gambro AB, Vattenteknik AB, Millipore AB, Hitachi Sales Scandinavia AB.

REFERENCES

1. Holm, E., Persson, B.R.B., Hallstadius, L., Aarkrog, A. and Dahlgaard, H., Radio-cesium and transuranium elements in the Greenland and Barents Seas. Oceanologica Acta, 1983, 6, 457-462.

2. Aarkrog, A., Dahlgaard, H., Hansen, H., Holm, E., Hallstadius, L., Rioseco, J. and Christensen G. Radioactive tracer studies in the surface waters of the northern North Atlantic including the Greenland, Norwegian and Barents Seas. Rit Fiskideildar, 1985, 9, 37-42.

3. Salo, A., Tuomainen, K. and Voipio, A., The inventories of certain long-lived radionuclides in the Baltic Sea. In Study of Radioactive Materials in the Baltic Sea, IAEA-TECDOC-362, IAEA, Vienna, 1986, pp. 52-62.

4. Erlenkeuser, H. and Balzer, W., Rapid appearance of Chernobyl radiocesium in deep Norwegian Sea sediments. Oceanologica Acta, 1988, 11, 101-106.

5. Pentreath, R.J., Harvey, B.R.,and Lovett, M.B., Chemical speciation of transuranium nuclides discharged into the marine environment. In Speciation of Fission products in the Environment, ed., R.A. Bulman and J.R. Cooper, Elsevier Applied Science Publishers, London, 1985, pp. 312-325.

6. Fukai, R., Yamato, A., Thein, M. and Bilinski, H., Speciation of plutonium in the Mediterranean environment. In Techniques for Identifying Transuranic Speciation in the Aquatic Environments, IAEA, Vienna, 1981, pp. 37-41.

7. Livingston, H.D., Bowen, V.T., Casso, S.A., Volchok, H.L., Noshkin, V.E., Wong, K.M. and Beasley, T.M. Fallout Nuclides in Atlantic and Pacific Water Columns: GEOSECS Data. Technical Report, Woods Hole Oceanographic Institution, May 1985.

8 . Krishnaswami, S., Lal, D. and Somayajulu, B.L.K. Investigations of gram quantities of Atlantic and Pacific surface particulates. Earth and Planetary Science Letters, 1976, 32, 403-419.

9. Krishnaswami, S. and Sarin, M.M. Surface particulates: Composition, settling rates and dissolution in the deep sea. Earth and Planetary Science letters, 1976, 32, 430-440.

10. IAEA methods for radiochemical analysis of plutonium, americium and curium. In Measurements of Radionuclides in Food and the Environment, 1986, technical Report Series No. 295, pp. 105-116.

11. UNSCEAR: United Nations Scientific Committee on the Effects of Atomic Radiation. Ionizing radiation: Sources and biological effects, United Nations, 1982

12. Fowler, S.W., Buat-Menard, P., Yokoyama, Y., Ballestra, S., Holm,E. and Van Nguyen, H., Rapid removal of Chernobyl fallout from Mediterranean surface waters by biological activity. Nature, 1987, 329, 56-58.

13 Holm, E., Ballestra, S. and Whitehead, N.E., Radionuclides in Macroalgae at Monaco following the Chernobyl Accident. In International Conference on Environmental Radioactivity in the Mediterranean Area., Sociedad Nuclear Espanola, 1986, pp.439-451.

14. Fowler, S.W., Ballestra, S., La Rosa, J. and Fukai, R., Vertical transport of particle-associated plutonium and americium in the upper water column of the Northeast Pacific. Deep-Sea Research, 1983, 12A, 1221-1233.

15. Thein, M., Ballestra, S., Yamato, A. and Fukai, R., Delivery of transuranic elements by rain to the Mediterranean Sea. Geochimica et Cosmochimica Acta, 1981, 3, 369-387.

RADIOACTIVE TRACERS IN THE GREENLAND SEA

H. DAHLGAARD, Q.J. CHEN[1], S.P. NIELSEN
Department of Environmental Science and Technology,
Risø National Laboratory, DK - 4000 Roskilde, Denmark

ABSTRACT

Under the Icelandic/Danish part of the international Greenland Sea Project, ^{99}Tc, ^{134}Cs, ^{137}Cs and ^{90}Sr have been measured in large volume water samples taken between Jan Mayen and East Greenland and across the Denmark Strait. A sampling in September 1988 has given the following results:

- Inputs of technetium-99 from European discharges during the 1970's and 1980's have resulted in Greenland Sea concentrations of 25 - 100 mBq m^{-3} by 1988. As the Irminger Water south of Iceland only contains approximately 10 mBq m^{-3}, ^{99}Tc is very promising as a tracer for the overflow of Greenland Sea water over the Denmark Strait sill. The results indicate a rapid exchange rate for the Greenland Sea water masses.

- Radiocaesium from the Chernobyl accident in 1986 appeared in 1988 in relatively high subsurface concentrations in the intermediate water at 80 - 300 meter. This is in accordance with a fast subsurface transport through the Greenland Sea of surface water originating in the Norwegian Coastal Current / West Spitzbergen Current.

- There are indications of high Chernobyl contributions in a water mass at 700-1400 meter. The measurements indicate that this water mass could contain a surface water contribution from a region receiving a relatively high amount of Chernobyl debris two years prior to the sampling. Alternatively, a scavenging of particulate radiocaesium from the Chernobyl accident immediately after the deposition in 1986 could also explain these measurements.

---

[1]. Guest scientist from the Institute of Atomic Energy, Beijing, China.

INTRODUCTION

Several water masses of different origin can be distinguished in the Greenland Sea. Knowledge of the origin of these water masses is of crucial importance for an understanding of the role of the Greenland Sea in the World Ocean circulation, including deep water formation. Different tracers, including a variety of man-made radionuclides of various origin, can provide an important contribution to this understanding. In this paper, we will concentrate on the nuclides ^{137}Cs, ^{134}Cs, and ^{99}Tc.

The main source of artificial radioactivity during the 1950's and 1960's was the atmospheric nuclear test explosions. During the 1970's, discharges (especially radiocaesium) from the nuclear fuel reprocessing plant, Windscale (now, Sellafield) in the UK rose to levels comparable to those from fallout in a large part of the North East Atlantic, including the Arctic waters. Finally, when radiocaesium discharges from Sellafield dropped dramatically in the mid 1980's, the Chernobyl Nuclear Power Plant accident contributed to atmospheric inputs in 1986. Technetium-99 has been discharged in significant quantities from Sellafield since 1970 [5], but it has not been possible to analyze the low concentrations in the Greenland Sea until now [9].

Several data sets from the Greenland Sea have been published [1-6] with specific reference to radiocaesium and ^{90}Sr. In our annual data reports, we have provided results from East Greenland Waters since 1962 [7-8]. In this paper, emphasis will be placed on water samples collected in the southern part of the Greenland Sea from the Icelandic fisheries research vessel BJARNI SÆMUNDSSON in September 1988 under the Danish/Icelandic part of the Greenland Sea Project.

MATERIALS AND METHODS

Deep seawater samples were collected with 270 ℓ Hydro Bios Large Volume Water Samplers. The depth was measured as the length of wire, and the accuracy was verified in several instances, when the sampler could be seen on sonar equipment. Surface water was pumped from approximately 5 m depth with the ship's pump.

The water was split into a 50 ℓ subsample, which was brought back to the laboratory for analysis of ^{137}Cs and ^{90}Sr, and a 200 ℓ subsample for ^{99}Tc and ^{134}Cs/^{137}Cs, which was partly processed onboard the ship. The 200 ℓ sample was spiked with $\approx$ 10 kBq ^{99m}Tc from

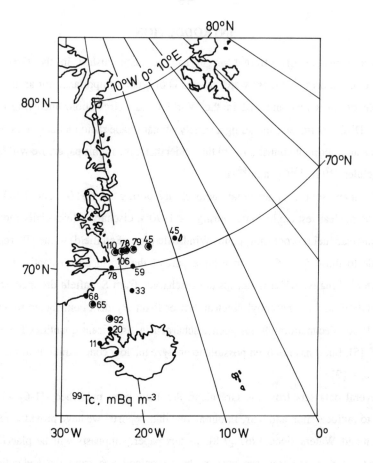

Figure 1. Technetium-99, mBq m^{-3}, in surface seawater September 1988. •: sampling location, ○: profile, see Figures 2-5.

a ^{99}Mo/^{99m}Tc source before it was pumped through a column with anionexchange resin to concentrate the technetium [9]. The yield of this first concentration was determined onboard by γ-counting of an acid eluate with a NaI detector. This eluate was brought home to the laboratory for the final technetium analysis [9]. In order to keep the ^{99}Tc contamination from the ^{99m}Tc spike at insignificant levels, the time between elutions of the ^{99m}Tc source was kept to a maximum of 24 hours. The effluent from the ion exchange column (200 ℓ) was collected in a container and caesium was coprecipitated with 100 g ammonium molybdo-phosphate (AMP) at pH 2.2. The AMP sample was counted for ^{134}Cs/^{137}Cs in the laboratory on a Ge detector. Caesium-137 and ^{134}Cs have physical half lives of 30.0 and 2.06 years, respectively. Technetium-99 is a pure β emitter with a 210,000 year halflife.

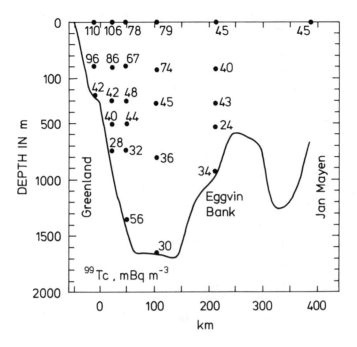

Figure 2. Technetium-99, mBq m^{-3}, in the 71°N section September 1988.

RESULTS

The sampling stations and depths were chosen as close as possible to the sites where the Greenland Sea Project's physical oceanographers also deployed current meters for one year [10]. Furthermore, a series of salinity and temperature profiles (CTD) were recorded. The samples were mainly taken in the southern part of the Greenland Sea on a section at 71°N between Jan Mayen and East Greenland and a section across the northern part of the Denmark Strait from 68°39'N, 26°20'W to 67°08'N, 22°54'W. The Greenland Sea section was taken 11-15 September, and the Denmark Strait section, 24-25 September 1988. Results for ^{99}Tc, ^{137}Cs and "Chernobyl ^{137}Cs" are given in Figures 1-5. Chernobyl ^{137}Cs is based on measurements of ^{134}Cs and application of the ^{134}Cs/^{137}Cs ratio of 0.52 measured in fallout from the Chernobyl accident during April and May, 1986 [11], which, after decay correction to September 1988, is 0.26. The primary data will be published in our annual data reports [8]. The names of places and currents used in the text are shown in Figure 6.

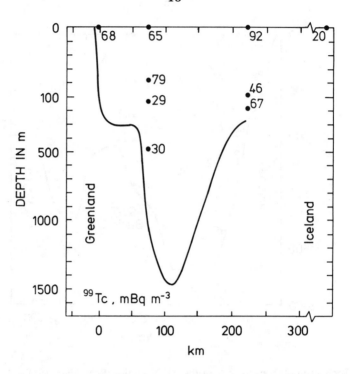

Figure 3. Technetium-99, mBq m^{-3}, in the Denmark Strait section September 1988.

DISCUSSION

The concentrations of both ^{99}Tc and other man-made radionuclides in the Greenland Sea and the Denmark Strait are influenced by the hydrographic conditions, both in the immediate area and in the North-East Atlantic in general. The main source of ^{99}Tc in the Greenland Sea is controlled discharges from Sellafield (formerly Windscale) to the Irish Sea and, to a lesser extent, La Hague discharges to the English Channel. The contribution from global fallout to the measured ^{99}Tc concentration in the surface waters of the East Greenland Current was estimated at only ~20% at the beginning of the 1980's [5]. The concentration levels measured in 1988 were of the same order of magnitude. The global fallout contribution can therefore now be estimated at 10-20%. The Chernobyl accident did not contribute to the ^{99}Tc concentration [12]. Figures 1 and 2 indicate, that the concentration of ^{99}Tc is highest in the part of the East Greenland Current closest to the Greenland coast, i.e. in the Arctic water. The discharges of ^{99}Tc from Sellafield peaked in 1978, and had dropped by an order of magnitude by 1981. We have earlier estimated the transport time from Sellafield to the East

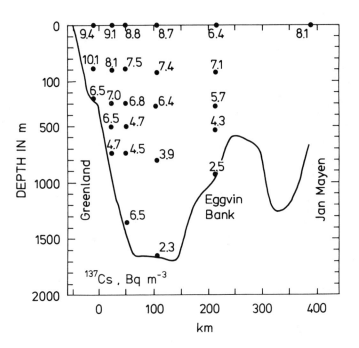

Figure 4. Caesium-137, Bq m^{-3}, in the 71°N section September 1988.

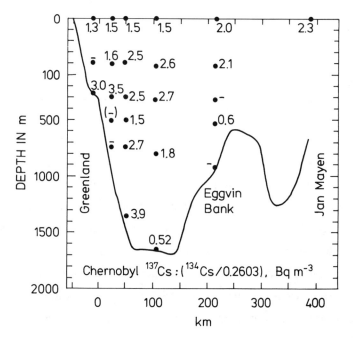

Figure 5. Chernobyl ^{137}Cs, (see text) Bq m^{-3}, in the 71°N section September 1988.

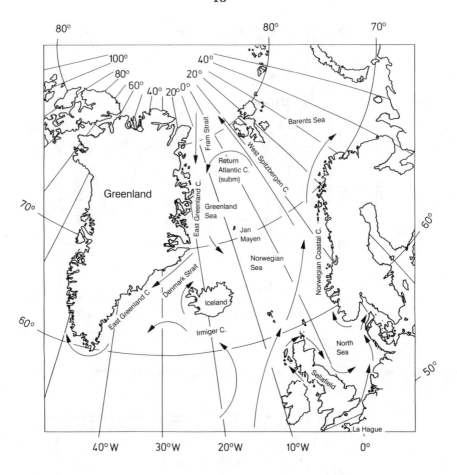

Figure 6. Selected currents of particular importance for the transport of European discharges to the Greenland Sea

Greenland Current to be ~7 years [4,5]. Presumably, the La Hague discharges get to the area two years faster, but Sellafield was still the dominant source in the Greenland Sea in 1988. We had therefore expected to see decreasing concentrations in 1988 in those parts of the Greenland Sea, that directly receive the contaminated European water masses, i.e. the submerged "Return Atlantic Current" which transports saline Atlantic water from the West Spitzbergen Current southwards in a layer below the polar water layer in the East Greenland Current [13]. As previous data for ^{99}Tc in the Greenland Sea nearly exclusively reflects surface water concentrations, it is not possible to make a direct assessment of concentration changes with time. It would, however, be expected, that a stable source in the Norwegian Coastal Current and the West Spitzbergen Current would result in peak concentrations in the intermediate water at 80 - 300 m depth on the Greenland Slope below the Polar Water in the East Greenland Current, as is seen for Chernobyl caesium (Figure 5). The observed maximum values in the Polar Water and the relatively uniform concentrations in the various remaining water masses of the Greenland Sea transect (Figure 2), can then be taken as an indication that the European ^{99}Tc, which has contaminated the Greenland Sea since 1977 [5], has been mixed throughout all water masses sampled in the 71°N transect (Figure 2), and that the transport with Atlantic Water has begun to decrease by 1988. The low values off the Iceland coast (Figure 1) indicate the low background levels of fallout having a negligible European contribution in the Irmiger Current, which strongly influences Icelandic coastal waters. The marked difference in concentration between the Irmiger water and the Greenland Sea water stresses the potential of ^{99}Tc as a tracer for the flow of Greenland Sea water over the Denmark Strait Sill into the deep water masses of the central North Atlantic.

For ^{137}Cs, the picture is more complicated. Assuming the mean East Greenland Current surface water (Polar water) value in the present study to be 9 Bq ^{137}Cs m^{-3} (Figure 4), and subtracting the Chernobyl contribution of $\sim$1.5 Bq ^{137}Cs m^{-3} (Fig. 5), the non-Chernobyl contamination of the polar water is estimated to be $\sim$7.5 Bq ^{137}Cs m^{-3} in 1988. Extrapolation of a time series (1973-1985) [14] of coastal water samples from various East Greenland stations, gives a value of $\sim$3.6 Bq m^{-3} for 1988. Presumably, this value represents mainly global fallout and other older sources in the Arctic Ocean such as direct fallout from the Novaya Zemlia nuclear tests and discharges through the Siberian rivers. The remaining $\sim$4 Bq ^{137}Cs m^{-3} can then be explained by European discharges: The ^{99}Tc/^{137}Cs activity ratios reported for the Sellafield discharges for the 7-10 year period prior to 1988 and the La Hague discharges 5-8 years before 1988 have an average value of 0.02. Multiplying this value by the above mentioned $\sim$4 Bq ^{137}Cs m^{-3} gives $\sim$80 mBq ^{99}Tc m^{-3}, which is exactly what was claimed above to be the European contribution to the polar water. Preferential precipitation of caesium relative to the fully conservative technetium in the sediment rich waters of the Irish Sea and the North Sea [5] could, however, increase the ^{99}Tc/^{137}Cs ratio of the European contribution by the time it has reached the East Greenland Current. This would allow for an additional ^{137}Cs source in the Arctic Ocean, e.g. more recent discharges from the USSR via the Siberian rivers.

In 1988 Chernobyl caesium was detected throughout the Greenland Sea section (Figure 5), although the contamination of the surface waters of the Greenland Sea and the Fram Strait was seen to be rather small in 1987 [8]. The level in the Polar water in the upper part of the East Greenland Current, ~ 1.5 Bq m^{-3}, may be consistent with the 1987 measurements in the Fram Strait. The somewhat higher levels in the "Atlantic water" at 80 - 300 m below the East Greenland Current can be explained by the higher Chernobyl contamination level of the Norwegian Coastal Current in 1986. The extent of the "Return Atlantic Current" according to Paquette et al [13] should, however, be much more limited than indicated here. The results in Figure 5 indicate that the entire intermediate water mass at 80 - 300 meters in 1988 was contaminated with Chernobyl activity, which was deposited on the sea surface two years earlier. As the exact concentrations of Chernobyl derived ^{137}Cs in the water masses in the Norwegian Coastal Current, the Barents Sea and the Arctic Ocean, which are the source of the intermediate 80 - 300 meter water mass in the Greenland Sea, are not known in sufficient detail, the percentage of 2 year old surface water cannot be quantified, although it must be rather large.

At depths of 700-1400 meters off the Greenland Shelf high values for the Chernobyl contamination were measured in a few samples (Figure 5). As this water mass is generally contaminated with lower levels of ^{137}Cs and ^{90}Sr from other sources (mainly global fallout) compared to the water masses noted above, the origin of these additional inputs is not clear. If it was mainly surface water less than two years old from any region in the Arctic Ocean or the Barents Sea, the background fallout levels should be higher. One explanation could be that the source water for this region includes a relatively small contribution of 2 year old surface water from an area where the Chernobyl contamination in 1986 was very high. This is plausible because the Chernobyl deposition was observed to be very patchy, and only minor parts of the relevant sea areas were monitored following the Chernobyl accident. Alternatively, a similar effect might be evident if a substantial part of the Chernobyl caesium deposited in 1986 was associated with particles, which did not undergo dissolution until they had reached the deep water masses. Although this process probably occurred it has not been demonstrated that such a precipitation of Chernobyl caesium would be sufficient to explain these results. A more unlikely explanation for the high levels of ^{134}Cs at 700-1400 meters might be "other sources", e.g. a leaking nuclear submarine. As the physical half life of ^{134}Cs is only 2.06 years, all older discharges including Sellafield can be ruled out.

CONCLUSIONS

Man-made radionuclides in the Greenland Sea can provide knowledge on the long term origin of water masses and water mixing rates. The most serious limitation in the use of man-made radionuclides as oceanographic tracers is the small quantity of routine monitoring results and the limited time series data for representative water masses.

Assuming that the Chernobyl accident was the only significant contributor to the concentration of ^{134}Cs in the Greenland Sea in 1988, the appearance of Chernobyl Caesium in the intermediate water at 80 - 300 meters indicates a significant contribution of surface water having an age of only two years. Inputs of technetium-99 from European discharges during the 1970's and 1980's have resulted in Greenland Sea concentrations of 25 - 100 mBq m^{-3} by 1988, approximately one decade after the first appearance of ^{99}Tc in the Greenland Sea. This is consistent with a very high water renewal rate. The high levels of ^{99}Tc from European discharges in the entire Greenland Sea transect and the low fallout background levels in the Irmiger Water indicate that the ^{99}Tc may be useful as a tracer for the transport of Greenland Sea water over the Denmark Strait Sill.

ACKNOWLEDGEMENTS

The project was partly sponsored by the Danish National Science Council, the Danish Council for Scientific and Industrial Research and the Commission for Scientific Research in Greenland via the Danish Greenland Sea Project. Support was furthermore received from the Radiation Protection Programme of the European Communities 1985-1989. We are specially indebted to the Icelandic crew and scientists onboard Bjarni Sæmundsson.

REFERENCES

1. Aarkrog, A., H. Dahlgaard, L. Hallstadius, H. Hansen, and E. Holm, Radiocaesium from Sellafield Effluents in Greenland Water. Nature, 1983, **304**, 49-51.

2. Casso, S. A. and H. D. Livingston. Radiocesium and Other Nuclides in the Norwegian-Greenland Seas 1981 - 1982. Woods Hole Oceanog. Inst. Tech. Rept. WHOI-84-40, 1984.

3. Aarkrog, A., H. Dahlgaard, H. Hansen, E. Holm, L. Hallstadius, J. Rioseco, and G. Christensen, Radioactive Tracer Studies in the Surface Waters of the Northern North Atlantic Including the Greenland, Norwegian and Barents Seas. Rit Fiskideildar, 1985, **9**, 37-42.

4. Dahlgaard, H., A. Aarkrog, L. Hallstadius, E. Holm, and J. Rioseco, Radiocaesium Transport from the Irish Sea via the North Sea and the Norwegian Coastal Current to East Greenland. Rapp. P.-v.Réun. Cons. Int. Explor. Mer., 1986, **186**, 70-79.

5. Aarkrog, A., Boelskifte, S., Dahlgaard, H., Duniec, S., Hallstadius, L., Holm, E., Smith, J.N.: Technetium-99 and Cesium-134 as long Distance Tracers in Arctic Waters. Estuarine, Coastal and Shelf Science, 1987, **24**, 637-647.

6. Ostlund, G.O., H. Craig, W.S. Broecker, D. Spencer, GEOSECS Atlantic, Pacific, and Indian Ocean Expeditions, Vol. 7, Shorebased Data and Graphics. U.S.Government Printing Office, Washington, 1987.

7. Aarkrog, A., J. Lippert, J. Petersen. Environmental Radioactivity in Greenland in 1962. Risø Report No. 65, 1963.

8. Aarkrog, A., E. Buch, Q.J. Chen, G.G. Christensen, H. Dahlgaard, H. Hansen, E. Holm, S.P. Nielsen. Environmental Radioactivity in the North Atlantic Region including the Faroe Islands and Greenland 1987. Risø-R-564, Risø National Laboratory, Roskilde, Denmark, 1989.

9. Chen Qing Jiang, Dahlgaard, H., Hansen, H.J.M., Aarkrog, A. Determination of ^{99}Tc in environmental samples by anion exchange and liquid-liquid extraction at controlled valency. Analytica Chimica Acta, 1990, **228**, 163-167.

10. Buch, E., S.-Aa. Malmberg. Joint Danish-Icelandic cruise to the Iceland Sea - Greenland Sea, September 1988. Greenland Sea Project, Internal Report **20**, March 1989.

11. Aarkrog, A., The Radiological Impact of the Chernobyl Debris Compared with that from Nuclear Weapons Fallout. J. Environ. Radioactivity, 1988, **6**, 151-162.

12. Aarkrog, A., Carlsson, L., Chen, Q.J., Dahlgaard, H., Holm, E., Huynh-Ngoc, L., Jensen, K.H., Nielsen, S.P., Nies, H., Origin of Technetium-99 and its use as a Marine Tracer. Nature, 1988, **335**, 338-340.

13. Paquette, R.G., R.H. Bourke, J.F. Newton, W.F. Perdue. The East Greenland Polar Front in Autumn. J. Geophys. Res., 1985, **90**, 4866-4882

14. Aarkrog, A., Chernobyl related monitoring and Comparison with fallout data. Proc. Sem. on The Radiological Exposure of the Population of the European Community from Radioactivity in North European Marine Waters, Project "MARINA", Bruges 14-16 June 1989. Commission of the European Communities, XI/4669/89-EN, 229-249.

SHORT- AND LONG-RANGE TRANSPORT OF CAESIUM AND PLUTONIUM IN SPANISH MEDITERRANEAN WATERS

J.A. SÁNCHEZ-CABEZA [1], J. MOLERO [1], A. MORÁN [1], M. SOLANO [1], A. VIDAL-QUADRAS [1], P.I. MITCHELL [2] AND M. BLANCO [3]

(1) Servico de Física de las Radiaciones, Universidad Autónoma de Barcelona, Bellaterra (Barcelona), Spain

(2) Laboratory of Radiation Physics, Department of Experimental Physics, University College, Dublin, Ireland

(3) Laboratorio de Química Analítica, Departmento de Química, Universidad Autónoma de Barcelona, Bellaterra (Barcelona), Spain

ABSTRACT

Surface sediments and large volume water samples were collected during May-June 1989 along the Spanish Mediterranean coast, from both inshore and offshore sites. Special attention was paid to the zone near the Vandellós I Nuclear Power Plant, the first Spanish NPP to be decommissioned, where a more intensive sampling was carried out which may serve as a baseline for further studies.

Water samples were analysed for Radiocaesium and Plutonium isotopes in both the soluble and particulate fractions (0.22 µm). Surface sediments were analysed for γ-emmitters and Plutonium isotopes. Results show average concentrations in water similar to those reported previously for the same sites in the period 1986-1988. However, the observed geographical distributions may be indicative of radionuclide transport from the Gulf of Lyons and Palomares. In the case of surface sediments, the use of Pb-210 as a normalization factor for artificial radionuclides is shown for the zone close to the Vandellós I NPP. The distortion of Pb-210 levels from other sources, such as the Ebro River, is also discussed. Results confirm the existence of a generalised north-to-south current which disperses, to different degrees, artificial radionuclides orginated at the Plant (i.e. Co-60, Cs-134, Cs-137, Pu-239,240, Pu-238). Finally, preliminary results on the speciation of global fallout plutonium in Mediterranean waters are presented.

RADIONUCLIDES IN WATER AND SUSPENDED PARTICULATE MATTER FROM THE NORTH SEA

H. Nies, H. Albrecht, and J. Herrmann
Bundesamt für Seeschiffahrt und Hydrographie
Bernhard-Nocht-Str. 78
D-2000 Hamburg 36

ABSTRACT

The reprocessing plants at Sellafield and La Hague are primarily the sources of artificial radionuclides in the North Sea. Due to the prevailing transport of water masses, the contamination of water and suspended particulate matter (SPM) with Cs-137, Cs-134, Co-60, Sb-125, and Ru-106 constitutes a characteristic pattern depending on the source of contamination.
Th-234 gives the highest activity of natural radionuclides in suspended matter. Its activity depends mainly on the suspension load.
Resuspension processes and biological activity are responsible for the specific activities in SPM of the North Sea. SPM was collected from several cubic meters of seawater by means of a continuous flow centrifuge. Results from several cruises to the North Sea are presented.

INTRODUCTION

Radioactivity in the North Sea and adjacent sea areas has been monitored for many years. This knowledge has been used to study transport processes of water masses and to derive dilution factors [1, 2, 3, 4]. Due to the two main sources of artificial radionuclides in the North Sea, the nuclear reprocessing plants at Sellafield (UK) and La Hague (F), the distribution of almost conservatively behaving nuclides such as Cs-137 and Sr-90 forms a characteristic pattern in the North Sea. The amounts of activity released from these facilities are reflected by the activity concentration in the North Sea with the time delay required for water transport from the source to the area investigated [5, 6]. The annual discharges from Sellafield decreased significantly during recent years and, consequently, the concentrations in the Irish and North Sea followed this decrease. Fig. 1 shows the distribution of the activity concentration of Cs-137 in the surface layer of the North Sea as measured in November 1989. The maximum concentration of Cs-137 is still to be found in the southern part. In the northern Skagerrak, some influence of Baltic Sea outflow can be ascertained which is the result of the high contamination of the Baltic Sea due to the Chernobyl fallout. The levels of radiocesium activity concentration in the Baltic have not yet decreased to concentrations as measured prior to this accident [7].

Much is known about the spatial and temporal distribution of artificial radionuclides in the water of the North Sea, but information about radionuclides in suspended particulate matter (SPM) of shelf sea areas is very limited. As far as sinking material is concerned, much scientific work has been performed on material collected from sediment traps located on continen-

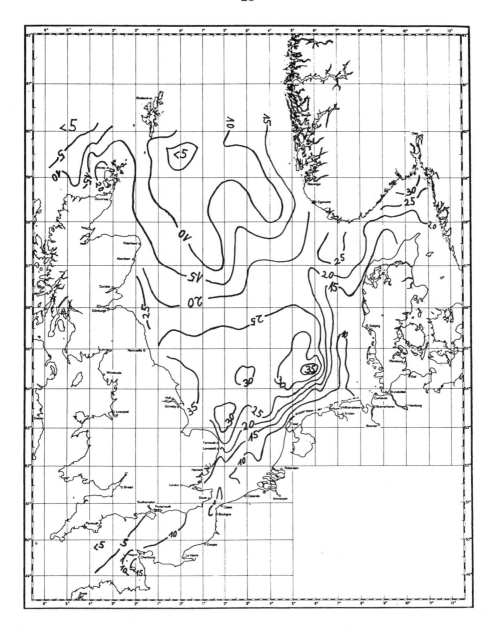

Figure 1
The distribution of Cs-137 (Bq m^{-3}) in the surface water layer of the North Sea in November 1989. The distribution is based on radiocesium determinations of 84 samples collected at stations with a separation of about 30 miles in the northern part and between 10 and 15 miles in the southern part.

tal margins [8, 9, 10]. These traps recorded the time series of the vertical flux. This paper is primarily focussed on radionuclides in SPM collected in the North Sea from 1986 to 1989 by means of continuous flow centrifugation of large water volumes.

MATERIALS AND METHODS

Seawater was collected on several cruises of the German RV GAUSS and RV VALDIVIA to the North Sea and, in January 1988, also to the Irish Sea. A volume of 50 l of seawater was processed for the determination of Cs-137 and Sr-90, and 100 l was used for the Transuranics determination without filtration, as described by Nies [6].

SPM was collected by means of a continuous flow centrifuge on board the research vessels mostly from a depth of 4.5 m. Separation rate was between 1.2 and 1.5 $m^3 h^{-1}$. The amount of suspended material which can be obtained by pressure filtration through 0.45 μm filters is not sufficient to carry out a gamma-spectrometric analysis. Depending on the actual particle concentration, water volumes between 3 000 and more than 10 000 l were centrifuged. Due to limited ship time, filtration was mostly carried out while the ship was moving. The separated material was frozen on board and freeze-dried before being analysed. The material was placed into plastic beakers calibrated in a geometry of 5, 10, 15, 20, 40 etc. ml volume in fixed position on the Germanium detectors and interpolated on the calibrated geometry. All data are given on a dry weight basis and are decay corrected for sampling date. The 1-sigma relative counting error (between 3 and 20 %) is not given, because it depends upon sample size and counting time. If data for Th 234 are given, the gamma-lines at 92.38 and 92.80 keV and 63.3 keV were evaluated.

The two coaxial intrinsic Germanium detectors used have a relative efficiency of 18.5 and 23 % and a resolution of 1.8 keV at 1332 keV. The detector with 23 % efficiency is an n-type low-level background detector with high sensitivity from 10 keV. The limited amount of the samples required counting times of normally more than 24 h.

Some of the suspension samples were also analysed for conventional parameters such as $CaCO_3$, Total Organic Carbon Content (TOC), Al, Fe and various heavy metals.

RESULTS

First samples for radionuclide analysis in SPM were taken from areas of the German Bight and estuaries of some German rivers in 1986. The data are given in Table 1. Numerous Chernobyl derived artificial radionuclides were detectable in these samples. During a cruise covering the entire North Sea from January to March 1987, samples of particulate material were obtained by centrifugal separation while underway along the track of RV GAUSS. Most of these data still show some influence by the Chernobyl fallout on the North Sea. The concentration of Cs-137 and the activity ratio Cs-137/Cs-134 in these samples is given in Figure 2. Ru-106 is the dominant artificial radionuclide due to both the discharges at La Hague and the accident at Chernobyl. However, nuclides such as Ru-103, Co-60, Sb-125 could be determined in the samples, as well. Due to the fact that most samples were not taken in nearshore waters, the particle concentration observed in most cases was less than 1 mg l^{-1}. In general, lower suspension concentrations were determined in the northern North Sea, where resuspension from sediment is limited due to larger water depth. Maximum suspension loads of 11.5 mg l^{-1} were determined in samples taken north of the Westfrisian Islands and from the Thames estuary (7.32 mg l^{-1}). High suspension concentrations were found mainly in the southern North Sea and the German Bight. On the other hand, maximum the Total Organic Carbon (TOC) in the samples were highest in the Skagerrak and the northeastern North Sea with values up to 25 % by weight. This is due to the fact that these samples were collected in March 1987, when

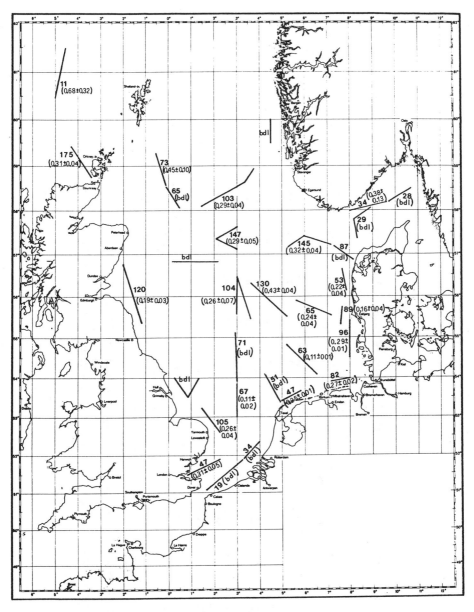

Figure 2
Concentration (Bq/kg) of Cs-137 in suspended particulate matter in the North Sea taken by centrifugal separation of surface water during the passage track of RV GAUSS No. 95, January to March 1987,
(In brackets: Cs-134/Cs-137 activity ratio with relative 1 sigma counting error)
bdl = Cs-137 and Cs-134 below limit of detection
(bdl) = Cs-134 below limit of detection

The theoretical Chernobyl fallout Cs-134/Cs-137 activity ratio is 0.38 at this time.

Table 1
Specific activity (Bq/kg) in suspended particulate matter collected in 1986 from the German Bight.

Location	Position	Date	Volume water (l)	Suspension load (mg/l)	Cs137	Cs134	Ru103	Ru106	Sb125	Ag110m	Co60	Zr95	Nb95	Ce144
	54° 03'N 08° 13'E	13.08.	8700	8.8	92.7	37.5	101.9	172.5	12.7	6.7	6.5	2.8	3.0	-
Alte Hever	54° 03'N 08° 13'E	11.12.	10880	9.74	65.5	20.4	7.6	125.8	14.4	3.7	9.3	<0.8	2.4	10.3
	54° 19'N 07° 06'E	12.12.	18155	1.73	82.2	24.7	7.8	163.2	19.0	<3.0	19.7	-	-	-
Ems	54° 24'N 08° 20'E	13.12.	12025	40.5	113.5	37.4	14.8	203.7	18.5	3.2	7.0	<0.5	1.6	8.0
Ems	53° 30'N 06° 47'E	15.12.	2345	77.2	57.3	17.3	6.5	119.2	14.8	3.2	8.9	<1.0	-	5.3
Jade	53° 30'N 06° 47'E	15.12.	5075	40.0	77.7	24.5	6.2	154.2	18.0	3.1	11.7	-	1.7	2.9
Jade	53° 44'N 08° 05'E	16.12.	2670	41.6	48.7	17.1	8.2	92.7	11.2	2.8	6.5	-	2.8	6.7
Büsum Reede	53° 44'N 08° 05'E	16.12.	5630	41.2	57.0	18.4	7.3	108.1	13.4	3.1	7.4	2.5	5.2	-
Büsum Reede	54° 07'N 08° 52'E	17.12.	2610	47.2	71.4	23.5	5.7	84.2	10.0	1.9	4.5	2.2	2.1	6.4
	54° 07'N 08° 52'E	17.12.	3785*C	56.8	45.0	14.0	3.9	57.0	8.3	1.5	2.8	-	-	2.8
			3785*F	20.7	111.3	27.5	14.0	163.8	20.9	<3.7	<7.8	-	-	16.9
Medemsand[§]	53° 51'N 08° 46'E	18.12.	1505	32.3	77.2	20.3	7.9	98.5	20.2	<2.2	<7.9	<2.1	-	-
Twielenfleet[§]	53° 37'N 09° 33'E	18.12.	2310	132.2	231.0	95.0	14.9	101.8	9.7	<4.8	3.3	-	-	15.9
Twielenfleet[§]	53° 37'N 09° 33'E	19.12.	1280	45.4	349.8	140.3	15.5	158.4	17.1	<3.3	-	<3.2	-	-

[§] Twielenfleet and Medemsand are located in the River Elbe

* This sample has been subdivided into a coarse-grain (C) and a fine-grain fraction (F).

Table 2 Suspended Particulate Matter Taken by Centrifugal Separation of Surface Water, October and November 1987, R/V GAUSS Cruise 103 and 105

Sea Area	Location from to	Date	Water-volume l	Suspens.-load mg/l	Cs137 Bq/kg	Cs134 Bq/kg	Ru106 Bq/kg	Sb125 Bq/kg	Co60 Bq/kg	Be7 Bq/kg	Th234 Bq/kg
Hornsref	55°43'N 07°29'E / 55°01'N 07°58'E	18.10.	6530	1.72	39	11	52	–	15	256	1438
R/Platf.	54°59'N 07°58'E	18.10.	7400	1.83	46	8	107	13	10	151	1272
Nordsee east	54°54'N 06°27'E / 55°29'N 00°59'W	22.10.	7940	0.35	21	–	133	–	–	587	6317
Newcastle east	55°00'N 00°58'E / 54°27'N 00°31'E	23.10.	7230	0.50	40	–	–	–	–	197	4179
Hull	54°00'N 02°05'E / 54°00'N 02°59'E	23.10.	10210	1.70	50	5	47	8	9	154	1614
Lowestoft north	53°30'N 02°00'E / 53°13'N 01°20'E	24.10.	7010	5.31	31	–	68	9	11	153	1254
Lowestoft east	53°00'N 02°59'E / 52°30'N 02°01'E	24.10.	3600	1.94	13	–	60	9	23	128	647
Lowestoft north	51°59'N 02°00'E / 51°16'N 02°16'E	25.10.	6600	3.92	20	–	215	20	31	207	1358
Dover Hoek van Holland	51°32'N 02°52'E / 51°59'N 03°57'E / 52°41'N 03°57'E	25.10.	8320	1.85	25	–	162	38	22	269	2906
Skagerrak	58°21'N 09°14'E	24.11.	24865*	0.14	53	24	182	< 38	19	220	64219
Skagerrak	57°55'N 10°30'E / 58°20'N 09°30'E	24.11.	4290	0.53	39	< 8	< 91	< 17	24	536	6893
Jyske Ref	57°00'N 08°02'E / 55°35'N 06°55'E	25.11.	9660	1.97	38	13	98	24	13	253	2886
Helgoland	54°25'N 07°39'E	27.11.	9440	2.98	41	7	93	23	11	308	1813
Hever	54°24'N 08°21'E	26.11.	1480	21.2	47	8	92	18	5	217	414
Cuxhaven	53°52'N 08°44'N	28.11.	890	22.7	59	13	55	15	9	114	278
Elbe / St. Margarethen		28.11.	890	70	101	33	52	7	3	100	133
Bützfleth Elbe Störmünd.		28.11.	1510	9.1	106	29	30	< 11	2	71	129
Lühe Elbe		28.11.	1800	30.2	126	35	32	< 5	5	102	278

*This sample was taken from a depth of 100m (total water depth about 600 m).

a spring diatom bloom was developing. TOC content in those samples was less than 5 % with a suspended load between 1.6 and 7.3 mg l^{-1} indicating a more lithogenic origin of this matter. The data are very similar to the situation found by Eisma & Kalf [11] and Nolting & Eisma [12]. However, the data of [12] are given for samples taken in January, when lower biological activity can be assumed resulting in lower TOC in the Skagerrak area. Other conventional parameters from our sampling campaign are given by Kersten et al. [13].

During two cruises of RV GAUSS in October and November 1987, suspended particulate matter was collected. The first covered the southern North Sea, the second the Skagerrak and the German Bight. The results are presented in Table 2. Noteworthy is the result of the sample taken at the 24 November from a depth of 100 m in the Skagerrak. A very low particle concentration was found at this position due to the halocline at about 50 m which is more or less stable throughout the year. This halocline generated by low salinity water outflow from the Baltic reduces the downward transport of small particles. A correlation matrix for the different parameters measured in 1987 is given in Table 3.

Table 3 Correlation Matrix of Radionuclide and Elementary Concentration in Suspended Particulate Matter
January to March 1987, October and November 1987,
RV GAUSS Cruise 103 and 105

	Cs137	Cs134	Ru106	Sb125	Co60	Be7	Th234	Po210	TOC	N	CaCO₃	Al
1/Cspm	.19	.34	.45*	.55*	.17	.56*	.84*	.70*	.15	.19	.11	-.28
Cs137		.65*	.35*	.60*	-.31	.39*	.15	.58*	-.17	-.15	-.02	.64*
Cs134			.51*	.36	.05	.43*	.26	.39*	.60*	.56*	-.33	.01
Ru106				.78*	.63*	.63*	.29	.57*	.02	.14	.38*	-.21
Sb125					.42*	.78*	.56*	.63*	.39	.47*	.57*	-.54
Co60						.20	.08	.07	.09	.00	.68*	-.37
Be7							.26	.59*	.26	.38*	.06	.10
Th234								.57*	.24	.23	.08	-.55*
Po210									.01	.06	.28	.20
TOC										.99*	-.47*	-.79*
N											-.42*	-.79*
CaCO₃												.39

* significance level less than 0.05
Cspm = Concentration of SPM

The results of a cruise with RV GAUSS in January 1988 to the southern North Sea and around the British Isles are given in Table 4; the data of a cruise with RV VALDIVIA in April 1989 from the Kattegat to the English Channel are given in Table 5.

DISCUSSION

The Chernobyl fallout introduced various artificial radionuclides into the marine environment. Therefore, besides the investigation of the contamination levels of the marine environment affected, it provided an opportunity to study processes of enrichment, partition, and vertical transport of the different phases of the sea, i.e. water, SPM, and sediment. The fact of suspended material representing the interphase between water and sediment is of special interest for the understanding of sedimentation processes.

The radionuclide concentration in SPM collected during various cruises to the North Sea and adjacent sea areas shows that the contamination by the major sources of radioactivity are distributed similarly as shown in Fig. 1 for Cs-137 in the water phase. Seston samples taken in areas mainly influ-

Table 4 Suspended Particulate Matter Taken by Centrifugal Separation of Surface Water, January and February 1988, R/V GAUSS Cruise 107

Sea Area	Location from to	Date	Water-volume l	Suspens.-load mg/l	Cs137 Bq/kg	Cs134 Bq/kg	Ru106 Bq/kg	Sb125 Bq/kg	Co60 Bq/kg	Be7 Bq/kg	Th234 Bq/kg
German Bight	54°10'N 08°05'E	12.01.	9410	8.08	50	10	64	20	8	223	1286
Yarmouth	53°58'N 07°24'E 53°21'N 02°28'E 52°31'N 02°02'E	13.01. 14.01.	10390	6.52	48	4	71	8	17	139	1751
Thames-estuary	52°30'N 02°00'E 51°29'N 01°34'E	14.01. 15.01.	9850	7.80	18	-	96	-	48	168	1402
Rhine-estuary	51°24'N 02°55'E 52°01'N 04°00'E	15.01.	8720	2.88	28	5	144	21	19	180	1573
Hoek van Holland	52°02'N 03°58'E 53°00'N 04°30'E	15.01. 16.01.	7530	2.38	39	-	230	38	27	325	3621
Dover	50°48'N 01°00'E 50°13'N 01°50'W	16.01. 17.01.	14560	1.33	39	-	458	40	217	454	7278
Plymouth	49°59'N 03°36'W 49°56'N 04°29'W	17.01.	6510	0.47	-	-	-	-	-	555	7932
Irish Sea	51°00'N 07°01'W 51°00'N 05°05'W	20.01.	8430	2.55	27	-	-	-	-	349	4471
Irish Sea	52°00'N 07°00'W 52°28'N 06°00'W	21.01.	8210	2.76	73	19	54	-	-	293	4432
Irish Sea	52°30'N 05°04'W 53°01'N 05°20'W	22.01.	11300	6.66	58	-	-	-	-	200	1839
Liverpool	53°30'N 04°00'W 53°00'N 04°01'W	23.01.	5190	3.61	390	16	64	22	8	364	3347
North-channel	54°30'N 05°00'W 56°01'N 07°00'W	27.01.	9940	1.87	474	26	172	18	13	229	5367
Minch	57°02'N 06°59'W 58°14'N 05°29'W	28.01.	12870	0.52	145	-	-	-	-	372	9411
Pentland Firth	58°57'N 03°51'W 58°39'N 04°27'W	29.01. 30.01.	17350	0.21	115	23	212	41	-	631	1972
Hohe Weg Roads	53°47'N 08°08'E	04.02.	7470	2.11	43	10	63	15	7	166	582
Neuwerk Roads	54°02'N 08°11'E	05.02. 06.02.	5350	1.12	25	6	56	10	7	119	762
Twielen-fleth	53°37'N 09°33'E	06.02.	6430	2.09	127	32	41	-	3	91	118

Table 5 Suspended Particulate Matter Taken by Centrifugal Separation of Surface Water, April 1989, R/V VALDIVIA Cruise 80B

Sea Area	Location from to	Date	Water-volume l	Suspens.-load mg/l	Cs137 Bq/kg	Cs134 Bq/kg	Ru106 Bq/kg	Sb125 Bq/kg	Co60 Bq/kg	Be7 Bq/kg	Th234 Bq/kg
Kattegat	56°30'N 12°00'E 57°22'N 11°30'E	16.04.	7250	0.31	67	9	< 50	< 21	< 9	700	3640
Skagerrak	58°00'N 10°00'E 58°28'N 10°13'E	17.04.	3670	0.24	< 24	< 51	264	< 57	< 30	684	3170
Skagerrak	58°28'N 10°13'E	17.04.	9380	0.30	23	< 21	< 70	< 28	12	841	3970
West Jutland	57°00'N 07°00'E 56°00'N 07°45'E	18.04.	6310	0.15	32	< 87	< 167	< 58	< 20	460	1906
West Sylt	55°00'N 08°00'E 54°30'N 08°00'E	19.04.	4980	2.38	28	< 3	50	17	7	190	983
West German B.	54°30'N 06°00'E 54°00'N 06°00'E	20.04.	4015	0.34	< 13	< 15	< 139	63	< 13	< 238	1090
Outer Well Bank	54°00'N 04°00'E 54°06'N 03°30'E	20.04.	8860	0.28	34	< 21	< 75	37	< 8	206	3230
Thames Estuary	52°00'N 02°00'E 51°13'N 01°43'E	21.04.	7010	2.85	14	< 2	138	< 5	58	117	2000
West Channel	50°25'N 01°00'W 49°35'N 03°59'W	22.04.	10205	0.81	< 8	< 9	47	9	39	260	3370
Rhine Estuary	51°30'N 02°54'E 52°00'N 03°45'E	24.04.	8010	1.62	9	< 2	86	8	14	139	1630
Hoek van Holland	52°02'N 03°40'E	24.04.	10485	0.91	17	< 2	69	19	10	153	1500
Elbe 1	54°00'N 08°00'E	26.04.	2360	1.63	26	< 10	97	< 15	14	157	707

enced by Channel water show Ru-106, Co-60 and Sb-125 as the dominating radionuclides. However, Cs-137 is the dominating nuclide in samples biased mainly by Sellafield discharges. The reprocessing plant at Cap de La Hague is not the only important source for Co-60 in Channel waters. The Winfrith Establishment (UK) can also be taken into consideration [14].

The distribution of the Cs-137 activity in SPM from Fig. 2 is mainly influenced by the relative deposition of the Chernobyl fallout over the North Sea. The radionuclide concentration in SPM from the central North Sea with larger water depth still shows the presence of Chernobyl activity, whereas the samples from coastal, shallow water are already depleted of Chernobyl material due to sedimentation, dilution in the sediment due to mixing processes with aged material and resuspension into the water column. The Skagerrak samples, however, contain only low concentrations of activity, which may be due to the commencement of a diatom bloom in this region.

Figure 3 shows the relation between Cs-137 and Ru-106 in samples taken in January 1988 with the corresponding location. Ru 106 is the dominating nuclide in samples taken in the Channel or southern North Sea. The same is valid for Co-60. The samples taken in the northern Irish Sea and the North Channel have higher Cs-137 activities. A surprising high Ru-106 concentration was measured in one sample in the North Channel which could be indicative for discontinuous peak releases of Ru-106 from Sellafield.

Activities in samples taken 1986 and 1987 from the German Bight are primarily determined by the Chernobyl fallout with an almost constant ratio between Cs-137, Cs-134 and Ru-106 plus an additional contribution of Ru-106 from La Hague, except for the samples taken in the river Elbe or its water plume in the German Bight which have higher Cs-137 activities. This is due to the higher Chernobyl deposition over the upper part of the Elbe in Czechoslovakia than over other drainage areas of large rivers running into the North Sea. Consequently, also suspended material was contaminated at higher concentrations and it was shown that for a longer period the contamination level in the River Elbe was higher than in other rivers [16].

Some of the nuclides determined in SPM samples are known to behave relatively conservative in seawater, such as Cs-137 or Ru-106 and Sb-125 which form anionic complexes. Co-60 is supposed to be mainly adsorbed onto particles. However, all of these nuclides could be detected in SPM. During two cruises (GAUSS 107, January 1988, and VALDIVIA 80B, April 1989) the Cs-137 activity concentration in seawater was determined at the same time when the SPM samples were collected. The concentration of the other nuclides in seawater were not measured. Assuming equilibrium conditions the distribution coefficient K_d for Cs 137 can be calculated. The calculated K_d-values during the GAUSS cruise range from $0.42 \cdot 10^3$ l kg^{-1} in the samples of Thames estuary to $6.2 \cdot 10^3$ in the Irish Sea. This last value might be overestimated, because it was calculated from an SPM sample taken during track passage of the vessel in an area, where a strong Cs-137 concentration gradient might have been present, whereas the water sample was taken on one location. An average K_d of $3.6 (\pm 1.7) \cdot 10^3$ l kg^{-1} was calculated from 14 values, which conforms sufficiently with a value of $3 \cdot 10^3$ l kg^{-1} given for coastal sediments by IAEA [15]. The K_d coefficients samples in April 1989 (VALDIVIA 80B) are somewhat lower. They range between $0.8 \cdot 10^3$ to $2.5 \cdot 10^3$ l kg^{-1} with an average K_d of $1.43 (\pm 0.63) \cdot 10^3$ l kg^{-1}.

The specific activities of the natural nuclides Th-234 and Be-7 are normally much larger in SPM than the artificial nuclides. Th-234 is used to study short term scavenging processes of suspended material in the marine environment. There is a strong negative correlation between the particle concentration and specific activity of Th-234 in SPM, which has already been shown by Kershaw and Young [17]. This is also valid for scavenging processes in the deep sea in the nepheloid layer [18]. Uranium-238 is the mother nuclide which can be regarded as a conservative constituent in seawater. With higher particle concentration, increased scavenging of Th-234 from the water column is observed as shown in Figure 4. However, available Th-234 is diluted by particles accessable in the water column, which leads to lower particulate Th 234 concentration in particle abundant coastal waters. The highest con-

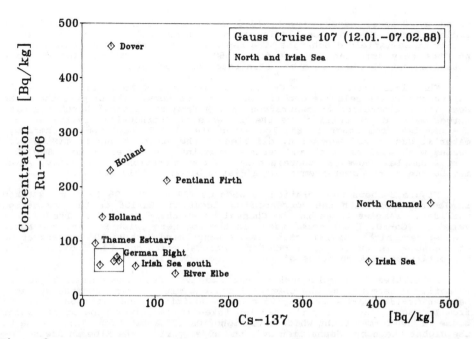

Figure 3
Correlation between the concentration (Bq kg^{-1}) of Cs-137 and Ru-106 in samples of suspended particulate matter (SPM) collected on RV GAUSS cruise 107 to the North Sea and around the British Isles.

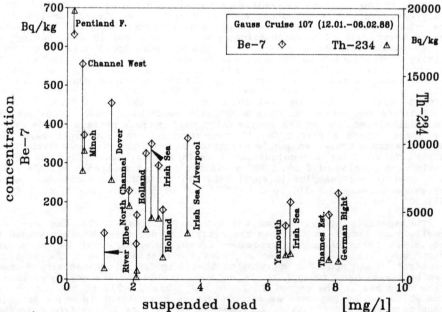

Figure 4
Correlation between the concentration of Be-7 and Th-234 (Bq kg^{-1}) and the particle concentration (mg l^{-1}) in samples of SPM collected on GAUSS cruise 107.

centration of Th-234 was found in particles from a depth of 100 m in the Skagerrak beneath the permanent halocline (Table 2) where an extremely low particle concentration is available.

A similar behavior is found for the 53 days half-life nuclide Be-7 which is deposited to the sea surface from the atmosphere (Fig. 4) and has a high tendency to be fixed to particles at the pH of seawater. The flux of Be-7 onto the sea surface, depending strongly on weather conditions and Be-7 production in the stratosphere, leads to a slightly lower correlation coefficient relative to the inverse particle concentration (Table 3) compared to Th-234.

CONCLUSION

The SPM samples collected in the North Sea and adjacent sea areas show a characteristic composition in their radionuclide concentrations depending upon the location where the samples were taken. Migration of particles by resuspension in the North Sea change their relative radionuclide composition. The concentration of the radionuclide in SPM is also depending on particle concentration, elementary composition of the particles, i.e. biogenic or lithogenic origin. The study of artificial and natural radionuclides in marine particles provides an excellent opportunity to understand sedimentation and resuspension processes. Thus, it appears to be possible to identify the sea area in the North Sea or adjacent seas, where a sample of particulate matter was collected and, to a certain degree, to derive the particle concentration of the sampling site.

ACKNOWLEDGEMENTS

The authors are indebted to the officers and crew of RV GAUSS and VALDIVIA, and they wish to record their thanks to I. Goroncy and V. Rechenberg and other laboratory staff for their assistence in sample collection and analysis.

REFERENCES

1. Kautsky, H., The distribution of the radionuclide ^{137}Cs as an indicator for the North Sea water mass transport. Dt. Hydrogr. Z., 1973, 26, 241-246.

2. Livingston, H.D., Bowen, V.T., and Kupferman, S.L., Radionuclides from Windscale discharges II: Their dispersion in Scottish and Norwegian coastal circulation. J. Mar. Res., 1982, 40, 1227-1258.

3. Dahlgaard, H., Aarkrog, A., Hallstadius, L., Holm, E., and Rioseco, J., Radiocaesium transport from the Irish Sea via the North Sea and the Norwegian Coastal Current to East Greenland. Rapp.P.-v. Réun. Cons. Int. Explor. Mer., 1986, 186, 70-79.

4. Guegueniat, P., Gandon, R., Baron, Y., Salomon, J.C., Pentreath, R. J., Brylinski, J.M., and Cabioch, L., Utilisation de radionucléides artificiels (^{125}Sb - ^{137}Cs - ^{134}Cs) pour l'observation (1983-1986) des déplacements de masses d'eau en Manche. In Radionuclides: A Tool for Oceanography, eds.: J.C. Guary, P. Guegueniat and R.J. Pentreath, Elsevier Applied Science Publishers, London, 1988, 260-270.

5. Nies, H., The Temporal Development of the Radioactive Contamination of the North Sea and Baltic Sea. In Proceedings of the Seminar on the Radiological Exposure of the Population of the European Community from Radioactivity in North European Marine Waters, Project MARINA, Bruges, Commission of the European Community, 1989, 155-170.

6. Nies, H., The Contamination of the North Sea by Artificial Radionuclides During the Year 1987. J. Environm. Radioactivity, 1990, 11, 55-70.

7. Baltic Marine Environment Protection Commission - Helsinki Commission - Three Years Observations of the Levels of Some Radionuclides in the Baltic Sea after the Chernobyl Accident - Seminar on Radionuclides in the Baltic Sea, 29 May 1989, Rostock-Warnemünde, Baltic Sea Environment Proceedings No.31, Helsinki, 1989.

8. Fowler, S.W., Buat-Menard, P., Yokoyama, Y., Ballestra, S, Holm, E., and Nguyen, H.V., Rapid removal of Chernobyl fallout from Mediterranean surface waters by biological activity. Nature, 1987, 329, 56-58.

9. Buesseler, K.O., Livingston, H.D., Honjo, S., Hay, B.J., Manganini, S.J., Degens, E., Ittekot, V., Izdar, E. and Konuk, T., Chernobyl radionuclides in a Black Sea sediment trap. Nature, 1987, 329, 825-828.

10. Kempe, S. and Nies, H., Chernobyl nuclide record from a North Sea sediment trap. Nature, 1987, 329, 828-831.

11. Eisma, D. and Kalf, J., Dispersal, concentration and deposition of suspended matter in the North Sea, J. Geol. Soc., London, 1987, 144, 161-178.

12. Nolting, R.F. and Eisma, D., Elementary Composition of Suspended Particulate Matter in the North Sea, Netherlands J. Sea Res., 1988, 22 (3), 219-236.

13. Kersten, M., Kienz, W., Koelling, S., Schröder, M., and Förstner, U., Schwermetallbelastung in Schwebstoffen und Sedimenten der Nordsee. Vom Wasser, 1990, 75, 245-272.

14. UKAEA, Annual Reports on Radioactive Discharges from Winfrith and Monitoring the Environment, Dorchester, UK.

15. IAEA, Sediment K_ds and Concentration Factors for Radionuclides in the Marine Environment. Vienna, 1985.

16. Mundschenk, H., Der Unfall im Kernkraftwerk Tschernobyl und seine Aus- und Folgewirkungen im Bereich Deutscher Gewässer. Bundesminister für Umwelt, Naturschutz und Reaktorsicherheit, 7. Fachgespräch: Überwachung der Umweltradioaktivität, 1987, 311-329, Ed: BMU, Bonn

17. Kershaw, P., and Young, A., Scavenging of ^{234}Th in the Eastern Irish Sea. J. Environm. Radioactivity, 1988, 6, 1-23.

18. Bacon, M.P., and Rutgers van der Loeff, M., Removal of thorium-234 by scavenging in the bottom nepheloid layer of the ocean. Earth Planet. Sci. Lett., 1989, 92, 157-164.

STUDIES ON THE SPECIATION OF PLUTONIUM AND AMERICIUM IN THE WESTERN IRISH SEA

P.I. Mitchell[1], J. Vives Batlle[1], T.P. Ryan[1], W.R. Schell[2], J.A. Sanchez-Cabeza[3] and A. Vidal-Quadras[3].

[1] Department of Experimental Physics, University College, Dublin 4, Ireland.
[2] Department of Radiation Health, University of Pittsburgh, U.S.A.
[3] Servicio de Fisica de las Radiaciones, Universidad Autonoma de Barcelona, Spain.

ABSTRACT

The distributions of plutonium and americium between filtrate (<0.22 μm) and suspended solids in sea water sampled throughout the western Irish Sea in September 1988 and September 1989 are examined. It is shown that typically 2-15% of the plutonium present in the surface water and 30-60% of the americium, is associated with the particulate fraction. Furthermore, measurements of the oxidation state distribution of plutonium in filtered water taken in December 1989 at six stations throughout the eastern and western Irish Sea showed little variation with some 85 ± 6% of the plutonium in the filtrate in an oxidized (Pu V,VI) form. Little or no distinction was observed between near-surface and near-bottom waters, reflecting the shallow nature of much of the Irish Sea and the importance of turbulent mixing in addition to advective transport. Differences in the ^{241}Am/^{239,240}Pu ratio in sea water, suspended particulate matter and surface sediments have been interpreted and evidence presented to show that the dominant source term for the western Irish Sea has become remobilised plutonium and americium from the fine sediments close to Sellafield. The ^{241}Pu/^{239,240}Pu ratio in filtered water and sediment has been examined and an estimate of 15 years deduced for the effective 'hold-up' time of plutonium in the sediments of the eastern Irish Sea. K_d coefficients for plutonium and americium in the western Irish Sea have also been determined and differences between these values and those observed in the eastern Irish Sea are discussed.

INTRODUCTION

One of the most striking features of the behaviour of plutonium and americium in natural water systems is their association with the particulate phase. The extent of this association is known to be influenced by many factors including the nature of the source term, the chemical speciation of both elements in the receiving water body, the salinity and pH of the water body, the concentration of organic and inorganic ligands, the presence of colloids and the size and composition of the particulate fraction. Of primary importance in assessing the long-term impact of these elements on the environment is an understanding of the species that are formed under different conditions together with the related biogeochemical cycles. Although, over the short term, the behaviour of both elements may be dominated by source-dependent species, releases from Sellafield being a case in point, the concentrations of such species will diminish with time and, ultimately, behaviour will be controlled by source-independent forms [1].

It is well known that plutonium can exist in aqueous solution in any of four oxidation states, III, IV, V or VI [2, 3]. In the Irish Sea, studies carried out in the late 1970's showed that between 70 and 90% of the plutonium present in the filtrate fraction was in an oxidized form, either Pu(V) or Pu(VI), while that adsorbed to the particulate fraction was almost entirely in a reduced form, Pu(III) or Pu(IV) [2, 4, 5]. Although there was considerable scatter in the data, little systematic variation in oxidation state distribution was evident over a concentration range of 0.05-50 Bq m^{-3} [1]. Other studies in shallow coastal waters such as the Bay of Fundy, the Gulf of Mexico, New York Bight and Eniwetok and Bikini lagoons have also shown that, on average, ~80% of the dissolved plutonium is in an oxidized form [1, 6]. In contrast to coastal zones, it has been reported that the two oxidation state groups are present in almost equal concentrations in the open ocean [7]. Little variation with depth is observed except near the bottom of the water column where the distributions change rapidly and become similar to those found in shallow waters. It is now generally accepted that the reduced category is mainly Pu(IV) [3, 8, 9], very likely as Pu(OH)$_4$ adsorbed on negatively charged [10], suspended sedimentary material, while the oxidized category is predominantly Pu(V), probably as PuO$_2^+$ and its complexes [9, 11, 12].

The affinity of americium for particulate matter is even greater than that of plutonium, as evidenced by their respective distribution coefficients [13, 14]. It has been shown that americium is present in eastern Irish Sea waters predominantly in the reduced form, Am(III), though it appears that there is also a small percentage in an oxidized form, Am(V) [12].

In the present study, the speciation of plutonium and americium in eastern and western Irish Sea waters is examined and comparisons are made with previous studies [2, 4, 5, 12] conducted mainly in the north-eastern Irish Sea. Of particular interest was the determination of representative K_d coefficients for the more particle-reactive reduced species in the western Irish Sea under equilibrium conditions - there is evidence that plutonium and americium concentrations in both the sea water and surface sediment compartments in the western Irish Sea are relatively constant at the present time. Moreover, it is not unreasonable to assume that the plutonium and americium released from Sellafield have 'lost' whatever source-dependent character they may have possessed by the time an admittedly small fraction reaches the western Irish Sea (the so-called intermediate zone). It was also of interest to compare the speciation of both elements in near-surface and near-bottom waters in order to determine whether any differences were apparent in the deeper waters of the western Irish Sea. Other objectives included a careful examination of the ^{238}Pu/^{239,240}Pu and ^{241}Am/^{239,240}Pu quotients in filtered water, suspended matter and surface sediments and an attempt to use the ^{241}Pu/^{239,240}Pu ratio to deduce an effective 'hold-up' time for plutonium in the sedimentary deposits close to the outfall.

SAMPLING AND ANALYSIS

Samples of sea water (100 l) and surface sediment (0-10 cm) were collected from the R.V. *Lough Beltra* at six stations in the western Irish Sea (Figure 1) in September 1988 and September 1989. Water samples were passed through 0.22 μm filters and the filtrate (acidified), suspended particulate and underlying sediment from each station were returned to the laboratory where they where analysed for plutonium and americium using standard radiochemical procedures based on co-precipitation with ferric hydroxide, followed by separation of plutonium from americium by solvent extraction (HDEHP) and/or anion and cation exchange, prior to electrodeposition and alpha spectrometry.

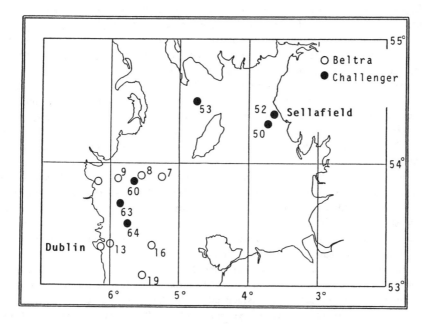

Figure 1. Location of sampling sites in the Irish Sea.

In a separate exercise, near-surface (3 m) and near-bottom (~3 m above seabed) water samples (45 l) were collected from the R.R.S. *Challenger* in December 1989 at six stations in the eastern and western Irish Sea (Figure 1) and analysed for plutonium and americium as above. Duplicate samples, taken at the same stations, were processed on-board ship to separate the two oxidation state groups of plutonium using the neodymium fluoride co-precipitation technique [5]. Carry-over between species was invariably small [Pu(III,IV) -> Pu(V,VI) = 0.0 - 6.0% (mean = 1.1%, n = 11) and Pu(V,VI) -> Pu(III,IV) = 2.2 - 3.0%, (mean = 2.5%, n = 11)] and was corrected for by assessing the cross-over between the ^{236}Pu(VI) and ^{242}Pu(IV) tracers added immediately upon sampling. Good agreement was observed between the sum of the oxidized and reduced plutonium measured in the filtrate using the latter technique and that determined in duplicate samples using the technique of co-precipitation with ferric hydroxide [mean difference = 1 ± 8%, n = 6]. The oxidation states of plutonium associated with suspended particulate in the western Irish Sea (Dublin Bay) were also examined and it was confirmed that the plutonium was almost completely in a reduced form (>97%), in excellent agreement with previous studies carried out elsewhere [2, 5].

RESULTS

R.V. Lough Beltra cruises

The concentrations of plutonium and americium, together with the relevant isotopic quotients, in sea water and surface sediment sampled from the R.V. *Lough Beltra* in 1988 and 1989, are summarized in Tables 1 and 2 (Concentrations on suspended particulate are given in units of mBq m^{-3}, whereas concentrations in surface sediments are given in mBq kg^{-1}). A number of features are evident from the data. Comparatively little change in transuranic concentrations in these compartments of the western Irish Sea appears to have occurred over this period. Preliminary analyses by this laboratory on samples collected from some of the same stations in September 1990 seem to confirm this observation. This would suggest that for the transuranics near-equilibrium conditions may presently prevail in the western Irish Sea, in contrast, for example, to the case of radiocaesium, where the concentrations continue to decline [15]. The latter is easily explained in terms of the progressive reduction in radiocaesium discharges from Sellafield, the solubility of this element in sea water and its consequent rapid removal from the Irish Sea via the North Channel with a half-time of about 200 days [16]. Although transuranic discharges have also been reduced to very low levels, it now appears that the dominant source term for the western Irish Sea has become remobilised plutonium and americium from the sedimentary deposits in the north-eastern Irish Sea close to the outfall, where it has been established that a significant 'hold-up' of transuranics occurs [17].

The data in Tables 1 and 2 show that some 2-15% (mean = 6 ± 5%, n = 11) of the plutonium and 30-63% (mean = 45 ± 13%, n = 7) of the americium in surface (3 m depth) waters sampled in mid-autumn are associated with the solid phase (>0.22 μm), the suspended load in these samples being typically 1-2 mg l^{-1}. Mean K_d values of (4 ± 3)x10^4 and (5 ± 3)x10^5 for plutonium and americium, respectively, were deduced from these data (Table 3) and appear to be significantly lower than total K_d values reported previously for eastern Irish Sea waters [1, 13, 14] and certain other shallow waters [1]. This point is discussed in detail below.

TABLE 1

Transuranic concentrations in sea water and surface sediment sampled in the western Irish Sea, September 1988 (Uncertainties quoted are ± 1 S.D.).

Location	Fraction	$mBq\,m^{-3}$ or $mBq\,kg^{-1}$ (dry weight)				Pu-238 / Pu-239,240	Am-241 / Pu-239,240	Am-241 / Cm-243,244
		Pu-238	Pu-239,240	Am-241	Cm-243,244			
Station 9 53°52′N 05°53′W	Filtrate	54 ± 3	257 ± 10			0.211 ± 0.007		
	Particulate	3.2 ± 0.4	16 ± 1			0.20 ± 0.03		
	Sediment	1620 ± 60	9300 ± 300	2050 ± 50	6 ± 1	0.175 ± 0.002	0.22 ± 0.01	320 ± 60
Station 8 53°53′N 05°33′W	Filtrate	100 ± 4	445 ± 16			0.225 ± 0.005		
	Particulate	2.4 ± 0.3	12.7 ± 0.9			0.19 ± 0.03		
	Sediment	2500 ± 100	15900 ± 700	15700 ± 400	34 ± 4	0.158 ± 0.002	0.99 ± 0.05	470 ± 50
Station 7 53°52′N 05°14′W	Filtrate	65 ± 3	317 ± 12			0.205 ± 0.007		
	Particulate	1.2 ± 0.3	8.0 ± 0.7			0.15 ± 0.04		
	Sediment	2070 ± 90	11400 ± 500	8900 ± 200	19 ± 2	0.182 ± 0.003	0.78 ± 0.04	470 ± 60
Station 16 53°21′N 05°22′W	Filtrate	35 ± 2	154 ± 6			0.23 ± 0.01		
	Particulate	1.4 ± 0.4	8.5 ± 0.7			0.16 ± 0.05		
	Sediment	174 ± 7	1110 ± 40	400 ± 20	< 5	0.156 ± 0.003	0.36 ± 0.02	
Station 19 53°04′N 05°31′W	Filtrate	17 ± 1	87 ± 4			0.19 ± 0.01		
	Particulate	2.6 ± 0.5	17 ± 1			0.15 ± 0.03		
	Sediment	33 ± 2	197 ± 7	67 ± 11	< 5	0.165 ± 0.004	0.34 ± 0.05	
Clogher Head	Filtrate	16 ± 1	90 ± 4			0.18 ± 0.01		
	Particulate	22 ± 1	197 ± 8			0.11 ± 0.01		
	Sediment	NM	NM	NM	NM	NM	NM	NM
Mean	Filtrate(n=6)					0.21 ± 0.02		
	Particulate(n=6)					0.17 ± 0.02		
	Sediment(n=5)					0.17 ± 0.01	0.5 ± 0.3	420 ± 90

TABLE 2

Transuranic concentrations in sea water and surface sediment sampled in the western Irish Sea, September 1989 (Uncertainties quoted are ± 1 S.D.).

Location	Fraction	$mBq\,m^{-3}$ or $mBq\,kg^{-1}$ (dry weight)				$Pu-238$ $Pu-239,240$	$Am-241$ $Pu-239,240$	$Am-241$ $Cm-243,244$
		Pu-238	Pu-239,240	Am-241	Cm-243,244			
Station 9 53°52′N 05°53′W	Filtrate Particulate Sediment	65 ± 3 1.0 ± 0.2 1550 ± 40	338 ± 11 5.6 ± 0.4 8500 ± 200	18 ± 1 9.0 ± 0.7 5300 ± 200	< 10	0.191 ± 0.008 0.17 ± 0.04 0.183 ± 0.003	0.053 ± 0.003 1.6 ± 0.2 0.63 ± 0.03	
Station 8 53°53′N 05°33′W	Filtrate Particulate Sediment	82 ± 3 1.7 ± 0.3 2760 ± 80	422 ± 13 12.8 ± 0.8 16600 ± 400	38 ± 2 16 ± 1 10600 ± 300	25 ± 5	0.195 ± 0.005 0.13 ± 0.02 0.166 ± 0.002	0.090 ± 0.005 1.2 ± 0.1 0.64 ± 0.02	420 ± 90
Station 7 53°52′N 05°14′W	Filtrate Particulate Sediment	69 ± 3 1.2 ± 0.3 2790 ± 90	368 ± 11 8.9 ± 0.6 14800 ± 500	21 ± 1 14 ± 1 11800 ± 400	50 ± 10	0.188 ± 0.005 0.14 ± 0.03 0.189 ± 0.002	0.056 ± 0.004 1.6 ± 0.2 0.80 ± 0.04	240 ± 50
Station 13 53°20′N 06°00′W	Filtrate Particulate Sediment	51 ± 3 9.3 ± 0.8 410 ± 20	248 ± 9 43 ± 2 2350 ± 60	17 ± 1 25 ± 1 700 ± 20	< 5	0.206 ± 0.009 0.22 ± 0.02 0.175 ± 0.006	0.069 ± 0.005 0.58 ± 0.04 0.33 ± 0.02	
Station 16 53°21′N 05°22′W	Filtrate Particulate Sediment	56 ± 3 2.8 ± 0.5 47 ± 4	327 ± 11 14 ± 1 340 ± 10	21 ± 2 17 ± 1 270 ± 40	< 5	0.172 ± 0.007 0.20 ± 0.04 0.14 ± 0.01	0.064 ± 0.006 1.2 ± 0.1 0.8 ± 0.1	
Station 19 53°04′N 05°31′W	Filtrate Particulate Sediment	55 ± 2 3.9 ± 0.6 74 ± 6	261 ± 8 27 ± 1 380 ± 14	15 ± 1 25 ± 1 190 ± 20	< 5	0.210 ± 0.008 0.15 ± 0.02 0.20 ± 0.02	0.059 ± 0.005 0.92 ± 0.06 0.50 ± 0.06	
Dublin Bay	Filtrate Particulate Sediment	31 ± 2 5.5 ± 0.6 NM	165 ± 6 29 ± 2 NM	10 ± 1 7.5 ± 0.9 NM	NM	0.19 ± 0.01 0.19 ± 0.02 NM	0.061 ± 0.006 0.26 ± 0.04 NM	NM
Mean	Filtrate(n=7) Particulate(n=7) Sediment(n=6)					0.19 ± 0.01 0.17 ± 0.04 0.18 ± 0.02	0.07 ± 0.01 1.2 ± 0.4 0.6 ± 0.2	300 ± 100

TABLE 3

Percentage of total plutonium and americium on suspended particulate in the western Irish Sea and associated (total) K_d values (Bq kg^{-1} suspended load/Bq l^{-1} of sea water filtered through a 0.22 μm filter), September 1989.

Location	% on suspended particulate			K_d(total)	
	Pu-238	Pu-239,240	Am-241	Pu-239,240	Am-241
Station 9 53°52'N, 05°53'W	1.5 ± 0.3	1.6 ± 0.1	33 ± 3	(1.4 ± 0.3) * 10^4	(4.2 ± 0.8) * 10^5
Station 8 53°53'N, 05°33'W	2.0 ± 0.4	2.9 ± 0.2	30 ± 2	(2.8 ± 0.5) * 10^4	(3.8 ± 0.8) * 10^5
Station 7 53°52'N, 05°14'W	1.7 ± 0.4	2.4 ± 0.2	40 ± 3	(1.1 ± 0.1) * 10^4	(3.0 ± 0.4) * 10^5
Station 13 53°20'N, 06°00'W	15 ± 1	14.8 ± 0.8	60 ± 3	(7.5 ± 0.8) * 10^4	(6.4 ± 0.7) * 10^5
Station 16 53°21'N, 05°22'W	4.8 ± 0.9	4.1 ± 0.3	45 ± 4	(2.9 ± 0.4) * 10^4	(5.4 ± 0.9) * 10^5
Station 19 53°04'N, 05°31'W	7 ± 1	9.4 ± 0.4	63 ± 3	(7 ± 2) * 10^4	(1.1 ± 0.3) * 10^6
Dublin Bay	15 ± 2	15 ± 1	43 ± 7	(2.8 ± 0.3) * 10^4	(1.2 ± 0.2) * 10^5
Mean(n=7)	7 ± 6	7 ± 6	45 ± 13	(4 ± 3) * 10^4	(5 ± 3) * 10^5

The ^{238}Pu/^{239,240}Pu quotients in filtered sea water, suspended particulate and surface sediment showed no change between 1988 and 1989 and are entirely consistent with those observed in other compartments of the western Irish Sea [18, 19]. The mean value for each category is clearly indicative of releases from Sellafield, being in good accord with those observed throughout the north-eastern Irish Sea [20]. The small difference observed in both years between the mean ratio in filtered sea water, 0.21 ± 0.02 and 0.19 ± 0.01, and that in suspended particulate (and sediment), 0.17 ± 0.02 and 0.17 ± 0.04, may not be statistically significant. Nevertheless, the likelihood that a considerable proportion of the plutonium associated with the particulate phase in the western Irish Sea is sourced by earlier rather than recent discharges and can, thus, be described as having 'aged', may account for the apparently lower ratio on particulate. There is firm evidence that the ^{238}Pu/^{239,240}Pu quotient has increased from about 0.05 in 1966 to about 0.2 or more since 1972 [21, 22].

As expected, there is a clear difference between the ^{241}Am/^{239,240}Pu quotient in filtered sea water (mean = 0.07 ± 0.01, n = 7) and that for the suspended load (mean = 1.2 ± 0.4, n = 6), reflecting the stronger particle reactivity of americium. Significantly, the ^{241}Am/^{239,240}Pu quotient in surface sediments from the western Irish Sea (mean = 0.58 ± 0.24, n = 11) was considerably lower than the latter, being almost identical to the mean value reported for mud samples from Carlingford Lough and Strangford Lough in 1989 [20] namely, 0.59 ± 0.16 (n = 6). The considerable scatter evident in these surface sediment ratios may be related to differences in particle-size distribution. However, the fact that a grab-sampler was used to take the samples may also have contributed to the variability as the precise depth from which the bulk of each sample was taken could not be controlled with any accuracy.

The concentrations of ^{241}Pu in filtered sea water and surface sediment at Stations 7, 8 and 9 (September 1988) were determined by low-level liquid scintillation counting. The ^{241}Pu/^{239,240}Pu ratio was found to be identical in both compartments, namely 18 ± 2. This value is almost half that reported for the source term during the period of peak discharges from Sellafield i.e., the mid-1970s, and implies that there is a 'hold-up' of plutonium between Sellafield and the western Irish Sea (mud bank) of about 15 years. Our estimate is similar to that proposed by Hunt [17] for the mean removal time for plutonium from the vicinity of Sellafield (excluding estuarine sediments).

Minute traces of ^{243,244}Cm were detected in surface sediments from the north-western zone. The mean ^{241}Am/^{243,244}Cm ratio was found to be 380 ± 100 (n = 5) and is in good accord with that measured in sediments from the West Cumbrian coast [20].

R.R.S. Challenger cruise

Plutonium and americium concentrations on suspended particulate and reduced and oxidized concentrations in filtered sea water sampled at near-surface and near-bottom water stations in both the eastern and western Irish Sea in December 1989 are given in Tables 4 and 5, respectively. The percentage of plutonium and americium on suspended particulate and the percentage of oxidized (V,VI) plutonium in the filtrate from each of the stations analysed, are summarized in Table 6.

It is evident from the data that proportionately more of the plutonium and americium in the water column close to the outfall is associated with suspended solid. The percentages (mean) of ^{239,240}Pu and ^{241}Am on suspended particulate in the eastern Irish Sea were 64 ± 19% (n = 6) and 92 ± 2% (n = 2), compared to 26 ± 8% (n = 7) and 64 ± 17% (n = 4) respectively, in the western Irish Sea. The latter are significantly higher than the values determined three months previously and, undoubtedly, reflect the influence of the observed increase in suspended load resulting from the severe gales which took place just prior to the *Challenger* cruise. In general, samples collected close to the settled sediment showed somewhat higher concentrations, though the difference was not always significant and, indeed, was reversed at Station 52 close to Sellafield.

In contrast to the systematic variation with increasing distance from Sellafield in the partition between suspended particulate and filtrate fractions, the percentage of Pu(V,VI) in the latter was essentially constant. No distinction between surface and bottom waters was detected, nor was there any difference between eastern and western Irish Sea stations. In fact, the agreement between these stations is remarkable, with little scatter in the data, and shows that throughout the open waters of the Irish Sea the percentage of ^{239,240}Pu(V,VI) in filtered water lies in the range 74 - 94% with an overall mean (± 1 S.D.) of 85 ± 6%. The equivalent figures for ^{238}Pu(V,VI) are 70 - 95% and 84 ± 7%, respectively.

Total K_d values for ^{239,240}Pu and ^{241}Am in both the eastern and western Irish Sea are given in Table 6. As is the case for the K_d values given in Table 3 (*Lough Beltra* western Irish Sea cruise, 9/89), the value for americium is consistently greater than that for plutonium by almost an order of magnitude. However, there appears to be a significant difference (factor of approximately 5) between the K_ds for both elements in the eastern and western Irish Sea zones. In the western zone, mean K_ds for plutonium and

TABLE 4

Plutonium and americium concentrations on suspended particulate and plutonium speciation in filtered (0.45 μm) sea water sampled at near-surface and near-bottom water stations in the eastern Irish Sea, December 1989 (Uncertainties quoted are ± 1 S.D.).

Location (Depth)	Sample	Fraction	Pu-238 ($mBq\,m^{-3}$)	Pu-239,240	Am-241	Pu-239,240 Pu-238	Pu-239,240 Pu-238	Am-241 Pu-239,240
54°25 00′N 03°34 00′W St. 52 (21 m)	Bottom	Particulate	4680 ± 120	20400 ± 500	27400 ± 500		0.230 ± 0.003	1.34 ± 0.04
		Filtrate(III,IV)	145 ± 9	640 ± 40			0.226 ± 0.007	
		Filtrate(V,VI)	600 ± 30	2700 ± 110			0.224 ± 0.004	
	Surface	Particulate	6110 ± 130	27100 ± 500	59300 ± 1100		0.226 ± 0.002	2.19 ± 0.06
		Filtrate(III,IV)	197 ± 12	770 ± 40			0.255 ± 0.007	
		Filtrate(V,VI)	670 ± 30	3000 ± 120			0.224 ± 0.004	
54°18 85′N 03°41 99′W St. 50 (36 m)	Bottom	Particulate	510 ± 30	2280 ± 130	2970 ± 70		0.225 ± 0.007	1.30 ± 0.08
		Filtrate(III,IV)	50 ± 5	255 ± 18			0.197 ± 0.020	
		Filtrate(V,VI)	320 ± 20	1500 ± 70			0.213 ± 0.007	
	Surface	Particulate	332 ± 22	1780 ± 100	2130 ± 70		0.187 ± 0.007	1.20 ± 0.08
		Filtrate(III,IV)	60 ± 4	286 ± 17			0.211 ± 0.009	
		Filtrate(V,VI)	350 ± 20	1630 ± 80			0.216 ± 0.009	
54°30 04′N 04°44 16′W St. 53 (37 m)	Bottom	Particulate	239 ± 13	1030 ± 30	1590 ± 30		0.232 ± 0.012	1.54 ± 0.05
		Filtrate(III,IV)	20 ± 3	52 ± 5	125 ± 4‡		0.39 ± 0.06	
		Filtrate(V,VI)	168 ± 7	780 ± 30			0.215 ± 0.005	
	Surface	Particulate	172 ± 8	816 ± 23	1262 ± 22		0.211 ± 0.010	1.55 ± 0.05
		Filtrate(III,IV)	21 ± 3	104 ± 8	144 ± 6‡		0.204 ± 0.024	
		Filtrate(V,VI)	183 ± 9	820 ± 40			0.222 ± 0.005	
	Mean (n=6)	Particulate					0.219 ± 0.017	1.5 ± 0.4
		Filtrate(III,IV)					0.219 ± 0.023	
		Filtrate(V,VI)					0.219 ± 0.005	

‡ Total Am-241 in filtrate

TABLE 5

Plutonium and americium concentrations on suspended particulate and plutonium speciation in filtered (<0.45 μm) sea water sampled at near-surface and near-bottom water stations in the western Irish Sea, December 1989 (Uncertainties quoted are ± 1 S.D.).

Location (Depth)	Sample	Fraction	Pu-238 ($mBq\,m^{-3}$)	Pu-239,240	Am-241	Pu-238 / Pu-239,240	Am-241 / Pu-239,240
53°50 15′N 05°39 87′W St. 60 (91 m)	Bottom	Particulate	18.3 ± 1.6	88 ± 5	283 ± 7	0.209 ± 0.016	3.22 ± 0.20
		Filtrate(III,IV)	26.0 ± 2.3	87 ± 6		0.299 ± 0.023	
		Filtrate(V,VI)	60 ± 4	252 ± 11		0.238 ± 0.012	
	Surface	Particulate	10.7 ± 1.3	57 ± 4	57.4 ± 2.3	0.189 ± 0.021	1.02 ± 0.08
		Filtrate(III,IV)	14.4 ± 1.5	62 ± 4		0.233 ± 0.024	
		Filtrate(V,VI)	66 ± 4	250 ± 11		0.262 ± 0.013	
53°40 18′N 05°50 27′W St. 63 (63 m)	Bottom	Particulate	16.4 ± 1.7	91 ± 6	76.7 ± 2.3	0.183 ± 0.017	0.84 ± 0.06
		Filtrate(III,IV)	5.1 ± 0.7	28.2 ± 2.1	41 ± 2†	0.180 ± 0.023	
		Filtrate(V,VI)	55 ± 3	252 ± 10		0.217 ± 0.008	
	Surface	Particulate	15.2 ± 1.2	77 ± 5	121 ± 3	0.198 ± 0.012	1.57 ± 0.11
		Filtrate(III,IV)	3.7 ± 0.5	17.5 ± 1.4	52 ± 3†	0.21 ± 0.03	
		Filtrate(V,VI)	46 ± 3	207 ± 8		0.221 ± 0.010	
53°30 07′N 05°44 88′W St. 64 (67 m)	Bottom	Particulate	31 ± 3	180 ± 11	136 ± 5	0.170 ± 0.011	0.76 ± 0.05
		Filtrate(III,IV)	8.0 ± 1.0	37 ± 3	34 ± 2†	0.22 ± 0.03	
		Filtrate(V,VI)	47 ± 3	217 ± 10		0.215 ± 0.013	
	Surface	Particulate	12.4 ± 1.3	68 ± 4	39.0 ± 1.8	0.183 ± 0.017	0.58 ± 0.05
		Filtrate(III,IV)	14.7 ± 1.8	58 ± 5	56 ± 2†	0.25 ± 0.03	
		Filtrate(V,VI)	64 ± 15	180 ± 30		0.3 ± 0.1	
Dublin Bay (5 m)	Surface	Particulate	12.9 ± 1.2	61 ± 3		0.211 ± 0.022	
		Filtrate(III,IV)	8.6 ± 0.8	43 ± 3		0.198 ± 0.017	
		Filtrate(V,VI)	19.1 ± 1.2	89 ± 4		0.215 ± 0.012	
	Mean (n=7)	Particulate				0.192 ± 0.015	1.33 ± 0.99
		Filtrate(III,IV)				0.23 ± 0.04	
		Filtrate(IV,V)				0.24 ± 0.03	

† Total Am-241 in filtrate

TABLE 6

Percentage of plutonium and americium on suspended particulate and percentage of Pu(V,VI) in filtered (<0.45 μm) sea water sampled throughout the Irish Sea, December 1989. Total K_ds for plutonium and americium and K_ds for Pu(III,IV) are also given.

Location	Fraction	% on suspended particulate			% Pu(V,VI) in filtrate		K_d		
		Pu-238	Pu-239,240	Am-241	Pu-238	Pu-239,240	Pu(total)	Pu(III,IV)	Am(total)
54°25 00'N	Bottom	86 ± 3	86 ± 3		81 ± 5	81 ± 4	$(6.8 ± 0.3) * 10^5$	$(3.5 ± 0.3) * 10^6$	
03°34 00'W	Surface	88 ± 3	88 ± 2		77 ± 5	80 ± 4	$(7.4 ± 0.3) * 10^5$	$(3.6 ± 0.2) * 10^6$	
54°18 85'N	Bottom	58 ± 4	57 ± 4		86 ± 7	85 ± 5	$(3.3 ± 0.3) * 10^5$	$(2.3 ± 0.2) * 10^6$	
03°41 99'W	Surface	45 ± 3	48 ± 3		85 ± 6	85 ± 6	$(2.6 ± 0.2) * 10^5$	$(1.7 ± 0.2) * 10^6$	
54°30 04'N	Bottom	56 ± 4	55 ± 2	93 ± 2	89 ± 5	94 ± 5	$(4.0 ± 0.3) * 10^5$	$(6.4 ± 0.8) * 10^6$	$(4.1 ± 0.3) * 10^6$
04°44 16'W	Surface	46 ± 3	47 ± 2	90 ± 2	90 ± 6	89 ± 6	$(2.7 ± 0.2) * 10^5$	$(2.4 ± 0.2) * 10^6$	$(2.7 ± 0.2) * 10^6$
53°50 15'N	Bottom	18 ± 2	21 ± 1		70 ± 6	74 ± 4	$(6.7 ± 0.6) * 10^4$	$(2.6 ± 0.3) * 10^5$	
05°39 87'W	Surface	12 ± 2	15 ± 1		82 ± 7	80 ± 5	$(1.1 ± 0.2) * 10^5$	$(5.4 ± 0.8) * 10^5$	
53°40 18'N	Bottom	21 ± 2	25 ± 2	65 ± 3	92 ± 7	90 ± 5	$(6.9 ± 0.6) * 10^4$	$(6.9 ± 0.7) * 10^5$	$(4.0 ± 0.3) * 10^5$
05°50 27'W	Surface	23 ± 2	26 ± 2	70 ± 2	93 ± 8	92 ± 5	$(1.1 ± 0.1) * 10^5$	$(1.4 ± 0.2) * 10^6$	$(7.5 ± 0.7) * 10^5$
53°30 07'N	Bottom	36 ± 4	41 ± 3	80 ± 4	85 ± 7	85 ± 5	$(6.5 ± 0.5) * 10^4$	$(4.5 ± 0.5) * 10^5$	$(3.7 ± 0.3) * 10^5$
05°44 88'W	Surface	14 ± 3	22 ± 3	41 ± 2	80 ± 30	80 ± 20	$(8.7 ± 1.3) * 10^4$	$(3.6 ± 0.4) * 10^5$	$(2.1 ± 0.2) * 10^5$
Dublin Bay	Surface	32 ± 2	32 ± 1		69 ± 2	67 ± 2	$(5.4 ± 0.5) * 10^4$	$(1.7 ± 0.2) * 10^5$	
Mean East.I.Sea	Bottom	70 ± 10	70 ± 10	93 ± 2	85 ± 3	87 ± 5	$(4 ± 2) * 10^5$	$(3 ± 2) * 10^6$	$(3 ± 1) * 10^6$
	Surface	60 ± 20	60 ± 20	90 ± 2	84 ± 5	85 ± 4	$(n = 6)$	$(n = 6)$	$(n = 2)$
Mean West.I.Sea	Bottom	25 ± 8	29 ± 9	73 ± 11	82 ± 9	83 ± 7	$(8 ± 2) * 10^4$	$(6 ± 4) * 10^5$	$(4 ± 2) * 10^5$
	Surface	16 ± 5	21 ± 5	56 ± 21	85 ± 7	83 ± 7	$(n = 7)$	$(n = 7)$	$(n = 4)$

americium of $(8 \pm 2) \times 10^4$ (n = 7) and $(4 \pm 2) \times 10^5$ (n = 4) respectively, were determined, whereas in the eastern zone the corresponding values were $(4 \pm 2) \times 10^5$ (n = 6) and $(3 \pm 1) \times 10^6$ (n = 2). The former are in good agreement with the corresponding values given in Table 3, while the latter are almost identical to the mean K_ds obtained from samples taken within the Irish Sea (mainly in the eastern zone) in 1977-79 by Pentreath et al. [13], namely $(3.2 \pm 0.5) \times 10^5$ (n = 64) and $(2.2 \pm 0.3) \times 10^6$ (n = 64), respectively.

The K_d for ^{239,240}Pu(III,IV) has also been determined and, again, a similar difference between the two zones is evident from our data. The mean K_d for ^{239,240}Pu(III,IV) in the eastern zone was found to be $(3 \pm 2) \times 10^6$ (n = 6), in excellent agreement with the mean of $(3.5 \pm 0.8) \times 10^6$ (n = 15) reported by Nelson and Lovett [4] for samples collected within 20 km offshore of Sellafield. In the waters of Dublin Bay the K_d for ^{239,240}Pu(V,VI), at $(2.5 \pm 0.5) \times 10^3$, was some two orders of magnitude smaller than the mean value for the reduced species namely, $(6 \pm 4) \times 10^5$ (n = 7).

In the eastern zone, the ^{238}Pu/^{239,240}Pu quotients on suspended particulate and filtrate (reduced and oxidised) were indistinguishable from one another, being identical to the mean value reported for the sediment (mud) compartment in the general vicinity of Cumbria [20] namely, 0.217 ± 0.005 (n = 16). Within the analytical uncertainties associated with considerably lower concentrations, similar quotients were observed in the western Irish Sea. The tendency, indicated above, for the 'filtrate' quotient to be marginally higher than the 'particulate', was again apparent. The ^{241}Am/^{239,240}Pu quotients on suspended particulate in both zones showed considerable scatter, though the respective mean values are consistent with the ratios reported for sediments near Sellafield [20].

DISCUSSION AND CONCLUSIONS

It is not surprising that the percentage of plutonium and americium on suspended solid in near-bottom samples is generally somewhat greater than that on near-surface samples and undoubtedly reflects measured differences in the suspended load between the two levels. However, differences in the suspended load in the eastern and western Irish Sea cannot account for the significant variation in these percentages as one moves away from Sellafield and suggests that other factors must be considered.

Interestingly, plutonium filtrate concentrations in near-surface waters of the eastern zone were all greater than in bottom waters, whereas the opposite was true in the western zone. In the case of americium, though the data are fewer, surface filtrate concentrations were invariably higher.

No systematic variation of K_d with depth at a given location was observed. However, a clear difference between the mean K_d values for plutonium and americium in the eastern and western zones was evident from our data. Furthermore, the concentration factors (CF) for surface sediments from the large mud bank in the western Irish Sea were indistinguishable from the latter, values of $(3.5 \pm 0.5) \times 10^4$ (n = 6) and $(4 \pm 2) \times 10^5$ (n = 3) being deduced for plutonium and americium, respectively.

Factors which could, conceivably, influence a K_d determination include suspended load particle-size distribution, the presence of significant concentrations of dissolved organic carbon (DOC), excessive silicon concentrations, the chemical speciation of the element in the receiving water body, the presence/absence of hot particles (i.e., particles of high specific activity) and experimental artifacts. As our results demonstrate, the oxidation state distribution of plutonium is identical in both the eastern and western Irish Sea zones and cannot account for the observed difference in the K_ds between the two zones. Organic carbon concentrations throughout the open waters of the Irish Sea are known to be much lower than the levels (>1 mg l^{-1}) required to alter measured K_ds [1, 23]. Further, although Irish coastal waters are enriched in silicon, while eastern waters are enriched with anthropogenically derived nitrogen and phosphorus [24, 25], these differences are small and very unlikely to give rise to a measurable perturbation in K_d. Moreover, we do not consider that the particle-size distribution or the composition of the suspended load is significantly different in the eastern and western Irish Sea, given the degree of turbulent mixing which takes place in these waters; nor do we believe that our data suffer from a systematic shift caused by an experimental artifact [14] given the fact that measurements made in the western Irish Sea on three separate occasions are in good agreement with one another.

We are, therefore, led to the conclusion that the variation with increasing distance from the source in the percentages of plutonium and americium on suspended particulate, together with the differences between the K_ds for both elements in the eastern and western zones, are related to the nature of the discharges themselves. It is known that only about 1% of the plutonium released from Sellafield is in an oxidized form, while all of the americium is in a reduced form [13]. Thus, the great preponderance of the particulate fraction in the effluent provides a ready explanation for the elevated percentages on suspended particulate close to the outlet. Indeed, this particular feature has been commented upon previously by other researchers. 'Hot' particles (less than 5 μm diameter) have also been identified in the effluent [26] and are known to persist in the environment for months before dissolving [27]. Moreover, it has been established that about 10% of the plutonium in surface sediments close to Sellafield is in the form of 'hot' particles [28]. The presence of these particles in the effluent from Sellafield may be one of the main reasons why K_ds for plutonium and americium in the eastern Irish Sea appear to be significantly higher than in the western Irish Sea, where 'hot' particles are less likely to be present in the same relative abundance and may, indeed, have become fully solubilised over the period of time taken to reach this zone.

Finally, the observation that the ^{241}Am/^{239,240}Pu quotient in suspended particulate in the north-western Irish Sea is higher than in surface sediment in the same zone, but similar to that found in suspended particulate and muddy sediments near the Cumbrian coast [20], lends weight to the view that the source term for the western Irish Sea zone is now, largely, the remobilisation of plutonium and americium from the sedimentary deposits in the general vicinity of Sellafield.

ACKNOWLEDGEMENTS

The support and encouragement provided by the Commission of the European Communities within the framework of the 1991-92 Radiation Protection research programme (Contract No. B17*0042-C) is gratefully acknowledged. The authors also wish to record their thanks to P.J. Kershaw and D.S. Woodhead at the MAFF Laboratory, Lowestoft for many helpful and stimulating discussions, and the captains and crews of the research vessels *Challenger* and *Lough Beltra*, without whose assistance this study could not have been completed.

REFERENCES

1. Nelson, D.M., Larsen, R.P. and Penrose, W.R., Chemical speciation of plutonium in natural waters. In *Proc. Symposium on Environmental Research on Actinide Elements*, Hilton Head, South Carolina, November 7-11, 1983. US Department of Energy, August 1987, CONF-841142, pp. 27-48.
2. Nelson, D.M. and Lovett, M.B., Oxidation state of plutonium in the Irish Sea. *Nature*, 1978, **276**, 599-601.
3. Aston, S.R., Evaluation of the chemical forms of plutonium in sea water. *Mar. Chem.*, 1980, **8**, 319-25.
4. Nelson, D.M. and Lovett, M.B., Measurements of the oxidation state and concentration of plutonium in interstitial waters of the Irish Sea. In *Impacts of Radionuclide Releases into the Marine Environment*, IAEA, Vienna, 1981, IAEA-SM-248/145, pp. 105-18.
5. Lovett, M.B. and Nelson, D.M., Determination of some oxidation states of plutonium in sea water and associated particulate matter. In *Techniques for Identifying Transuranic Speciation in Aquatic Environments*, Panel Proc. Ser., Ispra, IAEA, Vienna (STI/PUB/613), 1981, pp. 27-35.
6. Noshkin, V.E. and Wong, K.M., Plutonium mobilisation from sedimentary sources to solution in the marine environment. In *Marine Radioecology*, Proc. 3rd Nuclear Energy Agency Seminar, 1-5 October, 1979, Tokyo, OECD, Paris, 1980, pp. 165-78.
7. Nelson, D.M., Carey, A.E. and Bowen, V.T., Plutonium oxidation state distributions in the Pacific Ocean during 1980-1981. *Earth Planet. Sci. Lett.*, 1984, **68**, 422-30.
8. Silver, G.L., Comment on the evaluation of the chemical forms of plutonium in sea water and other aqueous solutions. *Mar. Chem.*, 1983, **12**, 91-6.
9. Edgington, D.N. and Nelson, D.M., The chemical behaviour of long-lived radionuclides in the marine environment. In *Proc. Symp. on the Behaviour of Long-Lived Radionuclides in the Environment*, 28-30 September 1983, La Spezia, Italy, Commission of the European Communities, Luxembourg, EUR 9214 en, 1984, pp. 19-69.
10. Neihoff, R.A. and Loeb, G.I., The surface charge of particulate matter in sea water. *Limnol. Oceanogr.*, 1972, **17**, 7-16.
11. Orlandini, K.A., Penrose, W.R. and Nelson, D.M., Pu(V) as the stable form of oxidized plutonium in natural waters. *Mar. Chem.*, 1986, **18**, 49-57.
12. Pentreath, R.J., Harvey, B.R. and Lovett, M.B., Chemical speciation of transuranium nuclides discharged into the marine environment. In *Speciation of Fission and Activation Products in the Environment* (Eds. R.A. Bulman and J.R. Cooper), Elsevier, 1986, pp. 312-25.
13. Pentreath, R.J., Woodhead, D.S., Kershaw, P.J., Jefferies, D.F. and Lovett, M.B., The behaviour of plutonium and americium in the Irish Sea. *Rapp. P.-v. Reun. Cons. int. Explor. Mer*, 1986, **186**, 60-9.
14. Kershaw, P.J., Pentreath, R.J., Harvey, B.R., Lovett, M.B. and Boggis, S.J., Apparent distribution coefficients of transuranium elements in UK coastal waters. In *Application of Distribution Coefficients to Radiological Assessment Models* (Eds. T.H. Sibley and C. Myttenaere), Elsevier, 1986, pp. 277-87.

15. Mitchell, P.I., Radionuclide monitoring in the Irish Sea. In *Proc. Seminar on the Radiological Exposure of the Population of the European Communities from Radioactivity in North European Marine Waters (Project Marina)*, Bruges, 14-16 June 1989, Commission of the European Communities, Luxembourg, XI/4669/89-EN, pp. 185-227.
16. Jefferies, D.F., Steele, A.K. and Preston, A., Further studies on the distribution of ^{137}Cs in British coastal waters. 1. The Irish Sea. *Deep-Sea Res.*, 1982, **29**, 713-38.
17. Hunt, G.J., Timscales for dilution and dispersion of transuranics in the Irish Sea near Sellafield. *Sci. Total Environ.*, 1985, **46**, 261-78.
18. Mitchell, P.I., Sanchez-Cabeza, J.A., Vidal-Quadras, A. and Font, J.L., Distribution of plutonium in inshore waters around Ireland using *Fucus vesiculosus* as a bio-indicator. In *Proc. CEC/CIEMAT Seminar on the Cycling of Long-Lived Radionuclides in the Biosphere: Observations and Models*, Madrid, 15-19 September 1986, Commission of the European Communities, Luxembourg, 1987, Vol. II, 13 pp.
19. Crowley, M., Mitchell, P.I., O'Grady, J., Vives, J., Sanchez-Cabeza, J.A., Vidal-Quadras, A. and Ryan, T.P., Radiocaesium and plutonium concentrations in *Mytilus edulis* (L.) and potential dose implications for Irish Critical Groups. *Ocean and Shoreline Management*, 1990, **13**, 149-61.
20. Hunt, G.J., Radioactivity in surface and coastal waters of the British Isles, 1989. Aquatic Environment Monitoring Report No. 23, Ministry of Agriculture, Fisheries and Food, Directorate of Fisheries Research, Lowestoft, 1990, 66 pp.
21. Hetherington, J.A., The uptake of plutonium nuclides by marine sediments. *Mar. Sci. Commun.*, 1978, **4**, 239-74.
22. Kershaw, P.J., Woodhead, D.S., Malcolm, S.J., Allington, D.J. and Lovett, M.B., A sediment history of Sellafield discharges. *J. Environ. Radioactivity*, 1990, **12**, 201-41.
23. Nelson, D.M., Penrose, W.R., Karttunen, J.O. and Mehlhaff, P., Effects of dissolved organic carbon on the adsorption properties of plutonium in natural waters. *Environ. Sci. Technol.*, 1985, **19**, 127-31.
24. Foster, P., Nutrient distributions in the winter regime of the northern Irish Sea. *Mar. Environ. Res.*, 1984, **13**, 81-95.
25. Irish Sea status report of the marine pollution monitoring management group (Ed. R.R. Dickson), Aquatic Environment Monitoring Report No. 17, Ministry of Agriculture, Fisheries and Food, Directorate of Fisheries Research, Lowestoft, 1987, 83 pp.
26. Pentreath, R.J., Lovett, M.B., Jefferies, D.F., Woodhead, D.S., Talbot, J.W. and Mitchell, N.T., The impact on public radiation exposure of transuranium nuclides discharged in liquid wastes from fuel element reprocessing at Sellafield, UK. In *Radioactive Waste Management*, IAEA, Vienna, 1984, IAEA-CN-43/32, Vol. 5, pp. 315-29.
27. Hamilton, E.I., Alpha-particle radioactivity of hot particles from the Esk estuary. *Nature*, 1981, **290**, 690-3.
28. Kershaw, P.J., Brealey, J.H., Woodhead, D.S. and Lovett, M.B., Alpha-emitting, hot particles in Irish Sea sediments. *Sci. Total Environ.*, 1986, **53**, 77-87.

ANTHROPOGENIC ^{14}C AS A TRACER IN WESTERN U.K. COASTAL WATERS

F H Begg*, M S Baxter†, G T Cook*, E M Scott** and M McCartney*

*Scottish Universities Research and Reactor Centre, East Kilbride, G75 0QU, Scotland
†IAEA International Laboratory of Marine Radioactivity,
Principality of Monaco, MC 98000.
**Department of Statistics, University of Glasgow, Glasgow,
G12 8QQ, Scotland

ABSTRACT

There is growing international awareness of the radiological significance of ^{14}C discharged by the nuclear industry. In addition, anthropogenic ^{14}C is a powerful tracer of partitioning and flux within the global carbon cycle. The research described here focuses on characterising and understanding the distribution of anthropogenic ^{14}C in the coastal marine environment of the U.K. where two of the known major sources are:-
1) the BNFL nuclear fuel reprocessing plant at Sellafield in Cumbria, where low-level effluents are discharged under authorization into the Eastern Irish Sea
2) the Amersham International radiochemical plant in Cardiff where waste from the production of radiotracers is released under licence into Cardiff Bay and ultimately into the Bristol Channel.

There are four reservoirs for carbon in the marine environment - the dissolved inorganic and organic (DIC and DOC) and the particulate inorganic and organic (PIC and POC) fractions. The DIC fraction, which is frequently used in tracer studies, has demonstrated the incorporation of ^{14}C from anthropogenic sources in both the Bristol Channel and the Eastern Irish Sea, with the measured activities reaching four times and twice the current ambient level respectively.

Investigation of the ^{14}C concentrations present in the water column, in conjunction with ^{137}Cs analysis, has indicated a similarity in the distribution pattern of the two nuclides allowing the modification of mathematical models, validated using ^{137}Cs, to describe the environmental transfer of ^{14}C within the marine system. In addition, accelerator mass spectrometry (AMS) has been used to assist in defining the partitioning and exchange of excess ^{14}C between the biogeochemical phases present in the water column.

INTRODUCTION

The levels of ^{14}C present in the various carbon reservoirs of the environment have in the past been, and in the future will be, determined by both natural processes and man's activities.

^{14}C ($t_{1/2}$ = 5730 years [1]) is produced in the upper atmosphere at an estimated annual rate of 1.0 - 1.4x10^{15}Bq [2,3] when cosmic ray-produced neutrons interact with naturally occurring ^{14}N nuclei (^{14}N + n → ^{14}C + p). Apart from cyclical modulations in this production rate due to changes in the strength of the earth's magnetic field (9000 year cycle), in conjunction with changes in the solar sun spot activity (11 year cycle), equilibrium conditions are believed to have prevailed in the upper atmosphere until the use of fossil fuels started in the 1860's. This introduced large amounts of stable carbon which brought about a decrease in the atmospheric specific activity of 2-3% by the 1950's before this was overshadowed by large-scale nuclear weapons testing which injected an estimated 2.2 - 3.5x10^{17}Bq [4,3] of ^{14}C into the atmosphere. This led to a doubling of the atmospheric specific activity by 1963 which has been slowly decreasing as excess ^{14}C has been incorporated into other carbon reservoirs - currently ≥ 80% of the produced excess can now be found in the surface oceans [4].

A further anthropogenic source of ^{14}C has been nuclear power and future global concentrations of ^{14}C will be dependent on the growth and waste management policies of this industry. Theoretical calculations indicate that 4.5x10^{12}Bq of ^{14}C are produced as an activation product in reactor components per GW(e), resulting in a global production of ~4x10^{14}Bq of ^{14}C per year using 1988 reactor distribution and capacity data [5]. Much of this will remain in the reactor until decommissioning with only ~40% being released directly to either the atmosphere or the marine environment.

^{14}C (both natural and anthropogenic) has been widely used as a tracer in marine processes - bomb-produced ^{14}C invasion rates into surface waters have been used to model the oceanic uptake of atmospheric CO_2 [6,7] whereas natural ^{14}C levels have been used to measure relative ages of deep water in the determination of the circulation pattern of the world's oceans [8].

Two of the known sources of anthropogenic ^{14}C in the U.K. are the BNFL nuclear fuel reprocessing plant at Sellafield in Cumbria which has been in operation since 1952 and the Amersham International radiochemical plant in Cardiff. These establishments release low-level effluents containing ^{14}C, under authorization, into the Eastern Irish Sea and Cardiff Bay (and ultimately the Bristol Channel) respectively, resulting in anthropogenic radionuclides entering the food chain due to their uptake by marine organisms. Although only ~10% of the ^{14}C discharged at Sellafield is released into the coastal waters, significant enhancements have been found in fish and other marine organisms [9,10]. In an attempt to determine the chemical form and behaviour of this excess ^{14}C, the study was widened to encompass the water column and analysis of the individual geochemical fractions present within it.

To this end, an extensive suite of water samples (~30) was collected, during a MAFF cruise in December 1989, for the analysis of the ^{14}C content of the DIC fraction, from both the Bristol Channel and the Irish Sea. These were collected along three working transects - one in the Bristol Channel to determine the influence of effluent from Amersham International and two in the Eastern Irish Sea (see Fig 1). ^{137}Cs analysis was also carried out along Transects 2 and 3.

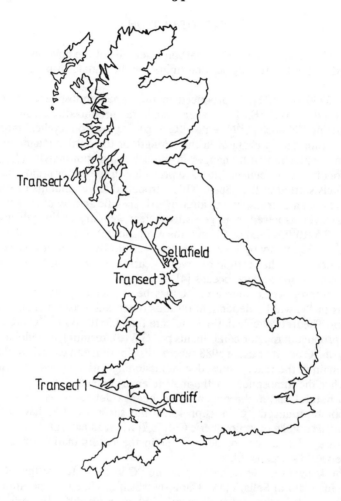

Figure 1. Location of the three transects sampled during MAFF cruise CH62B/89

At four selected sites further samples were collected for separation into the different biogeochemical phases - the dissolved and particulate inorganic carbon fractions (DIC and PIC) as well as the dissolved and particulate organic carbon fractions (DOC and POC). The sites were selected to show any fractionation effects in the Bristol Channel (one site) and also to monitor the distribution of ^{14}C in the various carbon pools at the entrance to (St Georges Channel) and the exit from (North Channel) the Irish Sea as well as close to the point of discharge from Sellafield.

MATERIALS AND METHODS

Analysis of the ^{14}C content of the dissolved inorganic carbon fraction of seawater requires the collection and filtration of ~200 l of surface water - the water is then acidified and purged with nitrogen with the evolved CO_2 bubbled through 4M NaOH in a closed system. The resultant sodium carbonate solution is then hydrolysed using 8M HCl in the laboratory to re-evolve the CO_2 which is converted to benzene via lithium carbide and acetylene and analysed by liquid scintillation spectrometry for a total counting period of 2000 minutes.

Due to the low concentrations of DOC, POC and PIC in seawater, the isolation procedures were carried out at SURRC but the actual analyses were carried out using the accelerator mass spectrometery (AMS) facility at Oxford as this requires only 1-2 mg of carbon whereas conventional radiometric ^{14}C analytical techniques generally require approximately 1 g of carbon. To obtain sufficient weights of particulate material 400 l of surface seawater were filtered through pre-combusted glass fibre paper (~0.45μm) - this yielded sufficient material to allow analysis of both the inorganic and the organic particulate fractions. Firstly, the filter papers were washed in ~2 l of distilled water to remove excess Cl$^-$ ions and then were hydrolysed with 50% H_3PO_4 under vacuum to isolate the inorganic fraction. The residue from this process was then reacted with chromic acid/potassium dichromate under vacuum and heated to encourage the wet oxidation of the organic fraction to take place. The CO_2 evolved from both fractions was collected and retained for analysis by AMS.

Dissolved organic carbon is present in seawater in concentrations of 0.3-3.0 mg (C) l^{-1} [11] (depending on the oxidation procedure used) hence, isolation procedures require an initial concentration step, this is achieved using XAD macroporous resins. XAD-2, 4 and 8 resins were used to concentrate the fulvic, hydrophilic and humic acids which make up 20 - 50% of the total DOC content [12], these are considered to be refractory and not easily broken down. 25 l filtered samples were put through 50cm^3 columns at a rate not exceeding 15 bed volumes per hour. The columns were set up in such a way that the seawater was passed through a column of XAD-4 at its natural pH, the eluent was then acidified to pH2 with conc. HCl before being passed through a column of XAD-2 followed by a similar one of XAD-8. The hydrophilic acids were eluted from the XAD-4 column using methanol whereas the fulvic and humic acids were eluted from XAD-2 and XAD-8 respectively using 0.1M NaOH. Once eluted, the samples were combined and oxidised under vacuum with hot chromic acid/potassium dichromate and the CO_2 retained for analysis by AMS. A total of 16 samples was analysed by AMS - consisting of 4 fractions (DIC, PIC, DOC and POC) from four sites.

RESULTS AND DISCUSSION

All ^{14}C results are reported relative to an international standard - NBS Oxalic Acid II - where 0.7459 of the observed count rate is assumed to be equivalent to a specific activity of 226 $Bqkg^{-1}C$ ie

$$\text{Specific Activity (Bqkg}^{-1}\text{C)} = A_{SN}/A_{OX} \times 226$$

where A_{SN} = net activity of the sample and A_{OX} = net activity of the standard.

The accepted present day atmospheric specific activity is 260 $Bqkg^{-1}C$ and this has been confirmed by analysis of samples from an area in north western Scotland where there are no known local inputs of anthropogenic ^{14}C.

Previous work has shown that concentrations of ^{14}C found in coastal biota samples collected around Sellafield appear to be dependent on the type of organism and the distance of the sampling site from the point of effluent release [10]. From the results obtained for activities found in the DIC phase of the water column (Table 1), distance again appears to be one of the determining factors.

Transect 1

Results from Transect 1 indicate that inputs from Amersham are significantly influencing the level of ^{14}C in the DIC pool within the Bristol Channel with measured concentrations reaching over four times the current ambient level. This enhanced ^{14}C activity steadily decreases with increasing distance, towards the mouth of the channel, where any remaining excess ^{14}C will be extensively diluted due to mixing with incoming Atlantic water. The observed concentration at the mouth of the channel (C6) is approximately equal to the accepted current ambient value, hence, it can be assumed that anthropogenic inputs of ^{14}C from Amersham International, Cardiff are not influencing the concentrations in the Irish Sea and that any enhancements observed in that area can be attributed to the Sellafield plant.

Transect 2

Of the two transects in the Eastern Irish Sea, Transect 2 again indicates decreasing activities of ^{14}C in the DIC of surface waters with increasing distance from the discharge location - directly analogous to the distribution found in coastal biota samples although the actual measured activities are considerably less.

In an attempt to empirically model this observed distribution of ^{14}C, to allow the prediction of ^{14}C levels at unsampled sites (in direct analogy to the Gaussian plume for atmospheric discharges), two parameters were considered - distance and direction from the discharge location. Almost 90% of the observed variation in the results can be explained empirically by an equation of the type:-

$$\ln {}^{14}C \text{ conc.} = x - y(DIST) + z(DIST)^2$$

which does not take into account the direction of the sampling site. The apparent irrelevance of direction may be in part due to the chosen transect - it approximately follows the direction of the predominant residual surface currents [13] and hence will reflect the rate of removal from the Eastern Irish Sea due to water movement.

TABLE 1
^{14}C and ^{137}Cs levels found in surface waters of the Bristol Channel and the Irish Sea

SAMPLE No	DISTANCE (km)	SPECIFIC ACTIVITY (Bgkg^{-1}C)	^{137}Cs (Bql^{-1})
TRANSECT 1			
C0	1.0	1082.73 ± 7.63	NA
C1	14.5	691.20 ± 5.99	NA
C2	21.0	638.58 ± 4.77	NA
C3	34.0	544.74 ± 3.56	NA
C5	115.0	274.55 ± 3.62	NA
C6	145.0	265.45 ± 1.94	NA
TRANSECT 2			
C08	4.8	573.06 ± 2.27	0.531 ± 0.003
C09	10.6	564.44 ± 1.94	0.460 ± 0.003
C10	26.5	543.67 ± 2.27	0.403 ± 0.002
C11	37.5	356.54 ± 1.60	0.215 ± 0.001
C12	63.1	354.62 ± 1.65	0.211 ± 0.002
C13	81.6	311.71 ± 1.82	0.138 ± 0.001
C14	112.4	329.88 ± 1.78	0.166 ± 0.001
C15	136.7	308.06 ± 1.15	0.130 ± 0.001
C16	152.6	303.77 ± 1.49	0.128 ± 0.001
C17	190.3	289.07 ± 0.64	0.102 ± 0.001
C18	262.4	268.96 ± 1.84	0.056 ± 0.001
TRANSECT 3			
C19	36.0 (south)	492.99 ± 4.83	0.386 ± 0.002
C20	23.9 (south)	894.75 ± 7.22	0.527 ± 0.002
C21	14.8 (south)	565.01 ± 7.20	0.470 ± 0.002
C22	7.4 (south)	551.17 ± 5.52	0.490 ± 0.003
C08	4.8	573.06 ± 2.27	0.531 ± 0.003
C23	9.0 (north)	574.58 ± 4.38	0.478 ± 0.003
C24	22.3 (north)	528.41 ± 4.42	0.420 ± 0.002
C25	37.6 (north)	448.41 ± 3.82	0.311 ± 0.002
C26	52.5 (north)	494.80 ± 4.24	0.393 ± 0.002

NA Not analysed

^{137}Cs is a well known and extensively used radioactive tracer of water movement due to its conservative behaviour in the water column. Sampling of both ^{14}C and ^{137}Cs at the same sites assisted in determining the behaviour of ^{14}C in the water column - a correlation of 0.989 between the two nuclides indicates a similar behaviour, hence, the ^{14}C found in the DIC fraction of surface water appears to behave in a conservative manner.

Transect 3

The concentrations of ^{14}C found along this transect are generally between 450 and 575 Bqkg^{-1} carbon with one point (C20) showing a much greater value than any of the others in the area. If this point is discounted, the maximum value is found in the immediate vicinity of the discharge location with decreasing values both to the north and south. When studied in conjunction with the ^{137}Cs data, in this case there appears not to be such a close relationship between the two nuclides (correlation factor of 0.688) which may just be a reflection of the relatively small variation in the observed concentrations.

Overall, there is a definite enhancement in the concentrations of ^{14}C in the DIC fraction of the surface water in both the Eastern Irish Sea and the Bristol Channel which appears to be directly attributable to the Sellafield and Amersham plants respectively. This enhancement is greatest closest to the discharge locations and decreases with increasing distance from the release point.

While Sellafield-derived ^{14}C can be used as a tracer of water movement in the Eastern Irish Sea (and Amersham-derived in the Bristol Channel) it may also assist in ascertaining the degree of partitioning between, and the relative fluxes of, the geochemical fractions of carbon within the coastal marine system.

Taking into account the four sites chosen for detailed analysis of the four biogeochemical fractions, the results have shown a wide variation in the specific activities between both fractions and sites (Table 2) with the inputs from Amersham and Sellafield being most clearly seen in the DIC fraction.

TABLE 2
Specific activity measured in the biogeochemical fractions of surface seawater

SITE	FRACTION			
	DIC	PIC	POC	DOC
BRISTOL CHANNEL	612.9±4.1	91.1±0.9	342.6±2.3	NA
ST.GEORGES CHANNEL	270.7±1.8	124.2±1.1	117.3±1.4	186.8±2.0
SELLAFIELD	549.9±3.8	108.3±1.1	165.3±1.6	143.3±1.8
NORTH CHANNEL	308.0±1.1	144.2±1.4	129.9±1.1	167.2±1.8

All results presented as specific activity, ie Bqkg^{-1}C. Those with values ≥260 Bqkg^{-1}C are enhanced relative to current ambient levels while those ≤260 Bqkg^{-1}C are depleted relative to current ambient levels.
NA Not Available.

From the results obtained it appears that the first carbon pool to be influenced by anthropogenic releases of ^{14}C is the DIC from where it can then be incorporated into other reservoirs within the carbon cycle, depending on the rate of transfer between each pool. At the Bristol Channel site the POC fraction is also enhanced and this may be indicative of the chemical form in which the excess ^{14}C is released, although in the Irish Sea the POC fraction appears slightly enhanced when compared to the activities of the POC material entering and leaving the area. This may indicate that the POC reservoir is linked to the DIC pool via its uptake by marine organisms which add organic matter to the water column during excretion of waste products in addition to that added on death of the organisms.

CONCLUSIONS

From the work carried out to date it appears that anthropogenic ^{14}C from both Sellafield and Amersham can be used as a tracer for coastal water movement due to the conservative behaviour of the DIC fraction.

The distribution of ^{137}Cs in the Eastern Irish Sea has been widely studied over the last 10 - 15 years and has culminated in the validation of a mathematical model which can reproduce measured ^{137}Cs concentrations and hence water movements over this time period. It is envisaged that a similar type of model will assist in determining the levels of ^{14}C present in coastal waters, given the Sellafield discharge data in conjunction with natural and nuclear weapons input levels.

As ^{14}C forms an integral part of living tissues, the model will need to take into account accumulation of ^{14}C by marine biota and its subsequent return to the water column - this will hopefully be achieved by the incorporation of the geochemical phases into the model structure.

The data collected to date only represent the distribution at the time of sampling, for a full study temporal sampling would need to be undertaken to ascertain whether the observed distribution is representative or whether some degree of seasonality occurs in the ^{14}C concentrations.

While the biogeochemical fractions analysed by AMS give some indication of the specific activities present, a more comprehensive sampling strategy is required to determine the validity of the results obtained and assist in the determination of the extent of partitioning, and the rate of transfer, between reservoirs.

ACKNOWLEDGEMENTS

The authors would like to thank Philip Naysmith for his technical assistance during the period of research, the officers and crew of RV Challenger, Alan Young and Dave Allington of MAFF for their assistance during the cruise and one of us (FHB) would like to thank the Natural Environment Research Council for funding for this project.

REFERENCES

1. Godwin, H., Half-life of radiocarbon. Nature, 1962, **195**, 984.

2. United Nations., Sources and effects of ionizing radiation. United Nations Scientific Committee on the Effects of Atomic Radiation 1977 Report to the General Assembly. United Nations, New York, 1977.

3.   Taylor, D.M., Moroni, J.P., Snihs, J.-O and Richmond, C.R., The metabolism of ^{3}H and ^{14}C with special reference to radiation protection. Radiation Protection Dosimetry, 1990, **30(2)**,87-93.

4.   Lassey, K.R., Manning, M.R. and O'Brien, B.J., Assessment of the inventory of carbon-14 in the oceans: an overview. In Inventories of selected radionuclides in the oceans. IAEA-TECDOC-481, 1988, IAEA, Vienna.

5.   IAEA, IAEA Bulletin, 1989,Vol 31 No.1, Quarterly Journal of the IAEA, Vienna.

6.   Broccker, W.S., Peng, T.H., Ostlund, G. and Stuiver, M., The distribution of bomb radiocarbon in the ocean. J. Geophys. Res., 1985, **90**, 6953-6970.

7.   Stuiver, M., ^{14}C distribution in the Atlantic Ocean. J. Geophys. Res., 1980, **85**, 2711-2718.

8.   Broecker, W.S., Geochemical tracers and oceanic circulation. In Evolution of physical oceanography, Eds. B.A. Warren and C. Wunsch, MIT Press, Cambridge, Mass., 1981. pp.434-60.

9.   Hunt, G.J., Aquatic Environment Monitoring Report No.23, Radioactivity in the surface and coastal waters of the British Isles, 1989, 1990, Ministry of Agriculture, Fisheries and Food, Lowestoft.

10. Begg, F.H., Cook, G.T., McCartney, M., Baxter, M.S.and Scott, E.M., Anthropogenic radiocarbon in the Eastern Irish Sea and Scottish Coastal Waters. Radiocarbon, 1992, **34**, In Press.

11. Sugimura, Y. and Suzuki, Y., A high-temperature catalytic oxidation method for the determination of the non-volatile dissolved organic carbon in seawater by the direct injection of a liquid sample. Marine Chemistry, 1988, **24**, 105-31.

12. Williams, P.M. and Druffel, E.R.M., Dissolved organic matter in the ocean: comments on a controversy. Oceanography Magazine, 1988, **1**, 14-17.

13. The status of current knowledge of anthropogenic influences in the Irish Sea. Eds. Dickson, R.R. and Boelens, R.G.V., March 1988, International council for the exploration of the sea (Denmark), Co-operative Research Report No.155.

RADIOCAESIUM IN LOCAL AND REGIONAL COASTAL WATER MODELLING EXERCISES

P E BRADLEY[1], E M SCOTT[2], M S BAXTER[3] AND D J ELLETT[4]
[1]SURRC, East Kilbride, Glasgow, G75 0QU
[2]University of Glasgow, Glasgow, G12 8QW
[3]IAEA, Monaco, MC98000
[4]Dunstaffnage Marine Laboratory, Argyll, PA34 4AD

ABSTRACT

In this paper we review briefly the last fifteen years of measurement of radiocaesium in Scottish coastal waters, present its geographical dispersal on both local and regional scales and consider the temporal variation in radiocaesium concentrations over the period of study.
Results from a comprehensive modelling exercise are interpreted in terms of the physical hydrography of the area. We conclude with a discussion of the advantages and disadvantages of the modelling approach taken in this work.

INTRODUCTION

Radiocaesium discharged into the Irish Sea acts as a valuable tracer in local and regional studies of the UK coastal water system. The particular area of study of this work comprises that region of the inner continental shelf between the North Channel and the Outer Hebrides of Scotland as shown in Figure 1, which also indicates sampling stations used throughout the period of study. Isopleths of ^{137}Cs concentrations from a representative cruise are also shown and indicate the extent of geographical dispersal.

Radiocaesium isotopes discharged from the Sellafield nuclear fuel reprocessing plant in Cumbria have been shown to pass through the region [1-8] and to provide a useful tracer system for the study of the rate and direction of the movement of water originally labelled within the Irish Sea. The radiocaesium isotopes of interest (^{137}Cs and ^{134}Cs) are of particular value as they exhibit relatively conservative behaviour and hence long residence times in coastal water [9]. They are present at easily detectable levels in Scottish coastal waters and their quantification is relatively straightforward using ion-exchange and γ

spectroscopy techniques [10].

In addition, since Sellafield discharges of radiocaesium have been seasonally pulsed, the short-term temporal trends in activity may be matched to observed trends in a number of geographical locations and a full compartmental model for the region developed. We

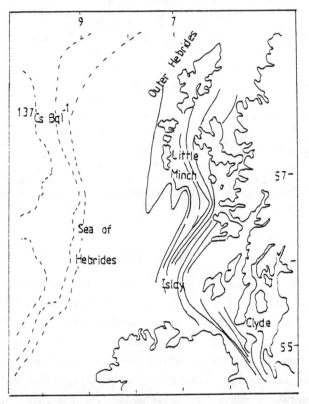

Figure 1. Survey area and geographical dispersal of ^{137}Cs, 1985

describe the development of such a compartmental model, its calibration, validation and interpretation for the region of study.

DATA

Radiocaesium distributions have been measured regularly over the previous fifteen years, and a total of 2000 radiometric measurements have been made by the Glasgow/SURRC laboratories. Sampling has been undertaken over a wide geographically dispersed grid, making use of transect sampling on cruises, in addition to sampling at identified locations on a regular basis. Thus in 12 locations, measurements have been made at different frequencies over the period of study, with the Clyde Sea Area and North Channel having the greatest number of measurements. Such a large and extensive data set, when used in conjunction with published radiocaesium discharge data allows exhaustive investigation of the dynamics of the region, on both temporal and geographical scales.

A conventional analysis of the data has involved mapping of the geographical dispersal of radiocaesium [11] in several years, and this has allowed a description of the hydrographic system to be prepared as input to the modelling work, providing a sound scientific basis for the compartments in the model.

COMPARTMENTAL MODELLING STUDIES

Compartmental models of widely varying complexity are commonly used as descriptive and inferential tools in environmental studies. Simple linear compartmental models are the most common type of model used in the study of the transport and exchange of material within a system. The structure of the real system is represented by a number of well defined functional units and the exchange of matter between them is described by corresponding transfer coefficients. One important assumption inherent in this formal structure is that the content of each compartment is considered perfectly homogeneous. This condition can of course be only approximately approached in any real system and depends on the initial choice of compartments. Excellent reviews of compartmental analysis are to be found in the literature [12-14] and the technique has successfully been used to describe various oceanographic systems [15-17]. The empirically derived model may be limited in its structure and hence considerably less complex than the natural system. Nevertheless, it may still prove useful in providing a general description of the system, and through its failure to reproduce the observed trends, highlight areas requiring further experimental and theoretical development.

The practical problems in any compartmental modelling exercise are typically dominated by the difficulties in defining numerical values for the rate coefficients. Given their values, the system response to any perturbations can be easily simulated. However, in this project, we have been concerned with estimation of the rate coefficients. One method which we have considered is that of inverse compartmental analysis (ICA).

ICA techniques have been used to determine optimised transfer coefficients for a model of the North Sea and adjacent areas and have resulted in an improved and validated model for the dispersion of effluent in the region [18]. In the light of the success of such methods and in view of the large data set associated with the Scottish coastal waters, we present here the results of an inverse compartmental modelling exercise applied to the coastal current system.

Choice of compartments and exchange pathways

The first step in any compartmental model study is to compartmentalize the region under study to allow the use of linear systems of equations to describe the transfer of ^{137}Cs and ^{134}Cs through the compartments. A detailed description of the system may require many compartments, but the critical process of model validation is of course, dependent on the availability of data for each of the model compartments. Figure 2 shows schematically the compartments assigned to the Scottish coastal current system, the allocation of compartments and exchange pathways being made on the basis of maps of geographic dispersal and known hydrographic features including depths and current information.

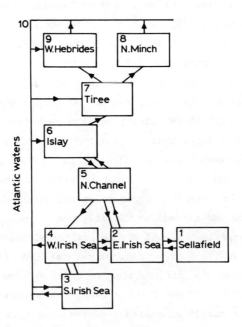

Figure 2. Compartments of the western U.K. coastal system

Experimental data and optimisation of transfer coefficients

The radiocaesium measurements used in this study form an extensive data set covering the period 1972 to 1988 and include, particularly for the Irish Sea, measurements made by M.A.F.F. (MAFF reports [19]). A large number of measured ^{137}Cs and ^{134}Cs activities are available for each of the compartments, (thus we have multiple time series of differing lengths and frequencies) which are linked through the model structure to the published discharges. The best characterised compartments are those representing the North Channel and Clyde Sea areas where measurements have been made on a routine basis. Figure 3 shows the observed data in the North Channel/Clyde Sea Area and the discharge data.

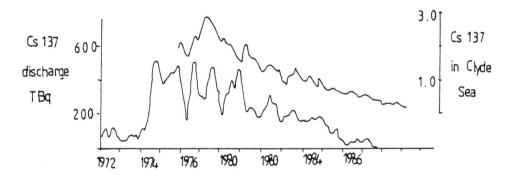

Figure 3. Sellafield discharge record and levels of ^{137}Cs in Clyde Sea area.

If the observed system can be described with sufficient accuracy by the compartmental model and the transfer coefficients are known, then the transfer of matter within the system is readily simulated using well established algorithms. However, in this situation, the opposite problem is encountered i.e. compartment contents have been measured over a long period of time and the corresponding functional properties of the system need to be determined. No straightforward algorithm exists for the numerically intensive calculations of this reverse operation. We have adopted an iterative procedure due to Hallstadius [20] in order to optimise the transfer coefficients for our compartmental model. A computer program for this optimization based on an iterative procedure has been developed and applied to a number of radiological investigations. The full details of the program are comprehensively documented in several publications [21-22] and are not reproduced here, but we describe the basis of the method for the estimation of transfer coefficients. The procedure depends on three steps:

1) Initial set of transfer coefficients defined
2) results for each compartment are simulated and
3) observed and simulated activities are compared, as a result of which a new set of transfer coefficients are defined and the procedure repeated. Convergence is achieved when the agreement between simulated and observed results is deemed acceptable.

RESULTS AND DISCUSSION

a) <u>Model calibration</u>

The procedure outlined above was applied to the radiocaesium data within the framework of the model described earlier and the model characteristics and the optimised transfer coefficients obtained after convergence are shown in Table 1. We have attempted to maximise the agreement between observed and modelled ^{137}Cs data. The quality of the fit is demonstrated in Figure 4. For the vast majority of the compartments the modelled data for compartmental inventories of ^{137}Cs shows a very good agreement with the observed data both in terms of the trend of the modelled data and in the absolute values of the inventories. A major discrepancy between the modelled and observed data values in the North Channel in 1977 has previously been reported and ascribed to a transient event, namely a change in the flow conditions. Given such a hypothesis, we have investigated recalibration of the model after the 1977 event; we find that post-1977, the agreement between observed and modelled values is much improved.

b) <u>Model validation</u>

The values of the optimised transfer coefficient in conjunction with compartment volumes allow calculation of the implied rates of volume transfer along the exchange pathways within the model and are also shown in Table 1. These values of volume transport are compared to similar estimates made by a number of other methods (such as salinity distribution methods, current meter data, radionuclide budgets and other modelling studies) in Table 2. The fact that the values obtained by inverse compartmental analysis are intermediate between the extremes of such values as exist and the fact that they are internally consistent gives considerable confidence in the framework of the model and allows credence to be put on those estimates of water flow for which no comparable data currently exists.

One of the major requirements of the model should be that the water mass is conserved. Although there are no constraints during the optimisation of transfer coefficients to force the flows into and out of the compartments to be balanced, we do observe that for this model structure, the resulting flows do exhibit a good balance (the maximum discrepancy between flows into and out of a compartment being 29%). Where there are discrepancies, such as between Box 5, the North Channel and Box 6, Islay, where an apparent loss of flow occurs, this may reflect either differences in Cs budgets between adjoining regions with differing stratification, a factor to which attention should be given in future work or that the behaviour of Cs is not conservative.

In general respects, the model confirms many of the features shown in the survey of July 1981, although this represented only a single summer month. The present work also demonstrates greater coastal current flow through the Minch (Box 8) than to the west of the Outer Hebrides (Box 9), but in the proportions 70:30%, rather than 84:16% as in July 1981. Similarly, a greater quantity of Atlantic water is entrained into the current by the time it passes Cape Wrath than issued from the North Channel, but again the proportion is larger (75:25%) than for the short term study (62:38%).

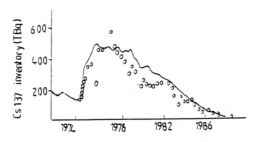

(a)   Compartment 6:   North Channel/Clyde Sea Area

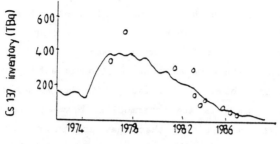

(b) Compartment 6: Islay Front Area

Figure 4. Comparison of observed and predicted levels

TABLE 1

Compartmental Box Model for Scottish Coastal Current System

Box	Volume (Km³)	Transfer (from/to)	Optimised Transfer Coefficient (months)	Implied Water flow (Km³/month)	(m³/sec) (x10⁻⁴)
1	65	1,2	0.46	30	1.2
2	340	2,1	0.09	30	1.2
		2,4	0.18	60	2.3
		2,5	0.18	60	2.3
3	1000	3,4	0.23	230	8.9
		3,10	0.01	10	0.4
4	800	4,2	0.08	60	2.3
		4,3	0.10	80	3.1
		4,5	0.21	170	6.6
5	200	5,2	0.30	60	2.3
		5,4	0.10	20	0.8
		5,6	0.85	170	6.6
6	360	6,5	0.28	100	3.8
		6,7	0.36	120	4.6
7	480	7,8	1.20	577	22.3
		7,9	0.52	247	9.5
8	590	8,10	0.94	450	17.4
9	480	9,10	0.59	349	13.5
10	5.0 x 10⁷	10,3	3.2×10^{-6}	160	6.2
		10,6	zero	-	-
		10,7	3.0×10^{-6}	451	17.4
		10,9	zero	-	-

TABLE 2

Volume Transport Estimates from Three Points in the Scottish Coastal System

Region	Period	Transport $km^{-3}\ day^{-1}$	Method	Reference
North Channel	1972-1986	4.8	ICA	This work
	long term mean	6.7	Salinity distribution	[23]
	long term mean	2.9	Salinity distribution	[24]
	Oct 71-Dec 75	4.5	^{137}Cs budget	[19]
	Jan 76-May 78	10.5	^{137}Cs budget	[19]
	Annual mean	6.9	Tide and wind model	[25]
	July 81	5.4	^{137}Cs budget	[7]
	June 84	5.8	$^{134}Cs, ^{137}Cs$ budget	[8]
Tiree passage	1972-1986	7.1	ICA	this work
	May 76	13.9	$^{134}Cs, ^{137}Cs$	[26]
	May 76	10.4	c/m 5days	[26]
	July 81	2.3	c/m 49 days	[27]
	July 83	4.1	c/m 31 days	[28]
	July 84	1.3	c/m 31 days	[28]
Minch	1972-1986	10.3 to 11.6	ICA	this work
	long term mean	5.3	Salinity distribution	[23]
	July 79	18.5	c/m 7 days	[26]
	July 81	10.4	^{137}Cs budget	[7]
W of outer Hebrides	1972-1986	4.5 to 5.2	ICA	this work
	long term mean	7.9	Salinity distribution	[26]
	July 81	1.4	^{137}Cs/cm	[7]

CONCLUSIONS

The process of model calibration and validation although appearing remarkably similar have considerable differences in their purpose. In model calibration, we wish to minimise the discrepancies between the observed and modelled values, and find estimated values for the model parameters which achieve this. We believe that the technique of ICA when combined with sufficient data, provides a useful calibration tool. In model validation, we are investigating the performance of the model, not in matching the observed data with which it was calibrated, but in predicting values at times and locations for which the model was not explicitly calibrated. The observed disagreements between model and data do not necessarily invalidate the model, they simply highlight potential processes and areas requiring further joint development between experimentalist and modeller. Thus for this model of the Scottish coastal system, we continue to make measurements of the ^{137}Cs in a number of locations and indeed are now extending the model to include other nuclides (in particular ^{14}C) [29].

This work clearly demonstrates the potential of applying inverse compartmental analysis techniques to an exploratory model and using an accurate and extensive data base of observations in the processes of model calibration and validation. The model has been shown to possess the following characteristics:

i) it exploits logistic sense in that the model's construction is based on known hydrographic flow patterns and distributions

ii) it demonstrates a good fit between observed and modelled data

iii) it has parameter values that bear a close resemblance to measured environmental parameters.

As a result of the above properties, the model may confidently be used to provide a description of the complex hydrography of the western Scottish coastal current system and has clear potential for the prediction of future radionuclide levels in Scottish waters in response to continuing Sellafield discharges for ^{137}Cs and for other radionuclides.

ACKNOWLEDGEMENT

We would like to acknowledge the help of Lars Hallstadius of the University of Lund, Sweden whilst one of us (PEB) thanks the UK Natural Environment Research Council for financial support.

The work has been undertaken in a tripartite arrangement between SURRC, DML and Glasgow University.

REFERENCES

1. Ellett, D.J., Temperature and salinity conditions in the Sea of Hebrides. Annales Biologiques, 1977, 33, 28-30.

2. Ellett, D.J., Some oceanographic features of Hebridean waters. Proc. Royal Soc. Edin., 1979, B77, 61-74.

3. Jefferies, D.F., Preston, A., Steele, A.K., Distribution of ^{137}Cs in British coastal waters. Marine Pollution Bulletin, 1973, 4, 118-122.

4. Kautsky, M., The distribution of the radionuclide caesium-137 as an indicator for North Sea water mass transport. Deutsche Hydrographs - Zeitschrift, 1976.

5. Livingston, M.I., Bowen, V.T., Windscale effluent in the waters and sediments of the Minch. Nature, 1977, 269, 586-588.

6. Baxter, M.S., McKinley, I.G., Radioactive species in sea water. Proc. Royal Soc. Edin., 1978, B76, 17-35.

7. McKay, W.A., Baxter, M.S., Ellett, D.J. and Meldrum, D.T., Radiocaesium and circulation patterns west of Scotland. J. Environ. Radioactivity, 1986, 4, 205-232.

8. Economides, B.E. Tracer applications of Sellafield radioactivity in British west coastal waters. Ph.D. Thesis, Glasgow University, 1986.

9. Swan, D.B., Baxter, M.S., McKinley, I.G. and Jack, W., Radiocaesium and Pb-210 in Clyde Sea Loch sediments. Est. Coast Shelf Sci., 1982, 15, 515-536.

10. McKenzie, A.B., Swan, D.S., McKinley, I.G., Baxter, M.S. and Jack, W., The determination of Cs-134, Cs-137, Pb-210, Ra-226 concentrations in nearshore marine sediments and sea water. J. Radioanalytical Chemistry, 1979, 48, 29-47.

11. Bradley, P.E., Economides, B.E., Baxter, M.S. and Ellet, D.J., Sellafield radiocaesium as a tracer of water movement in the Scottish coastal zone. In Radionuclides: a Tool for Oceanography, 1987 ed. J.C. Guary, P. Guegueniat, R.J. Pentreath. Elsevier.

12. Jacquez, J.A., Compartmental analysis in Biology and Medicine. 1972 Elsevier, Amsterdam.

13. Brown, R.F., Compartmental analysis: State of the Art. <u>IEEE Trans. Biomedical Eng.</u>, BME, 1980, **27**, 1-11.

14. Zievler, K., A critique of compartmental analysis. <u>Ann. Rev. Biophys. Bioeng.</u>, 1981, **10**, 531-62.

15. Bowden, K.F., Horizontal Mixing in the Sea due to a Shearing current. <u>J. Fluid Mech.</u>, 1965, Vol.21(2), 83-95.

16. Heinze, C., Schlosser, P and Koltermann, K.P., Deep water renewal in the European Seas as derived from a multi-tracer approach. CM 1986/C17, 1986 Hydrogr Committee 1-8.

17. Howarth, J., Kirby, C.R., Studies of Environmental Radioactivity in Cumbria, 11 Modelling the dispersion of radionuclides in the Irish Sea, 1988. DOE/RW/88.058.

18. Hallstadius, L., Garcia Montano, E., Nilsson, U., An improved and validated dispersion model for the North Sea and Adjacent waters. <u>J. Environ. Radioactivity</u>, 1987, **5**, 261-274.

19. Jefferies, D.F., Preston, A., Steele, A.K., Further studies on the distribution of ^{137}Cs in British coastal waters - Irish Sea. <u>Deep Sea Res.</u>, 1982, **29(6A)**, 713-738.

20. Hallstadius, L., A computer program for the determination of transfer coefficients in a compartment model. Radiation Physics Dept., U. of Lund, 1985a. LUNDFD6/INFRA 3059

21. Hallstadius, L., Compartment analysis. Radiation Physics Dept., U. of Lund, 1985b. LUNDFD6/INFRA-3060.

22. Hallstadius, L., Compartment modelling in nuclear medicine - a new program for the determination of transfer coefficients. <u>Nuclear Medicine Comms.</u>, 1986. **7**, 405-414.

23. Craig, R.E., Hydrography of Scottish Coastal Waters. <u>Marine Research 1958</u>, 1959, (2), pp.30.

24. Bowden, K.F., Physical and dynamical oceanography of the Irish Sea. In <u>The North-West European Shelf Sea: the sea bed and the sea in motion</u>. 1980, ed. F.T. Banner, M.B. Collins, K.S. Mansie, Vol. **2**, 391-413, Elsevier Oceanography Series 24B.

25. Prandle, D., A modelling study of the mixing of ^{137}Cs in the Seas of the European continental shelf. <u>Phil. Trans. R. Soc. Lond</u>, 1984, **A310**, pp. 407-436.

26. Ellett, D.J., Edwards, A., Oceanography and inshore hydrography of the Inner Hebrides. <u>Proc. Roy. Soc. Edinburgh</u>, 1983, **83B**, 143-160.

27. Ellett, D.J., Meldrum, D.T. and Edelsten, D.J., The Scottish coastal current in summer 1981. <u>Ann. Biol.</u>, 1984, Copenhagen, **38**, 39-42.

28. Economides, B.E., Baxter, M.S. and Ellett, D.J., Observations of radiocaesium and the coastal current west of Scotland during 1983-85. ICES-CM 1985, 1985, C:24, 10 pp.

29. Begg, F.H., Baxter, M.S., Cook, G.T., Scott, E.M. and McCartney, M. Anthropogenic ^{14}C as a tracer in western U.K. coastal waters. These proceedings.

MATHEMATICAL MODEL OF ^{125}Sb TRANSPORT AND DISPERSION IN THE CHANNEL

Jean Claude SALOMON*, Pierre GUEGUENIAT**, Marguerite BRETON*
* IFREMER, Centre de Brest, B.P. 70, 29280 PLOUZANÉ - F
** CEA/IPSN, B.P. 270, 50107 CHERBOURG F.

ABSTRACT

A mathematical model of the English Channel is used to describe the fate of radionuclides discharged by La Hague nuclear reprocessing plant. The modelling technique is described and the results discussed. A permanent and average wind-tide situation is used to determine mean trajectories and time scales of retention and evacuation to the North Sea. Comparison with *in situ* measurements are satisfactory.

Two additional simulations in the case of no wind and in the case of an easterly one are then used to investigate the variability linked to atmospheric conditions.

INTRODUCTION

Channel waters are known to originate in the Atlantic ocean and slowly proceed towards the North Sea. The time scale concerned is about a year, which means that some fast veins may travel along the Channel in a few months, while others away from the main movement, could remain there for a few years.

Until recently very little reliable information was available on this point, although the question is of great interest, as these waters carry innumerable physical, chemical and biological components. The inherent complexity of coastal dynamics makes this information very difficult to obtain.

Mathematical modelling and artificial radionuclide detection seem to be the best theoretical and experimental tools for progress. The purpose here, is to model the transfer of radionuclides released from the La Hague reprocessing plant throughout the Channel and validate computations by comparison with *in situ* measurements of ^{125}Sb activity. Our knowledge of transfer processes of these radionuclides will thus be increased, as for any other dissolved matter present in the Channel.

A) METHODOLOGICAL ASPECTS

1) General comments

In modelling currents and dispersion on the time scale of a year, present computational limitations still make it a rule to operate some low pass filtering at subtidal frequency. In the Channel, this frequency band is very energetic and great attention must be paid to non-linear effects which transfer energy to the part of the spectrum which is supposed to be resolved. In other words, determination of long term movement by some straightforward averaging of Navier-Stokes equations or of their result (Prandle, 1984, Becker, 1987) may lead to misleading results in the coastal zone. The tidal stress concept introduced by Ronday (1976) was an interesting attempt to retain most of the non-linearity effects in a residual calculation but applications have proved to be difficult. The barycentric method used here is another way to improve the procedure. It aims at combining eulerian residuals, stokes drift and lagrangian drift in a simple residual velocity field and then allows extension to long term simulations in an economic way.

The Channel being generally well mixed, computations will be done according to the 2D simplification. This approximation is no more valid in the western Channel during summer, which is of no importance as radionuclides do not reach this zone. It also fails in a thin band along the french coast, east of the Seine estuary, where fresh water discharge may accelerate the eastward movement and decrease transverse mixing. This will be a local limitation of the model.

2) The barycentric method

The method has been already described in previous papers (Salomon *et al.*, 1986 ; Orbi and Salomon, 1988). It starts with a classical computation of instantaneous velocities, then introduces water particle trajectories and makes an equivalence between the two initial parameters (point and instant of release), and the position of the center of gravity of each trajectory. The time parameter is thus eliminated and a unique velocity field which contains most of the above mentioned components is obtained. In doing this, the initial fixed coordinate system is exchanged for a moving one linked to the water mass.

Long term advection/dispersion calculations can then be done through a small number of those velocity fields taking care of two pecularities :

- For the reason mentioned earlier a fixed source or sink point is substituted by several others distributed along an apparent trajectory.

- Due to the numerical method of its elaboration, the residual velocity field is not fully conservative. Different possibilities to overcome the problem exist. One can use the advection-diffusion equation in its conventional non-conservative form or use the particle tracking and a Monte Carlo method. One can also slightly modify this "raw" velocity field to eliminate what is considered as a noise and obtain a fully conservative field nearly identical to the preceeding one.

This last possibility has been used here by a least square error minimization technique. The advection-dispersion equation is then solved in its conservative form :

$$\frac{\partial (HC)}{\partial t} + \frac{\partial (HUC)}{\partial x} + \frac{\partial (HVC)}{\partial y} - \frac{\partial \left(KH\frac{\partial C}{\partial x}\right)}{\partial x} - \frac{\partial \left(KH\frac{\partial C}{\partial y}\right)}{\partial y} + \alpha HC = 0 \quad (1)$$

U and V are velocity components, H is the water depth, K the dispersion coefficient and α the nuclear decay. The half-life of Antimony radioactivity is 2,7 years.

The dispersion coefficient, which is a very poorly known quantity in time averaged equations where its value is in the range of $10^3 m^2/s$, is no longer a problem here. As the main part of the tidal variability is present in the lagrangian velocity field, this coefficient has nearly the same significance, and the same value, as in real time computations. It is about 10^2 times smaller than the previous value. This makes the description of fine spatial structures possible.

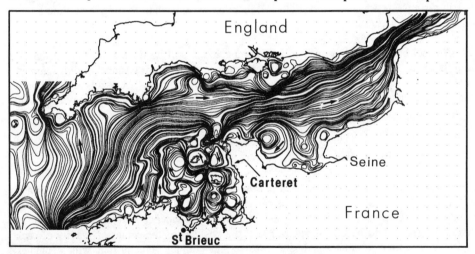

Figure 1. Long term trajectories (average tide – W.S.W. wind).

Application to the Channel

Velocities

The present 2D model is limited to 48°18'N and 51°20'N in latitude and to 6°28'W and 3° E in longitude. Its mesh size which has to be much smaller than tidal excursions is fixed to 1 852 m. Boundary conditions are provided by another general 2D model of the North European continental shelf which operates in spherical coordinates (Salomon and Breton, 1990). Hydrodynamic equations are solved by the well known A.D.I. method in finite differences.

Currentological results being already published (Salomon and Breton, 1991) a unique illustration is given in figure 1 of long term trajectories for an average tide and a moderate west–south–west wind (stress of 0.074 N/m^2).

Concentrations

Equation (1) is solved on the same grid defined above, combining a Kurihara expression (in Roache, 1982) of the temporal discretization and a Quick method (in Leonard, 1981) for spatial derivatives. Numerical stability is obtained for time steps greater than two hours, depending on hydrodynamic conditions.

The dispersion coefficients, spatially variable, are determined according to the following considerations :

- The subgrid dispersion, determined by the mesh size indicates values about 5 m^2/s (Lam et al., 1984).

- Empirical formulas, such as the Elder one indicate a relation between K, the depth and the velocity of the form :

$$K = \beta \ U.H$$

Coefficient ß is known to be about 0,25 in uniform flows but may be 2 or 3 times higher in transient ones.

Consequently, the following formula was adopted and at this stage, did not require any further adjustment :

$$K (x, y) = \text{maximum } (0{,}7 \ \overline{UH} \ (x, y), 5)$$

$\overline{UH}$ (x, y) is the local average of the product UH during the tidal cycle.

B) RESULTS

1) Permanent discharge

The first simulation refers to a theoretical scenario : starting from a situation of non-contamination in the Channel, a sudden and permanent discharge of 4 GBq/s of ^{125}Sb is supposed to take place indefinitely. The tide is an average one and the wind blows from the South West, inducing a stress of 0,074 N/m^2. Flow in the Dover strait is 119 000 m^3/s.

Results are presented in fig. (2) for different stages. From the beginning the plume forks into two distinct branches :

- One goes south along the west coast of Normandy, as far as Carteret (40 km), then suddenly turns west creating a sharp front.

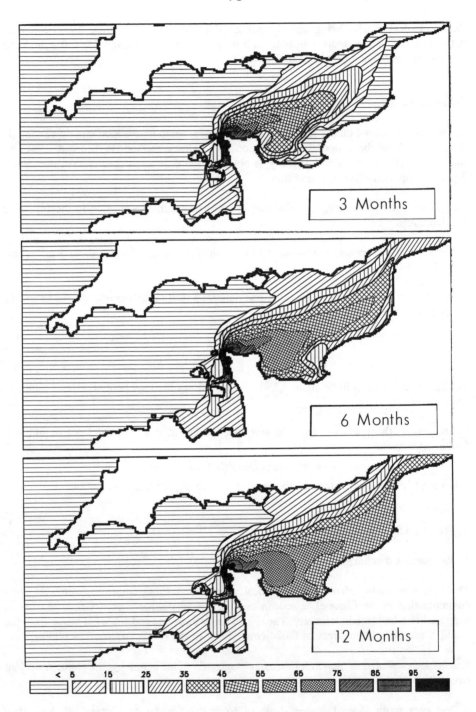

Figure 2. Simulation of a permanent discharge at La Hague (in m Bq/l)

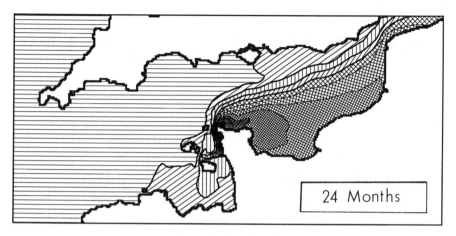

Figure 2. (continued)

– The other, which collects most of the effluent goes east in the direction of the Dover strait. On the way, it slowly diffuses laterally and first marks the western bay of Seine. Those radionuclides reach the strait in significant amounts near its center about three months later, while the Seine estuary and the coastal zone east of it remain unmarked.

From that moment concentrations near the center of the strait stay unchanged while they increase gradually towards the coast. Due to the mirror effect, 8 or 9 months after beginning, they are higher along the coast than near the center of the strait, but it takes about 6 months more to show the same pattern close to the bay of Somme.

A completely steady state is attained nearly 2 years after the start of the discharge at La Hague.

The abovementioned west vein forked again to the north west of Jersey. Part of it went north towards Alderney, the other part going south. Waters in the bays of St Michel and St Brieuc were marked near the fourth month, but radionuclides stopped a few kilometers west of Brehat. In this region the equilibrium was reached after a year.

The resulting situation may be compared to experimental indications in figure 3. Comparisons seem very good, taking account of the fact that conditions of simulations were purely theoretical.

Another confirmation could be found in Germain *et al.* (1988) who show the contamination of algae along the coast of France. Longitudinal distribution of radioactive tracers and concentrations computed along the coast are very similar.

If the injection stops (not illustrated here) concentrations decrease quickly around La Hague (about 30 km in one month) and then decontamination proceeds to the region of Dover in the central stream. Six months later the central part of the Channel is free of radionuclides, whereas the gulf of St Malo, the bay of Seine and the french littoral, east of the Seine estuary, still remain marked for other six months.

2) Meteorological effects

Wind forcing being in high proportion responsible for long term movements in the Channel simulations were done to test its role on radionuclide distribution.

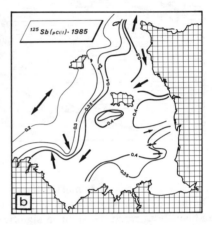
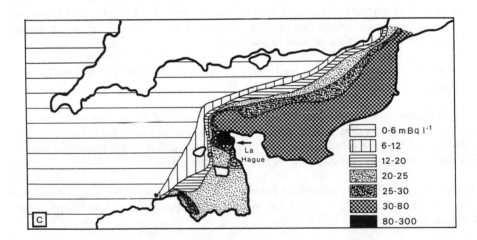

Figure 3.   Measured antimony concentrations
a) June 1984 b) June 1985 (from Salomon *et al.*, 1988)
c) June 1986 (from Guegueniat *et al.*, 1988)

Starting from the steady state obtained above, the model was operated for two months without any wind stress. The resulting situation is supposed to be representative of what may happen in summer, when wind velocities decrease and easterlies can cancel prevailing south-westerlies. The total flux in the strait of Dover is then 27 000 m^3/s.

Figure 4 shows the result. Higher concentrations are appearent near the Cap de la Hague, as well as a reduction of the east-west dissymetry.

The patch limit has moved north and west and the bay of Lannion is still uncontaminated.

In the Channel, isolines are less linear and gradients are smoother but the general aspects of the picture remain the same and similar to figure 3.

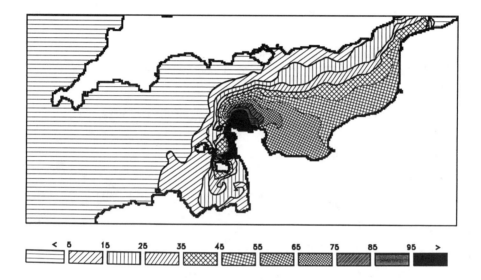

Figure 4. Antimony distribution after 2 months without any wind (in m Bq/l)

To go further we tried to find out the effect of strong easterly winds. Starting again from the equilibrium state, a constant wind stress of 0,074 N/m^2, from the east was imposed during a month (fig. 5). The flux in the Dover strait is then reduced to 10 000 m^3/s.

In that case discharged elements are not transported predominantly to the east. The flux oriented south increases greatly, marking Carteret's gyre and passing Jersey. Antimony is mainly accumulated near the headland. In the eastern Channel, the situation evolves slowly, the English coast being contaminated. The Isle of Wight seems to be quite a difficult limit to pass in usual meteorological conditions.

CONCLUSION

The present simulation of transport and mixing of radioactive tracers verifies the ability of the lagrangian barycentric method to elaborate long term velocity fields and make correct dispersion computations at time scales ranging from weeks to years. This result will probably be of some interest in the future, as the method applies to any other dissolved constituent (heat, nutrient, pollutant) or living passive cell.

It also confirms the great interest of artificial radionuclides to be used with long term mathematical models.

From the point of view of radionuclide dissemination numerous previous observations are going to be reconsidered in the light of model indications (Guegueniat et al., 1991). Present results already describe the main mechanisms responsible for their advection or retention and mixing :

- The central way from La Hague to the North Sea.
- The retention action of all gyres in the Normandy gulf or the bay of Seine.
- The very slow lateral diffusion towards the French coast.
- The sensitivity of the English coast to wind forcing.
- The usual limit of dissemination near the Isle of Wight or the bay of Lannion, etc.

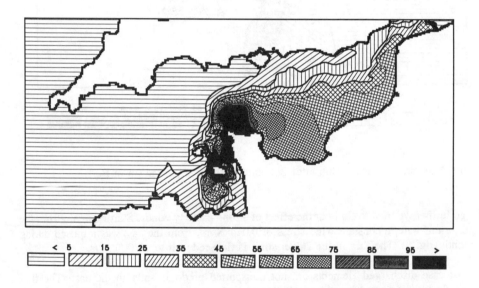

Figure 5. Antimony distribution after one month of easterly winds (in m Bq/l).

Work is still in progress. Next questions addressed to the model will be the effect of a varying injection function and the inclusion of heterogenous wind fields.

REFERENCES

1) Becker, G.A., Dispersion of substances in the North Sea. Deutsches Hydrographisches Institut, 1987, 61.

2) Germain, P., Baron, Y., Masson, M, Calmet, D., Répartition de deux traceurs radioactifs chez deux espèces indicatrices (*Fucus serratus, Mytilus edulis*) le long du littoral français de la Manche. *In* Radionuclides : a tool for oceanography. Elsevier Applied Science, 1988, pp. 312-320.

3) Guegueniat, P., Gandon, R., Baron, Y., Salomon, J.C., Pentreath, J., Brylinski, J.M., Cabioch, L., Utilisation de radionucléides artificiels (^{125}Sb - ^{137}Cs - ^{134}Cs) pour l'observation des déplacements des masses d'eau dans la Manche. *In* Radionuclides : a tool for oceanography. Elsevier Applied Science, 1988, pp. 260-270.

4) Guegueniat, P., Salomon, J.C., Wartel, M., Cabioch, L., Fraizier, A., Use of artificial radionuclides for the study of dissolved matter transfer mechanisms between the channel and the North Sea. Estuarine, coastal and shelf science (submitted).

5) Lam, D.C.L., Murthy, C.R., Simpson, R.B., Effluent transport and diffusion models for the coastal zone. Lecture notes on coastal and estuarine studies. Springer Verlag, 1984.

6) Leonard, B.P., A survey of finite differences with upwinding for numerical modelling of the incompressible correctives diffusion equation. *In* Computational techniques in transient and turbulent flows, Pineridge Press, 1981, **2**, pp. 1-36.

7) Orbi, A. and Salomon, J.C., Dynamique de marée dans le golfe Normand-Breton. Oceanologica Acta, 1988, **11**, 1, pp. 55-64.

8) Prandle, D., A modelling study of the mixing of ^{137}Cs in the seas of the european continental shelf. Phil. Trans. Roy. Soc. London. 1984, **A310**, pp. 407-436.

9) Roache, P., Computational fluid dynamics. Hermosa Publishers, Albuquerque, 1982, 446 p.

10) Ronday, F., Modèles hydrodynamiques, modélisation des systèmes marins, projet mer, rapport final. Services du 1er Ministre, Bruxelles, Vol. 3, 1976, 270 p.

11) Salomon, J.C., Guegueniat, P., Orbi, A., Baron, Y., A lagrangian model for long term tidally induced transport and mixing. Verification by artificial radionuclide concentrations. In Radionuclides : a tool for oceanography. Elsevier Applied science, 1988, pp. 384-394.

12) Salomon, J.C., Breton, M., Modèle général du plateau continental Nord européen. Rapport au programme CEE-MAST 89 0093, Fluxmanche, 1990, 20 p.

13) Salomon, J.C., Breton, M., Courants résiduels de marée dans la Manche. Oceanologica Acta, 1991 (in Press).

THE DISPERSION OF ^{137}Cs FROM SELLAFIELD AND CHERNOBYL IN THE N.W. EUROPEAN SHELF SEAS

D. Prandle & J. Beechey*
Proudman Oceanographic Laboratory, Bidston Observatory, Birkenhead
*Sandwich Student, University of Salford

ABSTRACT

In an earlier modelling study, the dispersion in the N.W. European Shelf Seas of ^{137}Cs discharged from Sellafield was simulated for the period 1964 to 1980. The present study extends this simulation up to 1988 and includes atmospheric deposition following the Chernobyl accident in April 1986. These model simulations are compared with annual observational surveys carried out in the North Sea from 1978 to 1988.

Atmospheric models have estimated a deposition in the North Sea of 1.4 PBq of ^{137}Cs from Chernobyl (and 3.7 PBq over the N.W. European Shelf Seas). An independent estimate of 0.9 PBq is made here using the Shelf Sea model to interpolate between marine surveys carried out in successive Septembers of 1985 and 1986. Immediately prior to the Chernobyl accident, the model indicated 2.3 PBq of ^{137}Cs in the North Sea and 8.2 PBq in the Shelf Seas. Thus, the Chernobyl deposition was of the order of half of the Sellafield material in both areas.

These Chernobyl fractions will subsequently decrease more rapidly due to their concentration towards the north-east of these sea areas. Meanwhile the concentrations due to Sellafield which peaked around 1980 in the North Sea will decline steadily at a rate determined by the flushing time of the combined Shelf Seas i.e. approximately 4.5 years. Estimations of the flushing time of the North Sea from both the model and observations indicate a value of the order of 1.5 years. These long flushing periods (relative to the 'annual' renewal period for many biological phenomena) may be important in maintaining the stability of various biological and chemical parameters.

INTRODUCTION

Prandle [1] used a numerical model to simulate the spread of ^{137}Cs, discharged from

Sellafield from 1964 to 1980. Results indicated a transport path northwards out of the Irish Sea thence clockwise around the Scottish coast followed by an anticlockwise circulation in the North Sea. The average timescale was 2 years for entry into the North Sea and up to 6 years for exiting along the Norwegian coast.

With enhanced interest in the water quality of the North Sea, estimation of its flushing time, F_T, is especially important. In the earlier modelling study, a value of F_T=530 days was calculated for the North Sea (51°-61°N, 4°W-9°E). Subsequent studies [2] have demonstrated how the mean residual flow field can vary significantly when hour-by-hour wind fields resolving storm events are incorporated in these models. Likewise studies by Jefferies et al [3] in the Irish Sea and Dickson et al [4] in the North Sea have suggested important inter-annual variability in the mean circulations.

The 35 km grid resolution of the present model is too coarse to resolve detailed dispersion processes in the Irish Sea (see Jefferies and Steele [5]), hence comparisons of model results against observations focus on the North Sea.

Atmospheric depositions of ^{137}Cs from the Chernobyl nuclear accident in April 1986 have been estimated by Ap Simon and Wilson [6]. Their values for inputs into the North Sea area are compared with new estimates made from the successive September marine surveys of 1985 and 1986. The subsequent dispersion of this Chernobyl material is simulated alongside that of the Sellafield releases. The absorption of ^{137}Cs on to sedimentary particles([7] and [8]) is not simulated.

SIMULATION OF SELLAFIELD ^{137}Cs RELEASES 1969-1988

Inputs of ^{137}Cs from the French re-processing site at Cap de la Hague and from Sellafield prior to 1969 are neglected in this simulation. Kautsky's [7] data are for ^{137}Cs + ^{134}Cs, Mitchell and Steele [9] show that by 1985 their ratio was 21:1 only 120km from Sellafield. River discharges are also neglected in this simulation, these reduce ^{137}Cs concentrations in the near-shore region especially in the German Bight. This study uses the same numerical model as described by Prandle [1] i.e. a two-dimensional depth-averaged formulation with a 35km square grid and an explicit finite difference scheme. The mixing model uses dispersion coefficients K_x=1000UR s^{-1} and K_y=1000VR s^{-1} where U and V are orthogonal M_2 tidal velocity amplitudes and $R=(U^2+V^2)^{1/2}$.

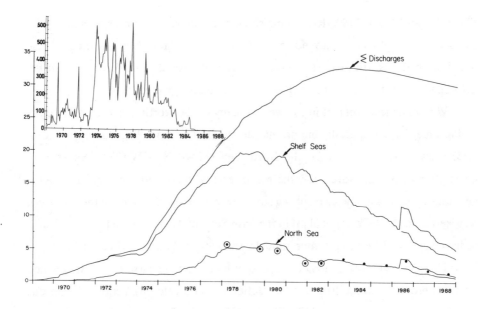

Figure 1 (insert) Monthly discharges of ^{137}Cs from Sellafield in T Bq
Figure 2 Net cumulative discharges (with radioactive decay)
Computed total in the North Sea and Shelf Seas with and without Chernobyl
Observed totals in the North Sea   Hunt [14].   Kautsky [7].

Advection is prescribed as the sum of tidal residual transports together with monthly-mean wind driven components. The latter were calculated by factoring simulated responses for unit stresses applied, respectively, for winds from the west and north.

In the earlier study, monthly-mean wind stresses were obtained from Thompson et al. [10] based on atmospheric pressure distributions. Here, winds at 6-hourly intervals, were obtained from the atmospheric model of the Norwegian Meteorological Office. Sea-surface stresses were obtained by assuming the relationship given by (Proctor and Wolf [11]). In correlating east and north components of this time series with Thompson's values, correlation values exceeding 0.95 are obtained. However to

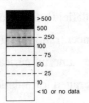

Figure 3:   Observed (left) and computed (right) levels of ^{137}Cs in
(next page)   the North Sea 1978-1988. Contours in Bq m^{-3}

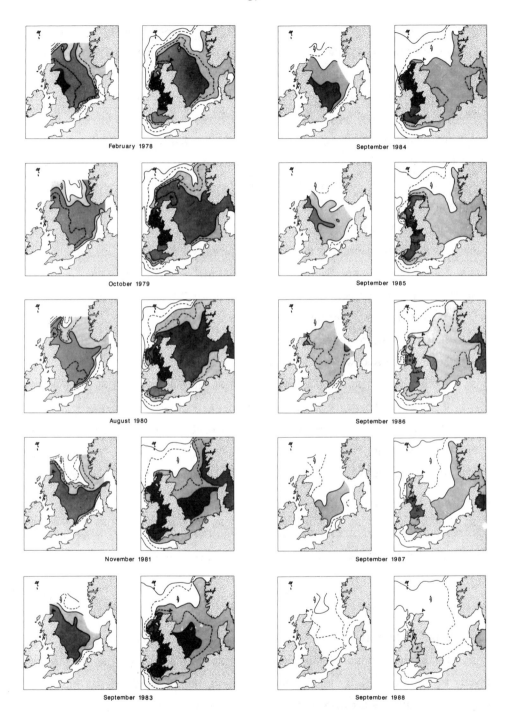

February 1978 September 1984

October 1979 September 1985

August 1980 September 1986

November 1981 September 1987

September 1983 September 1988

equate the variances, the Norwegian data had to be increased by a factor of 1.63.

Figure 1 shows the time-history of ^{137}Cs releases from Sellafield. Figure 2 shows: (i) cumulative time-releases; and the calculated net mass of ^{137}Cs in (ii) the North Sea and (iii) adjacent Shelf Seas (i.e. the total model area (47°-64°N, 15°W-12°E). Figure 3 shows observed and computed concentrations between 1978 and 1988.

To obtain the close agreement shown, the residual advective velocities were increased by multiplying the wind stress time-series by a 1.5 - equivalent to increasing wind speeds by 22%. This further adjustment of the wind time series might be attributed to the use of time-averaged responses with spatially uniform wind stresses.

Figure 2 shows that the major increase in Sellafield discharges (figure 1) which occurred in 1974 are evident in the North Sea by early 1976. While these discharges subsequently declined steadily until 1986 the North Sea concentrations remained reasonably constant between 1978 to 1980. The net wind stress for the 1980-81 season was the largest - amounting to 1.4 times the 20-year mean and was also more northerly. This resulted in a sharp reduction in the North Sea ^{137}Cs by about one-third, after which the decrease is slower, by only one-third between 1981 and 1986.

Precise comparisons are complicated since marine surveys are sensitive to localised conditions prevailing at the time of sampling and vertical averaging may be inaccurate. In general, the model exhibits less detailed variability with sharp features more dispersed. More ^{137}Cs moves southwards out of the Irish Sea in the model than in nature and the neglect of river flows along the Continental coast is also evident.

Characteristics of Model and Observed Dispersion

From Fischer et al. [12], the concentration C for a continuous input M, at a (large) distance X along a coastline with adjacent open ocean of depth D is

$$C = \frac{M}{D(\pi KUX)^{\frac{1}{2}}} \tag{1}$$

where U is a uniform residual velocity and K an isotropic dispersion coefficient.

This simple equation demonstrates that concentration C is proportional to the inverse square root of K, U and X. Thus, in general, both the advective velocities and dispersion coefficients prescribed in the dispersion model must be to the same order of accuracy as the resulting concentrations.

Interestingly, substitution of the following parameters for Sellafield dispersion, $M=1.5 \cdot 10^8 Bq/s^{-1}$ (equivalent to a monthly discharge of 400T Bq), $D=100m$, $K=500m^2 s^{-1}$, $U=0.02m\ s^{-1}$, $X=1500km$, gives $C \approx 200\ Bq\ m^{-3}$, in reasonable agreement with the concentrations along the Scottish east coast around 1980.

From figure 2 the total of 38.9 PBq of ^{137}Cs released from Sellafield between 1969 and April 1986 decreases to 31.7 PBq by radioactive decay (30 year half life). Some 0.26 of this latter amount remains within the model boundaries (8.2 PBq) and 0.07 within the North Sea (2.3 PBq).

A simple functional relationship between the discharge time series S(t) and computed total North Sea material NS(t) may be postulated of the form:

$$NS(t) = \sum_{i=1,N} a_i S(t-i) \qquad (2)$$

where $i=1,N$ represents N successive previous monthly discharges.

After some experimentation, the parameters a_i were simplified to annual parameters A_1, A_2 etc. with linear interpolation for intervening months. The response functions $A_1=0.04$, $A_2=0.46$, $A_3=0.56$, $A_4=0.34$ (with $A_o=A_5=0$) were calculated by least squares fitting. The corresponding time series for NS(t) accounts for over 90% of the variance of the simulated time-series. Thus, this simple model indicates that the total ^{137}Cs in the North Sea represents an integration of discharges over the preceding 5 years. Noting that the sum of a_i, i.e. $12\ (A_1+A_2+A_3+A_4)$ is 16.9, under steady state conditions the North Sea contains the equivalent of 16.9 monthly discharges i.e. $F_T \approx 500d$ (neglecting losses between Sellafield and the North Sea).

The peak concentrations in the North Sea in 1980 (figure 2) reflect both the transit time and the flushing time of the adjacent Shelf Seas. For the entire Shelf Seas the peak computed mass (M(t)max) in 1980 (≈ 20 PBq) can be related to an effective continuous steady discharge I from 1973 of 4.4 PBq y^{-1}. The theoretical time-history for which is [13] $M(t) = I.T_F(1-\exp(-t/T_F))$ indicating $T_F \approx 4.5$ years.

SIMULATION OF ^{137}Cs DISTRIBUTIONS FOLLOWING CHERNOBYL

The accident at the Chernobyl nuclear power plant occurred on 25th April 1986. Most subsequent radioactive deposition over the N.W. European Shelf Seas resulted from the movement of rain clouds over the following 10 days. The deposition of ^{137}Cs was specified from the data computed by Ap Simon and Wilson [6] (in a grid (approx) 1° latitude x 1.5° longitude), from the MESOS atmospheric model. Figure 4 shows these values in Bq m^{-2}, peak concentrations occurred over Scandinavia, decreasing towards the south west. Figure 4 shows these same values expressed as the increase in concentration in sea water in Bq m^{-3} (i.e. divided by water depth). The maximum increases are in the shallow waters off Denmark amounting to 60 Bq m^{-3} [9].

From these data, the total atmospheric input into the Shelf Seas from Chernobyl (to 62°N and 10°W) was 3.7 PBq of which 1.4 PBq enters the North Sea. Concurrent model calculations of Sellafield ^{137}Cs are 8.2 and 2.3 PBq.

The observed total of ^{137}Cs in the North Sea in September 1985 was 2.5 PBq. By April '86 the model reduction factor modifies this to 2.3 PBq. In the September

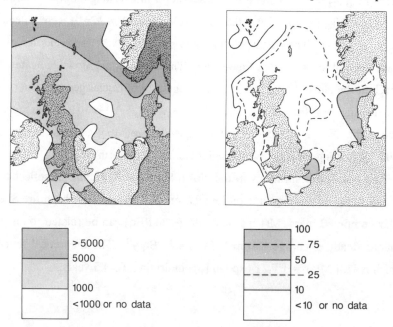

Figure 4: Chernobyl fallout of ^{137}Cs
(left) Activity per unit area Bq m^{-2}, (right) Equivalent concentration in sea water. Bq m^{-3}

1986 survey following Chernobyl the observed mass was 3.2 PBq, the model indicates only a minor loss between April and this survey, hence the marine surveys indicate a Chernobyl North Sea deposition of 0.9 PBq i.e. 0.64 of the atmospheric model value.

The observed and modelled distributions for September 1986 (figure 3) show reasonable agreement with peak contours of 100Bq m^{-3} off the Scottish coast and intruding out of the Kattegat. The agreement in 1987 is, perhaps, the closest in this 11-year sequence with only the Continental coastal region differing. In 1988 both indicate concentrations throughout the central North Sea close to 25Bq m^{-3}.

The simulation of the spread of the Chernobyl deposition in isolation (not shown) emphasises the more rapid loss of this material from the North Sea. By September 1986, only 4 months after the accident the peak depositions equivalent to 60Bq m^{-3} off the Danish coast are reduced to 40Bq m^{-3}, to 20Bq m^{-3} by September 1987 and to 7Bq m^{-3} by September 1988. For this last date, concentrations only exceed 10Bq m^{-3} adjacent to the Norwegian Coast and in the Kattegat and Skagerrak.

The more rapid depletion of the Chernobyl ^{137}Cs reflects its differing initial distribution, the large deposition off Norway being readily flushed. The net observed North Sea mass of September 1986 declines to 0.34 by September 1988. The corresponding remaining fraction of Chernobyl releases is 0.21.

With both Sellafield and Chernobyl material subsequently widely distributed, future decreases will depend on the flushing time of the Shelf Seas - estimated as 4.5 years. From the model simulation of Chernobyl input in isolation the ratio of Chernobyl:Sellafield ^{137}Cs in the North Sea in September '88 is 0.33, in reasonable agreement with the value of 40% obtained from the observed ratio of ^{134}Cs:^{137}Cs [14].

CONCLUSIONS

1. While the model calculates a North Sea flushing time of 530 days, North Sea concentrations of ^{137}Cs reflect an initial advective travel time from Sellafield of the order of 2 years and subsequent inputs over a further 3 years. Thus resulting concentrations reflect both losses en route and the dispersion pattern in the adjacent Shelf Seas. The flushing time of these adjacent Shelf Seas was estimated as 4.5 years. While Sellafield discharges declined steadily from an initial peak in 1974, maximum totals in both the North Sea and Shelf Seas occurred around 1980. By 1988 these

maximum totals had reduced by a factor of approximately 5.

This lengthy flushing time of the North Sea and surrounding seas may contribute to the long-term stability of many biological and chemical parameters.

2. The Chernobyl nuclear accident in April 1986 resulted in significant atmospheric deposition of ^{137}Cs directly into the Shelf Seas. Data taken from atmospheric models indicated net deposition into the Shelf Seas of 3.7 PBq of which 1.4 PBq was within the North Sea. An equivalent estimate of North Sea deposition based on marine surveys made in August/September of 1985 and 1986 indicated 0.9 PBq.

These depositions may be compared with estimates of concurrent Sellafield derived ^{137}Cs of: (a) in the Shelf Seas 8.2 PBq and (b) in the North Sea 2.3 PBq.

The Chernobyl material is flushed more readily than that from Sellafield as a consequence of its spatial distribution - with higher deposition close to Norway. However future discharges from the Baltic are likely to be of significance.

3. The estimated net Shelf Sea deposition of ^{137}Cs from Chernobyl of 3.7 PBq compares with 70 PBq estimated for the Northern Hemisphere [15]. Likewise the net Sellafield discharges ('69 - '88) of 39 PBq compares with estimated total global deposition by 1985 of 580 PBq [16]. Of the net Sellafield discharge of 39 PBq, 7.3 PBq has been lost by radio-active decay and 4.9 PBq remain within the Shelf Seas, indicating a net loss to the adjacent oceans [17] and bed sediments of approximately 27 PBq.

Acknowledgement. This study originated from an exchange visit to POL by Jose M. Abril of the Department of Physics at Seville University, Spain organised by J. Howarth of AERE, Harwell. The Norwegian Meteorological Institute provided the wind data.

REFERENCES

1. Prandle, D., A modelling study of the mixing of ^{137}Cs in the seas of the European Continental Shelf. Phil.Trans.R.Soc., A, 1984, **310**, 407-436.

2. Backhaus, J.O., On the atmospherically induced variability of the circulation of the northwest European Shelf Sea and related phenomena - a model experiment. In Modeling marine systems, Volume 1, ed. A.M. Davies, CRC Press, Florida, 1990, pp.93-134.

3. Jefferies, D.F., Steele, A.K. and Preston, A., Further studies on the distribution of ^{137}Cs in British coastal waters - 1. Irish Sea. Deep Sea Res., 1982, **A29**, 713-738.

4. Dickson, R.R., Kelly, P.M., Colebrook, J.M., Wooster, W.S. and Cushing, D.H., North winds and production in the eastern North Atlantic. J. of Plankton Res., 1988, **10**, 151-169.4.

5. Jefferies, D.F. and Steele, A.K., Observed and predicted concentrations of ^{137}Cs in seawater of the Irish Sea 1970-1985. J. of Environ. Radioactivity, 1989, **10**, 173-189.

6. Ap Simon, H.M. and Wilson, J.J.N., Modelling atmospheric dispersal of the Chernobyl release across Europe. Boundary Layer Met., 1987, **41**, 123-133.

7. Kautsky, H., Distribution and content of different artificial radio nuclides in the water of the North Sea during the years 1977 to 1981 (complemented with some results from 1982 to 1984). Dt. Hydrogr. Z., 1985, 38, 193-224.

8. Erlenkeuser, H. and Balzer, W., Rapid appearance of Chernobyl radiocesium in the deep Norwegian Sea sediments. Oceanol. Acta, 1988, **11**, 101-106.

9. Mitchell, N.T. and Steele, A.K., The massive impact of Caesium-134 and -137 from the Chernobyl reactor accident. J. of Environ. Radioactivity, 1988, **6**, 163-175.

10. Thompson, K.R., Marsden, R.F. and Wright, D.G., Estimation of low-frequency wind stress fluctuations over the open ocean. J. of Phys. Oceanogr., 1983, **13**, 1003-1011.

11. Proctor, R. and Wolf, J., An investigation of the storm surge of February 1, 1983 using numerical models. In Modeling marine systems, Volume 1, ed. A.M. Davies, CRC Press, Florida, 1990, pp.93-134.

12. Fischer, H.B., List, J.E., Koh, R.C.Y., Imberger, J. and Brooks, N.H., Mixing in inland and coastal waters. Academic Press, London, 1979, 483pp.

13. Prandle, D., Tides in estuaries and embayments. To appear in: Advances in Tidal Hydrodynamics, 1990, B. Parker (Ed.) John Wiley & Sons

14. Hunt, G.J., Radioactivity in surface and coastal waters of the British Isles: 1984 No. 13 46pp, 1985 No. 14 48pp, 1986 No. 18 62pp, 1987 No. 19 67pp, 1988 No. 21 69pp. Ministry of Agriculture Fisheries and Food, Directorate of Fisheries Research, Aquatic Environment Monitoring Reports.

15. Cambray, R.S., Cawse, P.A., Garland, J.A., Gibson, J.A.B., Johnson, P., Lewis, C.N.J., Newton, D., Salmon, L. and Wade, B.O., Observations on radioactivity from the Chernobyl accident. Nucl. Energy, 1987, **26**, 77-101.

16. Cambray, R.S., Playford, K., Lewis, G.N.J. and Carpenter, R.C., Radioactive fallout in air and rain, results to the end of 1987. Harwell Laboratory, Oxfordshire, UK. AERE R13226, 1989, 24pp.

17. Dahlgaard, H., Aarkrog, A., Hallstadius, L., Holm, E. and Rioseco, J., Radiocaesium transport from the Irish Sea via the North Sea and the Norwegian Coastal Current to East Greenland. Rapports et Proces-Verbaux des Reunions, ICES, 1986, **186**, 70-79.

GAS EXCHANGE AT THE AIR-SEA INTERFACE: A TECHNIQUE FOR RADON MEASUREMENTS IN SEAWATER

G. Queirazza*, M. Roveri*, R. Delfanti** and C. Papucci**

* ENEL (Italian Electricity Board), Thermal and Nuclear Research Centre, Via Monfalcone 15, 20132 Milan, Italy.

** ENEA (Italian Commission for Nuclear and Alternative Energy Sources), C.P. 316, La Spezia, Italy.

ABSTRACT

The rate of exchange of various gas species, such as O_2, CO_2 etc. across the air-water interface can be evaluated from the ^{222}Rn vertical profiles in the water column.
Radon profiles were measured in 4 stations in the NW Adriatic Sea, in September 1990, using solvent extraction and liquid scintillation counting techniques, directly on board the ship. The radiochemical procedure is described in detail. The lower limit of detection is approximately 0.4 mBq l^{-1}. The radon deficiency in the profiles gives estimates of the gas transfer rate across the air-sea interface ranging from 0.9 to 7.0 m d^{-1}.
The suitability of the radon deficiency method in shallow water, enclosed seas is briefly discussed.

INTRODUCTION

The greenhouse effect, related to the increasing average concentration of reactive gases such as CO_2 and CH_4 in the troposphere, causes the earth's surface as well as the entire troposphere, to heat up. The influence of the various atmospheric gases on radiative flux is also connected to the interactions of the different subsystems which constitute the climatic system of the earth. Among these, the air-sea subsystem deserves particular attention due to the exchange of various gases occuring at the interface.
The flux of gases across the air-sea interface can be calculated by the product of the concentration gradient which drives the exchange by the rate constant, termed the "transfer velocity" or "piston velocity" (1, 2, 3, 4, 5).

The variations in transfer velocity are governed both by wind velocity at the sea surface and the state of the sea. Information on the concentration gradients of the reactive gases and their transfer velocity is poor for enclosed seas; the experimental measurement of these parameters can improve our understanding of how the sea affects the global budget of a gas and how these parameters vary in space and time. Furthermore, the continuation of studies on the air-sea interface may permit prompt identification of possible anomalies in sea behaviour.

The present paper refers to the application of the ^{222}Rn deficiency method (5, 6, 7), to assess CO_2 exchange budgets in the Adriatic Sea, in a period (Summer 1990) marked by thermal stratification of the water. This approach assumes that the ^{222}Rn flux at the sea-air interface can be represented by a one-dimensional model in which the horizontal flux and that from the sediment are negligible (1).

In the Mediterranean Sea, characterized by large shallow water areas, ^{222}Rn concentration in the water column may be significantly influenced by relevant benthic fluxes (8). It is thus considered worthwhile to assess the applicability of ^{222}Rn deficiency measurements before planning spatial and temporal monitoring surveys in different areas of the Adriatic Sea.

In order to detect low levels of ^{222}Rn in seawater, a field, radiometric procedure was investigated as an alternative to the classic ^{222}Rn techniques (11, 12). The preliminary measurements indicate that a procedure based on a toluene liquid elution of the ^{222}Rn trapped on charcoal followed by liquid scintillation counting (LSC) is quite reliable. This paper also describes the extraction, elution and LSC procedures for determining ^{222}Rn and ^{226}Ra in seawater.

EXPERIMENTAL PROCEDURE

Study area, chemical and physical characterization and water sampling

The general surface circulation of the Adriatic Sea can be described as an elongated cyclonic gyre, with a northwestward flow along the Yugoslav coast and a return current flowing southeastward along the Italian coast (9). Due to the strong influence of the freshwater inflow from the rivers (mostly coming from the Po) the N Adriatic water masses are, in summer, warmer and less saline than the inflowing water masses from the SE Adriatic. During the summer, a well-stratified regime is present throughout the basin (10).

The samples were collected in September 1990 during a cruise aboard the R/V "BANNOCK" of the Italian National Research Council. The location of the sampling stations is shown in Fig.1: three (St. 52, 66 and 53) are in the northern and one (St. 74) in the central Adriatic Sea.

Vertical profiles of salinity, temperature, dissolved oxygen and pH were determined at each station, to identify the different water masses. A submersible pump was used to transfer seawater directly into the bottom of a 25 liter polyethylene extraction barrel. An overflow was used to obtain a

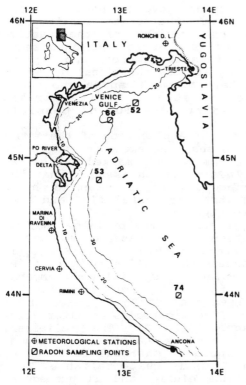

Figure 1. Water sampling points and meteorological stations.

representative water sample. Two 25 liter barrels were used to store the water samples from which ^{222}Rn was extracted to determine the total activities of ^{222}Rn and ^{226}Ra, respectively. One barrel was made of high density polyethylene (HDPE-0.2 cm wall thickness) and a second one of polyvinylchloride (PVC - 1 cm wall thickness). For ^{226}Ra analyses water samples were first degassed with helium and then kept sealed for 20 days, until equilibrium between ^{222}Rn and ^{226}Ra was reached.

Wind speed-measurements

The wind survey data from five coastal stations (Fig. 1) were processed in order to identify possible correlations between the radon profiles and sea surface wind. The measurements were carried out on the nearest coastline by the Italian Air Force Weather Service (Tab. 1) at a height of 10 m, according to W.M.O. standards. Two stations (16105,Venezia and 16108, Ronchi dei Legionari) were located near the water sampling sites 52 and 66, whereas three stations (16146, Marina di Ravenna ; 16148, Cervia and 16149, Rimini) were close to sampling sites 53 and 74. Wind data were taken at these stations every three hours at standard synoptic times.

TABLE 1
Sampling points and wind speed characterization

Station number	Date	Lat.	Long.	Max depth (m)	24-h mean wind (m/s)	3-week wind (m/s)
52	09/08	45°23'N	13°05'E	29	2.0	1.7
66	09/10	45°16'N	12°52'E	33	0.9	1.7
53	09/09	44°51'N	12°47'E	31	2.3	3.0
74	09/12	43°60'N	13°31'E	67	2.8	3.0

The measurements were processed to obtain the average wind intensity over the three weeks prior to the sampling period (from August 20 to September 12, 1990) as well as the average intensity during the 24 hours preceding each water sampling.

Sampling and extraction system

The radon extraction system was connected to the 25 l sample barrel (with ball valves R3, R4, and R5 open, and R1, R2 and R7 closed ; Fig.2). The gas present in the sample was recirculated through two traps (with ball valves R1, R2, R3 and R4 open, and R5 and R7 closed) filled with a bed of drying agent (80 ml drierite) and carbon dioxide absorber (40 ml ascarite), respectively. The gas was then passed through a U trap containing activated coconut charcoal (8-35 mesh in copper tubing 71 cm length and internal diameter 0.7 cm) to adsorb the radon. During the adsorption step the activated charcoal trap was cooled in a slurry of acetone-dry ice (-86 °C). Gas was allowed to recycle within the system for one hour at a flow rate of 1.6 l min^{-1}.

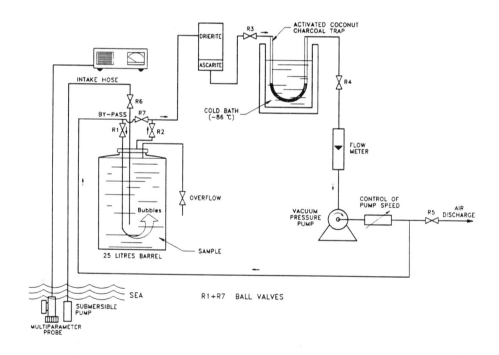

Figure 2. Schematic diagram of the sampling and extraction apparatus

Desorption procedures and liquid scintillation counting

The activated charcoal was transferred into a 70 ml separatory funnel, connected to a 50 ml separatory funnel containing reagent-grade toluene (Fig. 3). The stopcock in the latter funnel was then opened and the toluene permitted to flow down into the charcoal. Radon was desorbed from the charcoal and dissolved in the toluene with any radon escaping from the lower funnel being trapped in the upper funnel. After a few seconds all the toluene was gathered in the lower separatory funnel. Within two hours, all the radon was desorbed from the activated charcoal into the toluene and the separatory funnel was shaken.

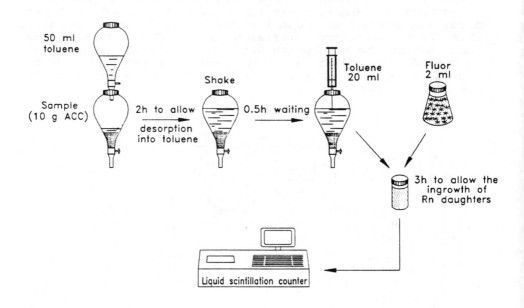

Figure 3. Flow scheme of radon desorption procedure from activated charcoal in toluene and LS counting.

After another 30 minutes, the settling of the fine activated charcoal particles was complete. Twenty ml of toluene were then collected with a syringe. Toluene was transferred into a glass vial and 2 ml of concentrated fluor solution added. After waiting three hours for the ingrowth of radon daughters, the sample was counted by LSC (Fig. 3). Due to the short half-life of ^{222}Rn (3.82 days), a portable LSC system was used for its direct measurement on board of the research vessel.

Calculation of gas exchange

The ^{222}Rn flux across the air-sea interface can be used to quantify the exchange rate between the near surface water and the atmosphere. In the sea ^{222}Rn is in secular equilibrium with its parent ^{226}Ra. However, in surface water there is a ^{222}Rn deficit caused by the flux of ^{222}Rn from the sea to the atmosphere. The gas exchange rate across the gas-water interface can be described in terms of a hypothetical boundary layer of water through which gases can pass only by molecular diffusion:

$$k_L = \frac{D}{Z} \quad (1)$$

where k_L is the gas transfer velocity, D is the molecular diffusivity of the gas in water, corrected for temperature, and Z is the depth in the stagnant boundary layer of water. The magnitude of Z characterizes the barrier limiting the rate of gas exchange and is a function of the state of the sea surface.

The transfer velocity k_L for radon can also be defined by:

$$k_L = \frac{F}{C_s - \alpha p} \quad (2)$$

where F is the flux of the gas across the sea-air interface, C_s is the concentration of the gas at the air-sea interface, α is the solubility of the gas and p is the partial pressure of the gas in the air at the interface.

As shown by Peng et al. (13) the transfer velocity can be calculated from radon profiles by:

$$k_L = \frac{\lambda \int_0^\infty (C_E - C_x) \, dx}{C_s - \alpha p} = \frac{\lambda I}{C_s - \alpha p} \quad (3)$$

where C_E is the concentration of ^{222}Rn at radioactive equilibrium with ^{226}Ra, C_x the concentration of ^{222}Rn at some depth x and λ is the radioactive decay constant (0.181 days^{-1}) of ^{222}Rn.

Since the atmospheric burden of radon is very small (i.e. $C_s \gg \alpha p$) eqs. (3) can be solved as:

$$k_L = \frac{\lambda I}{C_s} \quad (4)$$

The value of C_s is known for each radon profile on the basis of the best fit of the experimental data C_x. The integral I is evaluated numerically with the best fit function by using the trapezoid rule and the error in I is calculated by propagating the errors on each data point through the trapezoid rule calculations.

The molecular diffusivity of radon D is estimated from:

$$-\log D = (980/T) + 1.59 \quad (5)$$

where T is the absolute temperature (4).

In our case the mean temperature of surface water is 22.8°C and D = 110.6 x 10^{-6} m^2/day. The transfer velocity is calculated for each radon profile by using eq. (4) and the hypothetical boundary layer thickness Z from eq. (1). The results are shown in table 2.

TABLE 2
Radon Piston Velocity Calculations

Station number	C_S (mBq/l)	$\int_0^\infty (C_E - C_x) dx$ (Bq/m^2)	Rn transfer velocity (m/d)	Rn boundary layer thickness (μm)
52	0.68+0.10	26.2±1.2	7.0±1.8	16± 4
66	2.73±1.02	--	--	--
53	1.97±1.25	9.2±1.3	0.9±2.0	129±286
74	1.68±0.57	39.0±2.3	4.2±2.5	26± 15

* C_E= 2.78 ± 0.23 (mBq/l)

RESULTS AND DISCUSSION

^{226}Ra, ^{222}Rn concentrations and gas transfer velocity
Statistical analysis of the data (nine samples) indicates a mean ^{226}Ra concentration of 2.78±0.23 mBq/l (x ± 2σ/√N), i.e. within about 8%. Slight variations in the concentration of ^{226}Ra in Adriatic sea water, as reported by Giordani and Hammond (14), seem then to be confirmed. Since no significant dependence at the 80% confidence level (r=0.47, n=9) is evidentiated between this parameter and salinity (36-38.6‰) for Adriatic sea, we used the value of the reported mean ^{226}Ra concentration for C_E.

The calculated concentrations of ^{222}Rn at surface level (C_S) and related parameters are shown in Table 2 and Figure 4. Since the thermocline was well defined (Fig. 4), the radon deficit was exclusively calculated through the "equilibrium" concentration of ^{226}Ra in the wind mixed layer, ignoring the influence of other radon fluxes. Simple one-dimensional estimates of the Rn transfer velocity were made (1). These estimates, in some cases, suffer from uncertainties due to lateral and vertical transport from adjacent shallow water sediments and bottom sediments respectively. The ^{222}Rn values measured during the cruise in the vertical profile of the water column at stations 52 and 74 (Fig. 4) are compatible with the assumptions implicit in the deficit assessment method or radon disequilibrium. Station 66, on the other hand, shows average ^{222}Rn concentrations of 3.13 mBq/l in the mixed layer and 22.5 mBq/l in the layer below the thermocline. The radon contribution from the sea bed sediments seems to be significant and, in this shallow water area, possibly dominates gas transfer near the surface, producing C_S values similar to the equilibrium concentration C_E (Tab. 2). In the case of station 53, the high variability of the ^{222}Rn concentrations in the mixed layer results in a C_S estimate having a high associated error confirming that the contribution from the sediments cannot be ignored.

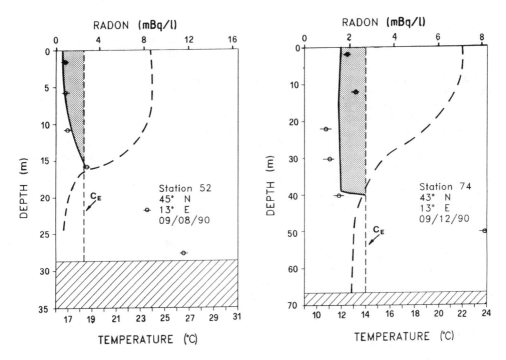

Figure 4. Radon profiles at the NW Adriatic Sea stations nos. 52 and 74. The dashed line is the temperature profile defining the depth of the wind mixed layer. The shaded area is a measure of the radon lost to the air.

The transfer velocities and the boundary layer thicknesses (Tab. 2) calculated for stations 52 and 74 are in the range of those observed in other marine environments (2, 15). However, when considering the environmental conditions expressed in terms of wind speed (Tab. 1), especially when data may be classified as belonging to the "smooth surface" regime (according to Liss, 3) it appears that the transfer velocity values obtained are higher than those resulting from laboratory or field surveys. It should, however, be emphasized that few field data are available for the "smooth surface" regime, characterized as it is by winds of less than 5 m/s. Attention should also be paid to the possible influence of horizontal fluxes of radon in the proper assessment of the gas exchange rates, so that in the future, advantage can be taken of the formalization of the mass balance of the radon gas. This may be possible by measuring the other sources and sinks and computing the exchange rate as a forced value (2)

^{222}Rn and ^{226}Ra procedure test

Tracer experiments were carried out to determine the leakage of radon from water kept both in HDPE and PVC barrels, using an aqueous ^{222}Rn solution from a commercial generator. Some screening tests of the leakage rate showed that the mean leakage time constant was 0.18 day^{-1} (i.e. T½ = 80 hours). Although radon loss occurs, it can be ignored with respect to the ^{222}Rn determination in sea water as the sample was treated with a low delay time (max. 3 hours). In the case of ^{226}Ra determination by radon emanation, PVC barrels having a measured leakage time constant of at least 0.033 day^{-1} (i.e. T½ = 500 hours) were selected.

Other tracer experiments were carried out to determine the separation yields of specific analytical stages and of the whole procedure. Solutions of ^{226}Ra were prepared to simulate conditions following sample collection. Trace experiments have shown that ^{222}Rn can be adsorbed (88±4%) on an activated coconut charcoal trap. The toluene desorption provides a relatively high recovery of radon (87±5%), while the whole procedure provides a mean recovery of 82±7%.

To check the reliability of the procedure, a comparison with other independent ^{226}Ra analytical procedures was undertaken. Values obtained with the methods employed by Lucas (11) and Mathieu et al. (12) are in agreement, within 4-20%, with the results obtained in the present procedure. An explanation for the differences may lie with the ^{222}Rn loss or gain during the storage period required for its ingrowth (5).

In order to eliminate this possible source of error in future, a procedure based on the use of Mn-impregnated acrylic fiber (8) will be used for the quantitative concentration of ^{226}Ra in the field. The fiber will be analyzed in the laboratory for ^{222}Rn, allowing ^{222}Rn ingrowth into a small, gas tight container.

Several commercial LSC systems have been compared in order to establish the lower achievable limit of detection (LLD). Table 3 shows the results obtained for samples transferred to 22 ml glass vials and containing toluene and fluor. The best LLD (17) for a 200 minute count at a confidence level of 95% (2σ) is 3.12 mBq, as obtained with the apparatus A directly on board the research vessel. In this case the LS counting permitted, with optimized discrimination, the detection of 4.6 counts per second (cps) for one Bq of ^{222}Rn in equilibrium with its progeny. The system blank, excluding the sampling step, obtained on board is 0.40 mBq/l, while in the laboratory it is 0.48 mBq/l. It should be noted that only a fraction (equal to 40%) of the toluene used for Rn extraction is counted by LSC. Choosing an optimum quantity of activated coconut charcoal and an appropriate volume of toluene may permit a reduction of the LLD (18) in the future.

TABLE 3
Background, counting efficiences and lower limit of detection (LLD) for several LSC instruments

LSC instrument	Window (keV)	Background (cps)	Efficiency (cps/dps)	LLD[a] (mBq)
A[b]	open	0.79	0.97	4.78
	29-1260	0.39	0.92	3.53
A[b] on board ship	open	0.64	0.97	4.30
	29-1260	0.30	0.92	3.12
B[b]	open	0.93	0.96	5.23
	29-1260	0.43	0.92	3.72
C[b]	open	0.55	0.94	4.10
	29-1260	0.35	0.89	3.47
D[c]	open	0.97	0.96	5.37
	optimum	0.22	0.77	3.17

[a] $LLD = \left[\dfrac{1 + (1+2bt)^{0.5}}{0.5 \; et} \; 1000 \right]$ where: b = background count rate, t = counting time, e = counting efficiency;
[b] This work;
[c] Published value in (16).

ACKNOWLEDGMENTS

We thank P. Bonelli (ENEL) for providing wind speed measurements and A. Moretti (ENEL) for technical work on board the R/V "BANNOCK".
This work was partially carried out under the contract B17-0042 of the Commission of the European Communities.

REFERENCES

1. Broecker, W.S., An application of natural radon to problems in ocean circulation. Symposium on Diffusion in Oceans and Freshwaters. Ed. T. Ichiye Lamont-Doherty, Geo. Observatory, New York., 1965, 116-145.

2. Hartman, B. and Hammond, D.E., Gas Exchange in San Francisco Bay. Hydrobiologia, 1985, **129**, 59-68.

3. Liss, P.S., Tracers of air-sea gas exchange .Phil. Trans. R. Soc. Lond. A, 1988, **325**, 93-103.

4. Peng, T.H., Takahashi, T., and Broecker, W.S., Surface Radon measurements in the north Pacific ocean station Papa. J. Geophys. Res, 1974, **79**, 12, 1772-1780.

5. Smethie, W.M. Jr., Takahashi, T., Chipman, D.W. and Ledwell, J.R., Gas exchange and CO_2 flux in the tropical Atlantic Ocean determined from ^{222}Rn and pCO_2 measurements. J. Geophys. Res., 1985, **90**, **C4**, 7005-7022.

6. Broecker, W.S., Peng, T.H., Mathieu, G., Hesslein, R. and Torgersen, T., Gas exchange rate measurements in natural systems. Radiocarbon, 1980, **22**, 676-683.

7. Thomas, F., Perigand, C., Merlivat, L. and Minster, J.F., World scale monthly mapping of the CO_2 ocean-atmosphere gas-transfer coefficient. Phil. Trans. Soc. Lond. A., 1988, **325**, 71-83.

8. Glover, D.M., Processes controlling radon-222 and radium-226 on the southeastern Bering Sea shelf. P.h.D Thesis, 1985, University of Alaska, Fairbanks.

9. Zore-Armanda, R., Les masses d'eau de la Mer Adriatique. Acta Adriatica, 1983, **10**, **3**, 1-87.

10. Malanotte Rizzoli, P. and Bergamasco, A., The dynamics of the coastal region of the Northern Adriatic Sea. J. Phys. Oceanography, 1983, **13**, **7**, 1105-1130.

11. Lucas, H.F., Improved low-level alpha-scintillation counter for radon. Rev. Sci. Instr., 1957, **28**, 680-683.

12. Mathieu, G.G., Biscaye, P.E., Lupton, R.A. and Hammond, D.E., System for measurement of ^{222}Rn at low levels in natural Waters. Health Physics, 1988, **55**, 989-992.

13. Broecker, W.S. and Peng, P.H., Gas exchange rates between air and sea. Tellus, 1974, **XXVI**,**1-2**, 21-35.

14. Giordani, P. and Hammond, D.E., Techniques for measuring benthic fluxes of ^{222}Rn and nutrients in coastal waters. Rapporto tecnico n. 20, CNR Istituto Geologia Marina, Bologna, 1985.

15. Peng, T.H., Broecker, W.S., Mathieu, G.G. and Li, Y.H., Radon evasion rates in the Atlantic and Pacific Oceans as determined during the Geosecs Program. J. Geophys. Res., 1979, **84**, **C5**, 2471-2486.

16. Prichard, H.M. and Mariën, K., Desorption of radon from activated carbon into a liquid scintillator. Analitycal Chemistry, 1983, **55**, 155-157.

17. Kahn, B., Rosson, R. and Cantrell, J., Analysis of ^{228}Ra and ^{226}Ra in public water supplies by a -ray spectrometer. Health Physics, 1990, **59**, **1**, 125-131.

18. Prichard, H.M. and Mariën, K., A passive diffusion ^{222}Rn sampler based on activated carbon adsorption. Health Physics, 1985, **48**,**6**, 797-803.

Scavenging and Particulate Transport Processes as Deduced from Radionuclide Behaviour

ARE THORIUM SCAVENGING AND PARTICLE FLUXES IN THE OCEAN REGULATED BY COAGULATION?

P H SANTSCHI AND B D HONEYMAN
Dept. of Marine Sciences, Texas A&M University, Galveston, TX 77553-1675)

ABSTRACT:

Trace metal scavenging in the ocean is classically explained by selective sorptive uptake of the metal by freshly produced particles, taking into account surface site and solute speciation, followed by settling of those particles. Kinetic models of oceanic trace metal scavenging rely heavily on the information gained from the application of Th nuclides as tracers of oceanic scavenging. The scavenging model of Honeyman and Santschi [1] links the formalism of coagulation/sedimentation models with that of the uranium/thorium decay series. It is based on the assumptions that scavenging rate constants reflect sorbed trace element removal from the water column *via* particle coagulation and sedimentation, and that the loss of particles by sedimentation is balanced by supply of colloidal particles to the settleable particle pool. Such model assumptions are verified here by a comparison of measured and calculated, depth-normalized sedimentation rates and by being able to predict the broad correlation of scavenging rate constants with particle fluxes, as observed by Bruland and coworkers. Our modeling results suggest that a physical step may be rate controlling, even though trace metal scavenging, overall, may still be driven by biological processes.

INTRODUCTION.

For two decades, Th nuclides have been used to study the self-cleaning capacity of seawater [2]. Oceanic residence times of Th which obtained from U/Th disequilibrium, range from days to decades, depending on the particular environment, and on the mean life of the isotope. This has been interpreted as reflecting different particle dynamical regimes [3]. The close linear correlation of the log of the total residence time of Th (log τ_T) with the log of the particle flux (log F_p) in the ocean [3], even though expected from a one box model approach, is still baffling since its slope is less than the expected -1.0. Another correlation, that of the scavenging or particle uptake rate constant with primary or new production (i.e., particle flux), was described by Bruland and co-workers [4-6]. Such a correlation is, however, not expected from any physical or chemical model or theory and thus remains to be explained from first principles.

Furthermore, the observations of Bruland and co-workers appear to be contradicted by yet another observation: the close correlation of the scavenging rate constant, λ_s, with particle concentration, C_p [7]. In this paper, we will show that these two observations, which are seemingly at odds with each other, can be produced by the same mechanism: coagulation of colloidal material in the ocean, which is a reactive intermediate in the scavenging process of many trace metals.

Colloids in the ocean are mostly composed of organic carbon [e.g., 8-10]. Total dissolved organic carbon in the ocean contains a mass of carbon equivalent to that in atmospheric CO_2 [11]. Only 10 % of this is well characterized. This fraction, which can be concentrated by solvent extraction, is composed of combined amino acids, carbohydrates and lipids [12,13].

High molecular weight components of DOC (such as humic compounds) are ubiquitous in seawater. They strongly and specifically complex metals [e.g., 14,15], form protective coatings on mineral surfaces [e.g., 16,17] and collect at the air-sea interface [18,16]. There is strong evidence that photolysis of high-molecular weight DOC compounds in the surface ocean is an important source of low molecular weight carbonyl compounds [19,20] and hydroxide radicals [21], which could be important intermediates in the polymerization of colloidal forms of DOC [22].

Th is taken up by non-living and living particles mainly by adsorption onto their outer surfaces rather than into their interior [23]. Chopin and coworkers have shown that Th in natural waters is probably strongly associated with humic material [24,25]. Humic acids, which, as macromolecular and hydrophobic compounds, are isolated by XAD resins at pH 2, make up only ≤ 10 % of DOC [26]. We hypothesize that they are the reactive colloidal intermediates which are posited in the scavenging model of Honeyman and Santschi [1].

The scavenging model of Honeyman and Santschi [1] is based on the following assumptions: 1) that scavenging rate constants reflect sorbed trace element removal from the water column *via* coagulation and sedimentation of particles of all sizes, whereby the rate of formation of settleable aggregates is controlled by individual coagulation mechanisms, and 2) the loss of particles by sedimentation is balanced by supply of colloidal particles to the settleable particle pool.

As will be shown in this paper, this model and its assumptions are supported by prediction of two seemingly unrelated observations:

1) A prediction of measured, depth-normalized sedimentation rates, Fp/z [with Fp=sediment accumulation rate (kg m^{-2}d^{-1}), z=mean depth of the water column (m)], for different oceanic regimes, using particle concentration, Cp (kg l^{-1}), as a master variable.

2) A prediction of the close correlation between the scavenging rate constant, $\lambda_s \cdot$(d^{-1}), for Th with particle flux (kg m^{-2}d^{-1}), as has been observed by Bruland and co-workers [4-6], using particle concentration, Cp (kg l^{-1}), as a master variable.

CONTROL OF PARTICLE MASS FLUX BY PARTICLE COAGULATION.

If the coagulation rate is in steady state with particle removal by settling, we can equate it to the depth-normalized particle flux, F_p/z, with z=mean depth (m). The coagulation-driven sedimentation rate can be calculated, according to Farley and Morel [27], as follows:

$$\partial C_p/\partial t = -B_{ds}C_p^{2.3} - B_{sh}C_p^{1.9} - B_bC_p^{1.3} \quad (kg\ l^{-1}\ d^{-1}) \quad (1)$$

with B_i = coagulation kernels for Brownian, shear, and differential settling, respectively, which individually include terms for particle sedimentation. Those terms contain the mixing depth (h), which we take as a scale length dependent on the time scale of the coagulation/sedimentation process; h is, however, not necessarily identical with z. This parameter, according to the authors, is important in defining the suspension time of each particle size and determines the likelihood of coagulation prior to removal, thus controlling the inital sedimentation behavior.

We can now compare the calculated value $\partial C_p/\partial t$ with measured values of F_p/z obtained from sediment trap measurements.

Figure 1 contains such a comparison. We examined and simulated the mass loss rate, F_p/z (kg $l^{-1}\ d^{-1}$), for three environments. These are: 1) the deep ocean and Sargasso Sea (F_p/z ca. $10^{-11.2}$ kg/l/d: [28]); 2) Funka Bay (F_p/z ca. $10^{-7.6}$ kg/l/d: [29]); and 3) MERL experiments to simulate Narragansett Bay, R.I. (F_p/z ca. $10^{-5.5}$ kg/l/d: [30]), and 4) the VERTEX site in the Eastern Pacific (F_p/z ca. $10^{-7.7}$ kg/l/d: [4]). Mass removal rates were calculated using the sedimentation model of Farley and Morel [27]. Calculations were made with α equal to 0.1 and 0.01, and an effective particle density, $\rho_p - \rho_{sw}$, of $1 g/cm^3$, with ρ_p = *in situ* particle density and ρ_{sw} = seawater density, and velocity gradient, G, of $1\ s^{-1}$. All other model parameters, except for z (water column depth), and C_p (particle concentration) are as given previously. For the sedimentation term in the coagulation/sedimentation equation of Farley and Morel [27] (eq. 1), we use a value of 6 m for the mixing depth, h, in all cases.

The agreement between simulated and measured particle fluxes in these very different hydrodynamic and particle dynamic regimes is encouraging. It needs to be further verified as more data becomes available.

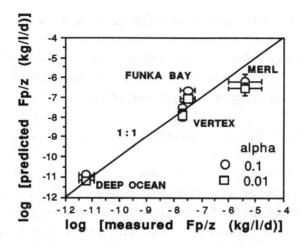

Figure 1 Correlation of the observed depth-normalized particle flux, Fp/z, with that predicted from Brownian and shear coagulation control, $\partial C_p/\partial t = B_b C_p^{1.3} + B_{sh} C_p^{1.9}$.

CORRELATION OF SCAVENGING RATE CONSTANT WITH PARTICLE FLUX (OR "NEW PRODUCTION").

Honeyman et al. [7] and Honeyman and Santschi [1] have shown that scavenging rates of ^{234}Th should be strong functions of particle concentration as coagulation rates are controlling the overall rates of removal from the water. As discussed before, coagulation rates, however, also control the mass flux of particles. Thus, both particle flux and metal scavenging rates are then functions of the rate of coagulation of colloids, which, at steady state, also would be equal to the rate of aggregation rate of suspended particles, as well as the production of colloidal sized particles. Their rates are then a function of the master variable particle concentration, C_p. As a consequence, the scavenging rate of Th would be expected to strongly correlate with particle flux (which is equivalent to new production). This is exactly what was observed by Bruland and co-workers [4-6].

An important factor is that "dissolved" Th activity is operationally-determined and usually indicates the less than ca. 0.4μm fraction of bulk solution. Thus, "dissolved" Th activity as generally described in the literature contains both truly dissolved and colloidal thorium. The discussion that follows is based on the concept that scavenging, as determined from the reference frame of 0.4μm or larger particle aggregates, is controlled by the rate of transfer of colloids to the macroparticle (≥0.4μm) size fraction.

The scavenging rate constant, λ_s, can be calculated from the activity ratios of particle-reactive ^{234}Th with its soluble mother nuclide, ^{238}U, as follows (e.g., [31,4]):

$$\lambda_s = \lambda_{234} [A_{238}/A_{234} - 1] \quad (2)$$

According to Honeyman and Santschi [1], we can use the coagulation-sedimentation rate, $\partial C_p/\partial t$ (defined by eq.1), to calculate the rate of uptake of Th into the macroparticle pool, as follows:

$$-\partial C_c/\partial t = \partial C_p/\partial t = -B_b C_p^{1.3} - B_{sh} C_p^{1.9} \quad (3a)$$
$$\approx -(B_b' + B_{sh}') C_p, \quad \text{with} \quad B_b' = B_b C_p^{0.3} \quad (3b)$$
$$B_{sh}' = B_{sh} C_p^{0.9} \quad (3c)$$
$$= -B C_p \quad (3d)$$

Equation 3a is based on the assumption that, at a quasi-steady-state, the sedimentation rate is balanced by the supply of material from the colloid, C_c, pool.

The uptake rate of the Th ions from the submicron sized pool onto macro-particles, $\partial[Me_{d+c}]/\partial t$, becomes then a function of the coagulation/sedimentation rate constant, B, and the fraction of Th$_{d+c}$ associated with colloids, $f_{c/d}$. The subscript "d+c" indicates that the Th fraction operationally considered as "dissolved" includes truly-dissolved and colloid-associated thorium. Thus,

$$\partial[Me_{d+c}]/\partial t = R_f[Me_{d+c}] = f_{c/d}B[Me_{d+c}] \quad (4a)$$
$$\text{with} \quad B = B_b' + B_{sh}' \quad (4b)$$

In order to predict λ_s (from U/Th), we make the following assumptions:
$$\lambda_s = B * f_{c/d} \approx R_f' \quad (5a)$$
$$\text{with} \quad f_{c/d} \approx 0.5 \pm 0.3 \text{ [32]} \quad (5b)$$
R_f' = scavenging rate constant (d^{-1}), calculated from [Th$_{part}$]/[Th$_{d+c}$] [7].
λ_s = scavenging rate constant (d^{-1}), calculated from eq. 2.

Eqs. 1-5 allow us to predict both λ_s and the particle flux, F_p, through a water column of depth z, F_p/z, using the particle concentration, C_p, as a master variable. The only data set where the parameters λ_s, F_p/z, and C_p were measured simultaneously is that of Coale and Bruland [4] which we used to test our hypothesis. Fig. 2 shows that the predicted correlation line is not significantly different from the one using measured values. However, even though the trend line can be correctly predicted, individual data points are sometimes not.

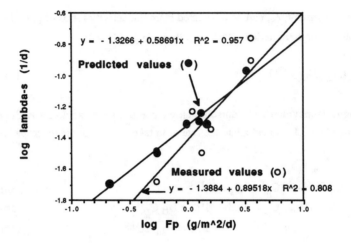

Figure 2. Comparison of the correlation of the scavenging rate constant, λ_s, with the particle flux, F_p, as observed by Coale and Bruland [4] with that predicted from particle coagulation. F_p was calculated assuming constant mass flux to primary productivity ratios.

There is another way to realize that such a correlation is to be expected from first principles for an open ocean environment where the coagulation rate is mostly given by the Brownian term. According to Honeyman and Santschi [1], we can set the scavenging rate constant, λ_s, equal to the coagulation/sedimentation rate times the fraction of Th associated with colloids, $f_{c/d}$:

$$\lambda_s = f_{c/d} \, (B_b' \, C_p^{0.3}) \tag{6}$$

The particle flux, $F_p = \partial C_p/\partial t$, can be expressed as a function of the scavenging rate constant, λ_s, as follows:

$$F_p = B_b' \, C_p^{1.3} \, z = [f_{c/d} \, B_b' \, C_p^{0.3}][C_p \, z/f_{c/d}] = \lambda_s \, C_p \, z/f_{c/d} \tag{7}$$

Rearranging:

$$\lambda_s = F_p \, f_{c/d} \, /[C_p \, z], \text{ or} \tag{8}$$
$$\log \lambda_s = \log F_p + \log[f_{c/d}/(C_p \, z)] \tag{9}$$

with z = mean water-column depth (m).

For environments with similar values of z, and where $f_{c/d}$ is itself a function of C_p [1] the second term in eq.9 would not be expected to change very much for the values of F_p observed by Coale and Bruland [4]. Ideally, we could expect then a linear correlation of log λ_s with log F_p, with a slope of 1. The fact that the slope is less than 1 indicates that C_p is likely a function of F_p as well.

Control of particle flux and trace metal scavenging in the ocean is known to be predominantly controlled by biological processes operating within the oceanic foodweb. A physical step, such as coagulation of colloidal sized particles as a rate controlling step of oceanic trace metal scavenging, which is suggested by our model calculations, is therefore unexpected from the point of view of biological processes. It is, however, not immediately obvious what this observed apparent correlation of particle flux with scavenging rate constant really means. The following ecological mechanism is suggested as a tentative explanation.

It is currently believed that a major portion (20-40 %) of the organic carbon produced in the ocean passes through a "microbial loop" to support bacterial production [33,34], resulting in the formation of marine detritus through aggregation aided by bacterial exudates. This view is also supported by the food chain model of Jackson [35] who concluded that approximately 40 % of the carbon transfer through the marine food chain passes through the detritus loop rather than to support secondary production. Furthermore, significant amounts of DOC in the deep ocean appear to adsorb onto POC as concluded from the $\Delta^{14}C$ value of POC decreasing with depth [36]. Passive uptake of the macromolecular (i.e., colloidal) fraction of DOC onto either suspended and sinking particles or directly onto bacteria could therefore regulate bacteria growth and detritus formation through coagulation, and thus constitute the physical mechanism causing the tight correlations of scavenging rate with both particle flux and particle concentration.

SUMMARY AND CONCLUSIONS:

This paper describes our attempts at predicting particle fluxes and their correlations with scavenging rate constants in the ocean, using particle concentration as a master variable. The agreement between simulated and measured particle fluxes for different hydrodynamic and particle dynamic regimes, even though encouraging, is puzzling for the following reasons: particle flux in the ocean is viewed as controlled by the supply of nutrients and composed of aggregates of fecal matter. It is thus controlled by biological productivity. We conclude that for both particle-flux (phytoplankton-detritus-pathway) and trace metal scavenging, a physical/chemical step is rate controlling, even though trace metal scavenging, overall, may still be driven by biological processes. Our findings are consistent with the view that a large fraction of organic carbon in the ocean passes through a microbial loop. Physical limitations on bacterial and detrital uptake of colloidal organic carbon could therefore constitute the required physical mechanism controlling particle and trace metal dynamics in the ocean. We note, however, that even though the arguments given in this paper show that we can correctly reproduce observed trend lines from first principles, we are as yet unable to predict individual data points. Our proposed trace metal scavenging model is just a first step towards the development of a comprehensive trace metal scavenging model. It should be further tested in oceanic environments where C_p and F_p vary independently. Furthermore, more work needs to be done to unravel the chemical and biochemical composition of colloids as well as their reactivity and age.

ACKNOWLEDGEMENTS:

This work was in part supported by grants of the National Science Foundation (OCE-9012103), and the Texas Institute of Oceanography. We acknowledge helpful criticisms by G. Benoit and M. Baskaran.

REFERENCES:

1. Honeyman, B.D. and Santschi, P.H. A Brownian-pumping model for oceanic trace metal scavenging: evidence from Th isotopes. Deep Sea Res., 1989, **47**, 951-992.
2. Broecker, W.S., and Peng, T.-H. Tracers in the Sea, Eldigio Press, Palisades, N.Y., 1982.
3. Santschi, P.H. Particle flux and trace metal residence times in natural waters. Limnol. and Oceanogr., 1984, **29(5)**, 1100-1108.
4. Coale, K.H. and Bruland, K.W.. ^{234}Th:^{238}U disequilibria within the California Current. Limnol. Oceanogr., 1985, **30**, 189-200.
5. Bruland, K.W., and Coale, K.H. In: Dynamic Processes in the Chemistry of the Upper Ocean, eds., J.D. Burton et al., Plenum Publ. Corp., 1986, 159-172.
6. Bruland, K.W. Scavenging of ^{234}Th in the Pacific: A tracer of surface water scavenging. In: Isotopic Tracers, U.S. Joint Global Ocean Flux Study Planning Report 12, July 1990, WHOI, Woods Hole, MA.,1990, pp. 53-58.
7. Honeyman, B.D., L.S. Balistrieri, L.S., and Murray, J.W. Oceanic trace metal scavenging: the importance of particle concentration. Deep Sea Res., 1988, **35**, 227-246.
8. Carlson, D.J., Brann, M.L., Mague, T.H., and Mayer, L.M. Molecular weight distribution of dissolved organic materials in seawater determined by ultrafiltration: A reexamination. Mar. Chem., 1985, **16**, 155-171.
9. Sigleo, A.C., and Helz, G.R. Composition of estuarine colloidal material: major and trace elements. Geochim. Cosmochim. Acta, 1981, **45**, 2501-2509.
10. Sugimura, Y. and Suzuki, Y. A high temperature catalytic oxidation method for the determination of non-volatile dissolved organic carbon in seawater by direct injection of a volatile sample. Mar. Chem., 1988, **24**, 105-131.
11. Williams, P.M., and Druffel, E.R.M. Radiocarbon in dissolved organic carbon in the Central North Pacific Ocean. Nature, 1987, **330**, 246-248.
12. Williams, P.M. Chemistry of the dissolved and particulate phases in the water column. In: Plankton Dynamics on the Southern California Bight, ed. R.W. Eppley, Springer Verlag, New York, 1986, pp. 53-82.

13. Williams, P.M., and Druffel, E.R.M. Dissolved organic matter in the ocean: comments on a controversy. Oceanogr. Rev., 1988, **1**, 14-17.
14. Bruland, K.W. Complexation of zinc by natural organic ligands in the central North Pacific. Limnol. Oceanogr. 1989, **34**, 269-285.
15. Coale, K.H., and Bruland, K.W. Copper complexation in the northeast Pacific. Linol. Oceanogr., 1988, **33**, 1084-1101.
16. Neihof, R.A., and Loeb, G.I. Dissolved organic matter in seawater and the electric charge of immersed surfaces, J. Mar. Res., 1974, **32**, 5-12.
17. Hunter, K.A., and Liss, P.S. Organic matter and the surface charge of suspended particles in estuarine waters. Limnol. Oceanogr., 1982, **27**, 322-335.
18. Williams, P.M. The distribution and cycling of organic matter in the ocean. In: S.J. Faust and J.V. Hunter, eds. Organic Compounds in Aquatic Environments. Marcel Decker, 1971, pp. 145-163.
19. Mopper, K., and Stahovec, W.L. Sources and sinks of low molecular weight organic carbonyl compounds in seawater. Mar. Chem, 1986, **19**, 305-321.
20. Kieber, D.J., McDaniel, J., and Mopper, K. Photochemical source of biological substrates in seawater: Implications for carbon cycling. Nature, 1989, **341**, 637-639.
21. Mopper et al., 1990.
22. Harvey, G.R., and Boran, D.A. Geochemistry of humic substances in seawater. In Humic Substances in Soils, Sediment and Water, eds. G.R. Aiken et al., Wiley-Interscience, New York, 1985, pp. 233-249.
23. Fisher, N.S., Teyssie, J.-L., Krishnaswami, S., and Baskaran, M. Accumulation of Th, Pb, U, and Ra in marine phytoplankton and its geochemical significance. Limnol. Oceanogr., 1987, **32/1**, 131 - 142.
24. Nash, K.L. and Choppin, G.R Interaction of humic and fulvic acids with Th(IV). J. Inorg. Nucl. Chem., 1980, **42**, 1045-1050.
25. Cacheris, W.P. and Choppin G.R. Dissociation kinetics of thorium-humate complex. Radiochemica Acta , 1987, **42**, 185-190.
26. Stuermer, D.H., and Harvey, G.R. The isolation of humic substances and alcohol-soluble organic matter from seawater. Deep-Sea Res.,1977, **24**, 303-309
27. Farley, K.J. and F.M.M. Morel Role of coagulation in sedimentation kinetics. Environ. Sci. Technol., 1986, **20**, 187-195.
28. Anderson, R.F., Bacon, M.P., and Brewer, P. Removal of ^{230}Th and ^{231}Pa from the open ocean. Earth and Planet. Sci. Lett., 1983, **62**, 7-23.
29. Minagawa, M. and Tsunogai, S. Removal of ^{234}Th from a coastal sea: Funka Bay, Japan. Earth Planet. Sci. Lett., 1980, **47**, 51-64.
30. Santschi, P.H., Adler, D., Amdurer, M., Li, Y.-H., and Bell, J. Thorium isotopes as analogues for "particle-reactive" pollutants in coastal marine environments. Earth and Planet. Sci. Lett., 1980, **47**, 327-335.
31. Santschi, P.H., Li, Y.-H., Nyffeler, U.P., and O'Hara, P. Radionuclide cycling in natural waters: relevance of scavenging kinetics. In Sediments and Water Interactions, ed. Sly, P.G., Springer-Verlag: New York, 1986, Chapter 17.
32. Baskaran, M., Benoit, B., and Santschi, P.H.. The role of sub-micron sized particles in the removal of thorium isotopes from the oceanic water column. Ms in preparation, 1991.
33. Azam, F., Fenchel, T., Field, J.G., Gray, J.S., Meyer-Reil, L.A., and Thingstad, F. The ecological role of water-column microbes in the sea. Mar. Ecol. Progr. Ser., 1983., **10**, 257-263.
34. Ducklow, H.W. Production and fate of bacteria in the oceans. BioSci., 1983, **33**, 494-501.
35. Druffel, E.R., Williams, P.M., and Suzuki, Y. Concentrations and radiocarbon signatures of dissolved organic matter in the Pacific Ocean. Geophys. Res. Lett., 1990, **16/9**, 991 - 994.
36. Jackson, G.A., and Eldridge, P.M. Filling in the flows of a marine planktonic food web off Southern California. EOS ,1990, **71/43**, 1385.

^{234}Th : AN AMBIGUOUS TRACER OF BIOGENIC PARTICLE EXPORT FROM NORTHWESTERN MEDITERRANEAN SURFACE WATERS

C.E. Lambert[*], S. Fowler[+], J.C. Miquel[+], P. Buat-Ménard[*], F. Dulac[*], H. V. Nguyen[*], S. Schmidt[*], J.L. Reyss[*], J. La Rosa[+]

[*] Centre des Faibles Radioactivités, CFR, Laboratoire mixte CNRS-CEA, F91198 Gif-sur-Yvette CEDEX France

[+] International Laboratory of Marine Radioactivity, IAEA, 19 av des Castellans, MC98000 Monaco

ABSTRACT

As part of the DYFAMED programme in the Ligurian sea, a record of ^{234}Th particulate fluxes has been obtained with sediment traps moored at 200 m depth from February 1988 through 1989. This site, situated in the central zone of the Ligurian sea, was chosen so that advective inputs from the coast would be minimal. Each sediment trap sample integrated a collection period of 9 to 15 days and was analyzed for radionuclides by non-destructive gamma ray spectrometry. The temporal variation of ^{234}Th flux is significantly correlated with that of the total mass flux, the highest flux generally being observed in winter and early spring. These variations are most likely related to biological activity in the upper layers of the water column. However, the relationship between total carbon fluxes and ^{234}Th fluxes is not clear. When two large mass fluxes occur two weeks apart, the C/Th ratio of the latter can be unusually high. This could be due to variations in the biological transport processes. There is also evidence that the lack of a clear relationship is due to the high proportion of clay in many sediment trap samples. The Mediterranean is indeed subject to frequent inputs of Saharan dust via the atmosphere. ^{234}Th adsorbs both on biological and inorganic detrital material. Consequently, in regions where the flux contains a high terrigenous component, the prediction of biogenic fluxes from ^{234}Th fluxes must be considered with caution.

INTRODUCTION

It is now recognized that the vertical flux of radionuclides and metals is closely linked to the vertical flux of matter leaving marine surface waters [1]. The radionuclide fluxes are thought to be be predictable from knowledge of primary productivity and "new production" leaving the euphotic zone [2], since it is well established that these nuclides can be transferred from dissolved to particulate forms through active and passive uptake by phytoplankton. Of special interest is ^{234}Th, a gamma-emitting nuclide with a convenient half-life of 24.1 d. produced by the decay of ^{238}U in seawater. Its susceptibility to scavenging has lead to numerous studies of removal rates, residence times and transport processes of marine particles. It has also been used to predict carbon fluxes out of the euphotic zone or "new production" [3]. Indeed, there is strong evidence that the scavenging of thorium from dissolved to particulate form varies as a function of primary production [4]. More recently, ^{234}Th has been used to assess and eventually question the efficiency of sediment traps to collect settling organic matter, for it seems that the use of ^{234}Th yields a prediction of higher carbon fluxes than are actually measured in sediment traps [5]. In this paper, we describe thorium scavenging processes in the northwestern Mediterranean and examine the assumption that ^{234}Th fluxes allow the prediction of fluxes of biogenic matter.

For the northwestern Mediterranean Sea, it has been shown [6,7] that i) there is a large seasonal variability in thorium fluxes, and ii) the flux of ^{234}Th correlates with the total mass flux. These observations were made on a series of sediment trap samples collected at 200 m depth from a site 11 nautical miles off Calvi, northwestern Corsica, in 1987. In 1988 and 1989, several additional cruises were made at a site 28 nautical miles off Villefranche-sur-mer (Station 1), 43°36'N 07°29'E, in a 2300 m water column. Data obtained from these studies are compared to those reported previously [6,7].

Sampling and analytical protocols have been described in detail elsewhere [6,7]. Briefly, samples were collected with an automated time-series sediment trap of 0.125 m^2 opening during intervals ranging from 9 to 15 days. After retrieval of the traps, samples were processed immediately, swimmers removed, and the sediment freeze-dried, weighed and analyzed for radioactivity by non destructive gamma spectrometry within 3 months following retrieval. Precision on ^{234}Th determination is 10% on average.

Dried aliquots of the homogenized sediment trap samples were also analyzed for total carbon using a Hereaus CHN-O-Rapid analyzer. Precision and reproducibility of the measurements range from 2 to 5%.

RESULTS

At station 1, in 1988 and during six months of 1989, we observed a variability of ^{234}Th particulate fluxes of the same order as was found at the Corsican site. This was demonstrated by comparing minimal and maximal fluxes and concentrations at both sites (Table 1). There was a significant correlation (r = 0.79, n=41) between the flux of ^{234}Th and total mass flux which was very similar to that observed at the Corsican site (Figure 1). The regression line in Figure 1 is shown for all samples ; earlier data from the Corsican site are presented as black dots.

TABLE 1

Minimal and maximal fluxes and concentrations of ^{234}Th

^{234}Th min flux Bq m^{-2} d^{-1}	^{234}Th max flux Bq m^{-2} d^{-1}	^{234}Th min conc. Bq g^{-1}	^{234}Th max conc. Bq g^{-1}
1987 (Calvi) 0.92	11.33	13.33	40.92
1988 (st.1) 0.17	11.17	2.50	101.67
1989 (st.1) 1.67	4.33	24.17	51.67

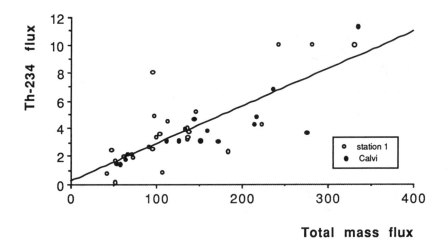

Figure 1 ^{234}Th fluxes (Bq m^{-2} d^{-1}) vs total mass fluxes (mg m^{-2} d^{-1}) at both sites.

Surprisingly, the correlation between ^{234}Th fluxes (Bq m^{-2} d^{-1}) and total carbon fluxes (mg m^{-2} d^{-1}) is less significant (r=0.59 for

41 samples). There is a slightly better correlation with Al flux, which represents the clay fraction (r=0.67 for 38 samples); this will be discussed later. The relatively poor correlation between ^{234}Th and total carbon is not due to measurement uncertainties but represents a real phenomenon as far as our data are concerned. We have observed, however, that there are obvious co-variations in a number of cases between ^{234}Th fluxes and total carbon fluxes (Figs. 2-4). Nevertheless, the amplitude of the variations in the Th/C ratio and some important exceptions blur the correlation (see e.g. generally low flux of C in 1989 compared to 1988, Fig. 3 and 4, in contrast to relatively similar thorium fluxes in 1988 and 1989).

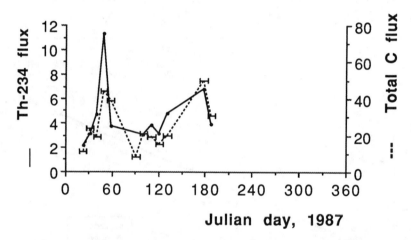

Figure 2. ^{234}Th fluxes (Bq m^{-2} d^{-1}) and total carbon fluxes (mg m^{-2} d^{-1}) at the Corsican site. The duration of collection (9 to 15 days) is indicated by bars on the C flux.

DISCUSSION

Possible explanations for such observations are numerous and difficult to verify. For example, there may be different adsorption capacities of thorium depending upon the nature of the particles, which can vary over short time scales. Furthermore, coastal advection could transport additional ^{234}Th into surface waters, although there is greater probability for that occurring at the Corsican site than at station 1, as suggested from ^{228}Ra measurements [8]. As a first attempt, we have tried to account for

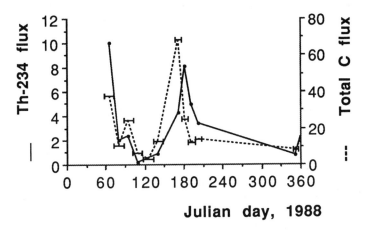

Figure 3. ^{234}Th fluxes (Bq m^{-2} d^{-1}) and total carbon fluxes (mg m^{-2} d^{-1}) at station 1 in 1988. The duration of collection (9 to 15 days) is indicated by bars on the C flux.

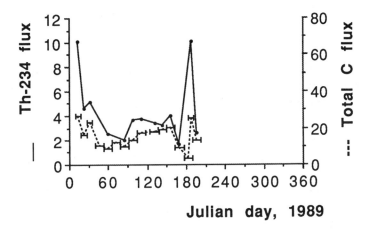

Figure 4. ^{234}Th fluxes (Bq m^{-2} d^{-1}) and total carbon fluxes (mg m^{-2} d^{-1}) at station 1 in 1989. Note that the thorium fluxes are of the same order as in Figure 2, whereas the carbon fluxes are lower. The duration of collection (9 to 15 days) is indicated by bars on the C flux.

the observed variability in the Th/C ratio by pooling samples from each site on a monthly basis. However, this approach neither smooths the data nor enhances the correlation (Figure 5). It is obvious from this figure that very similar carbon fluxes can be observed together with large variations of the thorium fluxes.

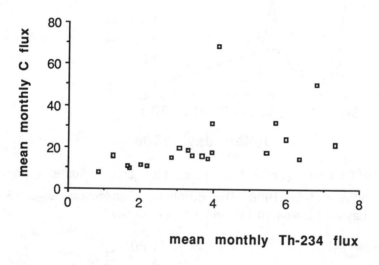

Figure 5. Relationship between the monthly average of total carbon flux (mg m^{-2} d^{-1}) and ^{234}Th flux (Bq m^{-2} d^{-1}) for each year of observation. Note that the variability of the ^{234}Th/C ratio is not reduced when pooling the samples together.

If these results are confirmed, one can conclude that in the northwestern Mediterranean, ^{234}Th flux gives, at most, only a very crude picture of the fluxes of biogenic particulates which leave surface waters. The ratio of ^{234}Th fluxes (Bq m^{-2} d^{-1}) / total carbon fluxes (mg m^{-2} d^{-1}) range between 5 and 15 in 65% of the cases, which indicates that the prediction of carbon fluxes can be made, at best, within a factor of 3, leaving 35% of the cases unexplained. This variability is likely not due to any variability in total carbon/organic carbon ratios, since in this region the total carbon/organic carbon ratio appears to be rather constant [9].

Another possible explanation is that temporal variability of the quality of organic material available to transfer the adsorbed thorium could play a role ; however, such variations are not sufficiently well-documented for the period considered. The C/N mass ratio of these samples varies by less than a factor of two (7 to 11, median = 8) and there is evidence that the material is generally fresh. There is also some evidence that among fecal pellet producers, copepods and salps may play a major role in packaging particulate material containing scavenged thorium, but their packaging efficiency might also differ from one species to another [10]. If, indeed, this is the case, such a temporal variability could exert an influence on the variation of C/Th ratios. We have an indication of this ; for example, five cases correspond to a similar situation : some samples which exhibit a very large thorium flux in the time series were followed by a low thorium flux (e.g. day 60, Fig. 2, Table 2). In some of these suites of cases, marked with an asterisk in Table 2, the variation in the C/Th ratio could only be explained if the particulate thorium is not in equilibrium with the carbon reservoir. We think that this may be an indication of some biogenic process that precludes the incorporation of thorium in the biogenic flux. It could happen in a period of low primary productivity whereas the zooplankton are still very active, and induce an important vertical flux. In this case, the fine particles would be in relatively low concentration, and therefore would be able to scavenge little thorium, resulting in a low thorium flux relative to the carbon flux. The result would be the same if the herbivore activity was low ; i.e. the carnivores would transfer a carbon poor in thorium (high proportion of meat diet). The biomass data provided by Baker [10] on station 1 in summer 1988 could be interpreted as follows : when the herbivore/carnivore ratio is low, the transfer of thorium is low because there is no transfer from the primary producers, i.e. the phytoplankton.

TABLE 2

Examples of variations of the C/Th ratio in time series

Month	Mass	^{234}Th	C	Al	C/Th			
	(1)	(2)	(1)	(1)			(3)	(4)
02/87	335	11.3	44	11.7	3.9			
	276	3.7	39	2.4	10.5	*		
03/88	331	10.0	37.7	7.1	3.8			
	73	2.0	10.2	1.8	5.2	*		
	183	2.3	24.4	2.0	10.5	*		
04/88	19	7.8	2.8	0.2	0.4			
	107	0.8	12.6	nd	15.8	*		
06-07/88	223	4.3	68.2	2.2	15.9	*	1.4	2800
	95	8.0	25.2	1.8	3.2		7.7	8300
	97	4.9	12.4	4.0	2.5		12.5	4500
	100	3.4	14.0	3.3	4.1		12.5	2200
01/89	243	10.1	26.4	17.8	2.6			
	113	4.6	16.2	10.5	3.5			

(1) mg m^{-2} d^{-1}
(2) Bq m^{-2}d^{-1}
(3) Ratio of the biomass of herbivores/ biomass of carnivores (in mg dry weight per m^2 on 6/23, 6/30, 7/8 and 7/14/88 at station 1 [10].
(4) Total zooplankton biomass in mg dry weight per m^2 [10].

More biological data are needed to confirm and refine such

hypotheses. Whatever the cause of these variations in the Th/C ratio, discarding them will, of course, result in a better correlation between thorium and mass fluxes (r = 0.84), and between thorium and C fluxes (r = 0.83).

We have further considered the possibility that the masking of the correlation between carbon and thorium fluxes could be caused by dust inputs to the Mediterranean. It is well known that dust plumes from the Sahara occur frequently over the western Mediterranean. Therefore some of the available thorium could adsorb on dust particles present in surface seawater. The residence time of these dust particles, primarily aluminosilicates, can be very short in surface waters, where they are efficiently grazed and incorporated within fast-sinking fecal pellets [11]. In such a case, their contribution to the total mass flux in the Mediterranean can be as high as 40% [11]. Thus it is likely that such dust inputs would alter the correlation between thorium and carbon, because the fate of terrigenous material is not only linked to the biogenic cycle, but also to atmospheric transport and deposition patterns. This is confirmed by examining the thorium/carbon relationship when all samples rich in Al are deleted. The correlation is enhanced only if all samples with an Al content of less than 1.8% are considered ; however, this is not a realistic treatment, because it necessitates rejecting half of the samples.

We have attempted to test the hypothesis that the particulate thorium flux was composed of two components : one biogenic and the other abiogenic, each having a constant relative abundance of thorium. If this is the case, the thorium flux could be expressed by the following equation :

^{234}Th flux = A Total carbon flux + B Al flux   (1)

Solving the equation, we obtain :

^{234}Th flux = 0.182 (± 0.024) C flux + 0.184 (± 0.113) Al flux where ^{234}Th is in Bq m^{-2} d^{-1}; C and Al in mg m^{-2} d^{-1}; n= 31 (all cases where Al, C, and Th were measured together minus the exceptions explained above); r = 0.95.

We can translate Al into the detritic fraction, considering Al = 7% of aluminosilicates [12], and C into a biogenic fraction, considering it equals C x 2.15, i.e. taking into account 70% of the C as organic matter and 30% of C as $CaCO_3$. This biogenic fraction does not include opal.

Then:

^{234}Th flux = 0.085 (± 0.011) biog. + 0.013 (± 0.008) detr.

According to this equation, thorium adsorbs more readily on biogenic particles than on detritic particles. This may be specific to the Mediterranean, as the detritic fraction and the fraction linked to total carbon are on average of equal importance (30%). Using equation (1) gives an average thorium/carbon ratio of 0.18 Bq/mg of total carbon (0.12 dpm ^{234}Th /µmol of organic carbon), very close to that estimated by Coale and Bruland [4],.

In conclusion, we estimate that in the Mediterranean, and probably in all regions including coastal waters where large detrital fluxes exist, caution has to be used when trying to predict biogenic fluxes from ^{234}Th fluxes or scavenging rates in surface waters, since in such cases, there is a second controlling variable. Only by taking into account the Al flux can the biogenic flux be predicted with, at best, 50% uncertainty.

Our data also highlight the presence of very high C/Th ratios which could be related to the successive stages of development of the ecosystem.

Acknowledgements

This research was supported in part by the Centre National de la Recherche scientifique, the Commissariat à l'Energie Atomique and grants from the Institut des Sciences de l'Univers (Programme flux océaniques). The International Laboratory of Marine Radioactivity operates under an agreement between the International Atomic Energy Agency (IAEA) and the Government of the Principality of Monaco. This is CFR contribution 1188.

References

[1] Fowler S.W. and G.A. Knauer, Role of large biogenic particles in the transport of elements and organic compouds through the oceanic water column, Prog. Oceanogr.,1986, **16**,147-194.

[2] Fisher N., J.K. Cochran, S. Krishnaswami, H.D. Livingston, Predicting the oceanic flux of radionuclides on sinking biogenic debris, Nature, 1988, **335**, 621-625.

[3] Coale K.H. and K.W. Bruland, Oceanic stratified euphotic zone as elucidated by ^{234}Th : ^{238}U disequilibria, Limnol. Oceanogr.,1987, **32**, 189-200.

[4] Coale K.H. and K.W. Bruland, ^{234}Th : ^{238}U disequilibria within the California Current, Limnol. Oceanogr.,1985, **30**, 22-33.

[5] Buessler K.O., M.P.Bacon, .K. Cochran, H.D. Livingston, Carbon and nitrogen export during the JGOFS North Atlantic bloom experiment estimated from ^{234}Th:^{238}U disequilibria, Deep-Sea Res., in press.

[6] Buat-Ménard P., H.V. Nguyen, J.L. Reyss, S. Schmidt, Y. Yokoyama, J. La Rosa, S. Heussner, S.W. Fowler, Temporal changes in concentrations and fluxes in the northwestern Mediterranean, in : Radionuclides : a tool for oceanography, J.C. Guary et al.eds, Elsevier applied science, 1988,121-130.

[7] Schmidt S., J.L. Reyss, H.V. Nguyen, P. Buat-Ménard, ^{234}Th cycling in the upper water column of the northwestern Mediterranean sea, Palaeogeography, pamaeoclimatology, palaeoecology, Global and

planetary change section, 1990, **8 9**, 25-33.

[8] Schmidt S., Cycle des radionucléides naturels dans l'océan : exemples d'application à l'étude des flux de matières en Méditerranée occidentale et dans l'Atlantique Nord,1991, Thèse, University of Paris VI.

[9] Fowler S.W., L.F. Small, J. La Rosa, Seasonal particulate carbon flux in the coastal northwestern Mediterranean sea and the role of zooplankton fecal matter, Oceanologica Acta, 1991,**1 4**, 77-85.

[10] Baker M., Estimation du bilan de matière dans l'écosystème pélagique en mer de Ligure (Méditerranée occidentale). Application aux missions DYPHAMED et TROPHOS II. Etude expérimentale sur les salpes, 1990, Thèse, University of Paris VI.

[11] Buat-Ménard P., J. Davies, E. Remoudaki, J.C. Miquel, G. Bergametti, C.E. Lambert, U. Ezat, C. Quetel, J. La Rosa, S.W. Fowler, Non-steady state removal of atmospheric particles from Mediterranean surface waters, Nature, 1989, **3 4 0**,131-134.

[12] Bowen, H. J. M., Trace elements in Biochemistry, Ac. Press, New York, 1966.

THE INTERACTION BETWEEN HYDROGRAPHY AND THE SCAVENGING OF ^{230}TH and ^{231}PA AROUND THE POLAR FRONT, ANTARCTICA

MICHIEL M. RUTGERS VAN DER LOEFF
Alfred-Wegener Institute for Polar and Marine Research,
Columbusstrasse, 2850 Bremerhaven, FRG.

ABSTRACT

The removal rate of particle-reactive radionuclides from the water column depends on particle concentration and on particle flux. The high export productivity that is associated with the southern Polar Front is therefore characterized by high scavenging rates. The distribution of natural radionuclides from the Uranium decay series in the water column is however not only determined by scavenging patterns, but also by transport of water masses. A major parameter controlling this distribution is the ratio between the scavenging residence time of the particular isotope and the residence time of the water mass in a particular rain rate regime. It will be shown how the distribution of natural radionuclides with appropriate scavenging residence times can be used to obtain information on both time scales.

PARTICLE/SOLUTION PARTITIONING OF THORIUM ISOTOPES IN FRAMVAREN FJORD: INSIGHTS INTO SORPTION KINETICS IN A SUPER-ANOXIC ENVIRONMENT

BRENT A. McKEE
Louisiana Universities Marine Consortium
Louisiana Universities Marine Center
Cocodrie, LA 70344

JAMES F. TODD
National Oceanic and Atmospheric Administration
Office of Global Programs
Silver Spring, MD 20910

WILLARD S. MOORE
Department of Geological Sciences
University of South Carolina
Columbia, SC 29208

ABSTRACT

Water samples were collected in a detailed depth profile (32 samples) of Framvaren Fjord (water depth: 183 m) and examined for dissolved (< 0.22 μm) and particulate (> 0.22 μm) phase ^{234}Th ($t_{1/2}$ = 24.1 days) concentrations. The particulate thorium fraction is relatively constant (50-60% of total) in the oxic surface waters of Framvaren Fjord. Below the oxic-anoxic interface (approximately 18 m), the particulate thorium fraction increases dramatically, reaching a maximum (96%) at 24 m, coincident with the maximum in total suspended matter concentrations (TSM). In this environment, the incorporation of thorium into the particulate phase seems to be more closely tied to the presence of bacteria than to the formation of particulate Fe and Mn oxyhydroxide phases. Distribution coefficients (K_d) for thorium in surface waters are 10^6-10^7 L/kg, typical for low particulate marine environments. Below the TSM maxima, K_d values are much lower (10^4-10^5 L/kg) and remain consistently low in the anoxic bottom waters. These are among the lowest K_d values for thorium ever observed for a marine environment. Equilibrium and kinetic approaches to quantifying the flux of thorium into the particulate phase indicate that, in the anoxic bottom waters of Framvaren Fjord, thorium sorption reactions become less dominant relative to reverse reactions such as desorption and particulate remineralization.

INTRODUCTION

From a global perspective, anoxic marine systems account for only a small percentage of the world ocean volume; ranging in size from the Black Sea - the largest present day permanently anoxic marine basin - to small fjords, such as the intermittently anoxic Saanich Inlet (British Columbia) and the permanently anoxic Framvaren Fjord (Norway). While not representative of waters comprising the majority of the world ocean, anoxic basins are nonetheless important geological features that have been of interest to chemical oceanographers for a number of years, mainly because of the redox cycling of chemical elements that occurs in the vicinity of the oxic-anoxic boundary in these environments. In particular, the geochemistries of manganese and iron have received considerable attention [1-5], largely because of the importance of Fe and Mn oxyhydroxides in oceanic trace element scavenging, and hence, global trace element cycles.

Thorium is a highly-reactive chemical element that exists only as the tetravalent cation, Th(IV), in nature [6]. The naturally-occurring isotopes of thorium (e.g., ^{228}Th, ^{230}Th, ^{234}Th) have been applied widely as analogues of particle-reactive elements in the marine environment [7-13]. Thorium isotopes are particularly useful tools for such applications since they have known sources, are strongly sorbed onto particle surfaces and can be measured at very low concentrations. The behavior of thorium isotopes in anoxic systems has not been sufficiently documented, however, and it was with the anticipation that such environments might offer unique insights into the aquatic geochemistry of thorium that the present study was initiated.

Study Area

Our investigation (June 3-4, 1989) was carried out at a central, mid-axis station in Framvaren Fjord (Figure 1), a small (5.8 km^2 area), deep (183 m), permanently anoxic basin in southern Norway (58° 10'N, 6° 45'W). Framvaren Fjord was chosen as an ideal location for the study because of its well-studied hydrography [14] and previously reported distributions of several chemical constituents that suggest the system has achieved an approximate steady-state condition. Framvaren Fjord receives runoff from a drainage area (31 km^2) comprised predominantly of farsundite, a rock having a monzonitic-granitic composition [15]. Freshwater runoff entering the fjord is small, averaging 1 m^3/sec annually. Water exchange between Framvaren Fjord and the Skagerrak Sea is severely restricted by four shallow (< 18 m) sills, with a 2 m sill lying at the immediate entrance to the fjord. Due to the restricted circulation imposed by these bathymetric features, a permanent oxic-anoxic interface exists at 18 ± 1 m, which is in the euphotic zone. Sorensen [16] has reported an abundance of the phototrophic sulfur-oxidizing bacteria, *Chromatium sp.* (purple colored) and *Chlorobium sp.* (green colored) below the interface.

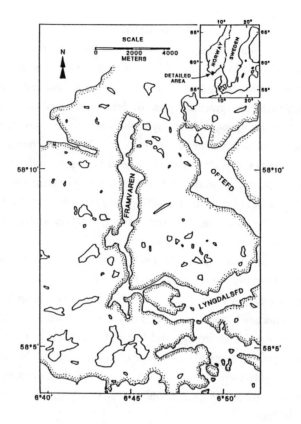

Figure 1. Location map of Framvaren Fjord, Norway

The depth profile of salinity obtained in Framvaren Fjord revealed: (a) a low salinity, isohaline surface layer between 0-4 m (9.70-9.80 ⁰/oo); (b) a steep halocline from 5-18 m (10.80-20.30 ⁰/oo); and, (c) gradually increasing salinity from 19 m to the bottom (20.60-23.06 ⁰/oo). The density (σ_θ)-depth gradient through the oxygen-hydrogen sulfide (O_2/H_2S) interface was sharp (0.13 σ_θ/m) when compared to other anoxic marine basins (e.g., Black Sea, Cariaco Trench). The anoxic waters below 18 m account for most (76%) of the total volume of Framvaren Fjord (0.32 km^3). The bottom waters are highly sulfidic, although sulfate is present at concentrations of about 10 mM [17], with dissolved H_2S concentrations up to 6 mM [18]. These concentrations are approximately 20 times greater than those reported for the bottom waters of the Black Sea and are more typical of dissolved H_2S concentrations in the porewaters of reducing marine sediments. Additionally, total alkalinity concentrations in the bottom waters exceed 20 mM [19], a factor of 8 greater than those encountered in normal seawater.

Depth profiles of O_2, H_2S and total suspended matter (TSM) are shown in Figure 2. Dissolved O_2 drop sharply from about 380 mM at 8 m to undetectable concentrations at

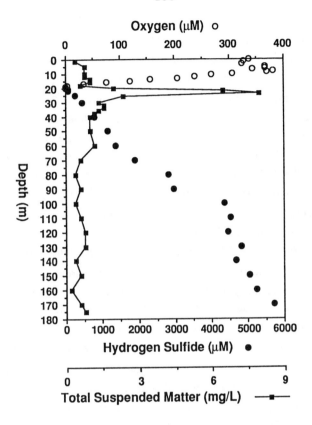

Figure 2. Oxygen, Hydrogen Sulfide and Total Suspended Matter in Framvaren Fjord

18 m, the depth of the O_2/H_2S interface. Dissolved H_2S concentrations increases rapidly with depth, attaining Black Sea bottom water concentrations (350 mM [20]) by 27 m and concentrations of 5.7 mM by 170 m.

A highly-resolved particle maximum is located between 20-26 m (Figure 2) with TSM concentrations reaching 8 mg/L at 24 m. The TSM filters for the 22 m and 24 m samples were visibly covered with purple bacteria. Sorensen [16] reported that this bacterial layer is evident as a sharp peak in light scattering (l = 430 nm) profiles, easily separated from the chlorophyll fluorescence (l = 685 nm) peak just above the interface. Within a few meters of the particle maximum, TSM concentrations drop back to 1-1.5 mg/L, averaging 0.6 mg/L at depths below 70 m. From our TSM profile, it is clear that the zone near the O_2/H_2S interface is an area of intense particle formation and destruction; a region where abundant bacterial growth is occurring, organic matter is being remineralized, and particulate sulfide phases are being formed. The partitioning behavior of ^{234}Th ($t_{1/2}$ = 24.1 days) between the dissolved and particulate phases in this remarkably unique environment is discussed below.

METHODS

Water-column samples were collected using a 30-liter Niskin bottle and transferred into 20-liter cubitainers. Samples were filtered through 0.22 mm Millipak^{TM} filters (Millipore Corporation) using a teflon-headed diaphragm pump. A separate 250 to 500 ml aliquot was filtered through a 0.2 mm Nuclepore^{TM} filter which was subsequently rinsed with deionized water to remove salts, and stored for TSM determinations. Filtrates from the 20-liter samples were acidified to pH 2.0 with concentrated HCl, spiked with known activities of ^{230}Th and ^{236}U for yield determination and with $FeCl_2$ as a carrier, then allowed to equilibrate. Concentrated NH_4OH was used to adjust the pH to 8.0, co-precipitating thorium and uranium with $Fe(OH)_3$. The precipitates were collected in settling jars connected to the cubitainers, then decanted, capped and returned to the laboratory.

In the laboratory, Millipak^{TM} filters were leached and back-flushed sequentially with NH_4OH, deionized water, warm HF-$HClO_4$ and HNO_3-HCl mixtures. Prior testing of this experimental procedure yielded near quantitative removal of thorium and uranium from Millipak^{TM} filters. The resulting leachates were combined, spiked with known activities of ^{230}Th and ^{236}U yield tracers, taken to dryness, and finally redissolved in 7N HNO_3. Iron hydroxide precipitates (dissolved thorium and uranium samples) were centrifuged, decanted and dissolved in 7N HNO_3. Uranium and thorium were separated and purified by ion exchange techniques modified from Aller and Cochran [21]. Samples were extracted into thenoyltrifluoroacetone (TTA), and evaporated onto separate stainless steel planchets. ^{234}Th activities were measured (immediately) using a low background (< 0.3 cpm) proportional beta counter, and activities of ^{230}Th, ^{236}U and ^{238}U were quantified by alpha spectrometry.

RESULTS AND DISCUSSION

The partitioning of ^{234}Th between dissolved (< 0.22 µm) and particulate (> 0.22 µm) phases varies dramatically with depth in Framvaren Fjord (Figure 3). With respect to the partitioning behavior of ^{234}Th, the water column can be divided into four distinct regions. In the surface 20 meters, ^{234}Th is relatively balanced between dissolved and particulate phases with the particulate fraction being slightly dominant (50-60%). The fraction of particulate thorium increases sharply at the oxic-anoxic interface; and, in the 20 meters below this interface, particulate ^{234}Th is the dominant phase (85-96%) reaching a maximum at 24 meters. Maxima for particulate iron and manganese are at, or just above, the oxic-anoxic interface in the oxygenated waters [3, 14]; 4 to 5 meters above the particulate thorium maximum located in the anoxic waters. Therefore, it is unlikely that co-precipitation of thorium with Fe and Mn oxyhydroxides is responsible for the high particulate thorium values observed in the 22-32 meter region. The particulate thorium maximum is coincident with the TSM maximum and with noted dense populations of purple bacteria.

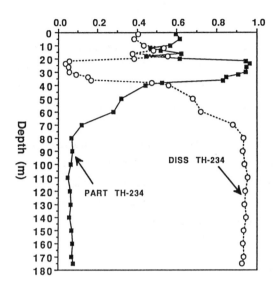

Figure 3. Fraction of thorium in particulate and dissolved phases in Framvaren Fjord

The efficient scavenging of thorium from the dissolved phase in this region appears to be associated with the abundant bacterial biomass. Whether the accumulation of ^{234}Th by bacteria is the result of passive or active mechanisms warrants further study.

In the 40 to 80 meter region, the particulate thorium fraction continues its steady decrease from the 24-meter maximum, with dissolved ^{234}Th becoming the dominant phase below 40 meters. Thorium partitioning is relatively constant below 80 meters, where the dissolved phase constitutes 92-95% of total ^{234}Th.

Distribution coefficients for ^{234}Th (K_d) range over two orders of magnitude in Framvaren Fjord, with the highest values observed in the region just below the oxic-anoxic interface. Values for log K_d range from 4.9 to 7.0 L/kg, exhibiting uniformly lower values in the anoxic bottom waters. This range of values is consistent with log K_d's from the Black Sea [22] which range from 5.4 to 6.9 L/kg; however, the anoxic bottom waters in Framvaren Fjord have considerably lower K_d values for ^{234}Th than are observed in the Black Sea or any other marine environment cited in the literature. These low K_d values are possibly due to the presence of particles with relatively unreactive surface sites for thorium sorption, changes in surface complexation chemistry resulting from high carbonate concentrations [23], and/or the rapid remineralization of thorium-bearing particulate matter.

Dissolved ^{234}Th/^{238}U activity ratios (Figure 4) exhibit large disequilibria in the surface 10 meters. Relatively high activity ratios are observed just above the oxic-anoxic interface, associated with the steep halocline. The extent of ^{234}Th/^{238}U disequilibrium (and the variability of activity ratios in the upper 18 meters) indicates the presence of very active scavenging and release processes, perhaps tied to primary production and regeneration of organic matter. In the 10 meters below the oxic-anoxic interface, large ^{234}Th/^{238}U disequilibria are again observed (similar to the activity ratios in surface waters). As was noted above, this region below the oxic-anoxic interface appears to be a zone of efficient scavenging associated with dense populations of bacteria. ^{234}Th/^{238}U activity ratios steadily increase below 30 meters, achieving complete equilibrium below 80 meters. The increase in ^{234}Th/^{238}U ratios suggests a decrease in sorption rates relative to rates of desorption and particle remineralization.

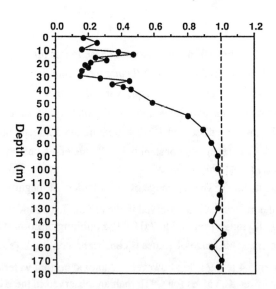

Figure 4. Dissolved Th-234/U-238 activity ratios in Framvaren Fjord

When thorium sorption is modeled as an irreversible, steady-state process [8], the removal residence time of dissolved ^{234}Th, τ_{Th}, is given by:

$$\tau_{Th} = \frac{R}{(1-R)\lambda_{Th}}$$

where R is the dissolved ^{234}Th/^{238}U activity ratio and λ_{Th} is the radioactive decay constant for ^{234}Th (0.029/day). Using this formalism, representative residence times (Table 1) are on the order of 30 days in surface waters, decreasing to 9 days below the oxic-anoxic interface at the TSM maxima, and increasing steadily with depth to values approaching 10 years in the bottom waters. In oxic environments with TSM values comparable to Framvaren Fjord, characteristic dissolved ^{234}Th residence times range from 5 to 35 days [24]. However, it is obvious from Figures 3 and 4 that desorption/particle remineralization is important and the assumption of irreversible sorption does not hold. Therefore, the traditional way of estimating residence time of dissolved thorium is not completely adequate in this environment and other means of quantifying the relative flux of ^{234}Th into the particulate phase is needed.

Table 1
Model parameters for ^{234}Th in Framvaren Fjord: June 1989

Depth (m)	R Dissolved ^{234}Th/^{238}U	^{234}Th log K_d (L/kg)	Apparent* Residence Time (days)	K_f/K_r	F_p (dpm/L/d)
14	0.473	6.02	31	1.05	0.0139
24	0.200	6.51	9	25.89	0.0081
70	0.888	5.36	273	0.14	0.0032
120	0.991	4.91	3797	0.07	0.0003

* The uncertainty associated with these values becomes high as R approaches unity

One steady-state approach is to balance dissolved ^{234}Th sources and sinks. Sources include production from dissolved ^{238}U ($\lambda_{Th}U_D$) and desorption/remineralization from particles ($K_r Th_P$); thorium removal from the dissolved phase includes ^{234}Th decay ($\lambda_{Th}Th_D$) and sorption into the particulate phase ($K_f Th_D$), where K_f is the forward rate constant (sorption), and K_r is the reverse rate constant (desorption and remineralization). The net flux of ^{234}Th into the particulate phase (F_p) is sorption minus desorption/remineralization; therefore, at steady state:

$$F_p = K_f Th_D - K_f Th_P = \lambda_{Th} U_D - \lambda_{Th} Th_D. \quad (2)$$

The absolute values of F_p decrease with depth in Framvaren Fjord (Table 1), indicative of increasing control by reverse reactions (desorption/remineralization). F_p fluxes in the anoxic bottom waters are almost two orders of magnitude lower than observed in the oxic surface waters. This indicates either: (a) decreasing sorption rates in bottom waters (perhaps related to changes in particle surface chemistry); (b) increasing rates of desorption and/or remineralization of thorium-bearing particles; or, (c) a combination of the two.

An alternative kinetic approach to quantifying the relative flux of ^{234}Th into the particulate phase is by expressing K_d in terms of rate constant ratios [25]:

$$K_d = \frac{K_f (TSM)^\phi}{K_r} \cdot \frac{1}{TSM} = \frac{K_f (TSM)^{\phi-1}}{K_r} \quad (3)$$

Data from Framvaren Fjord indicates little or no correlation between K_d (L/kg) and TSM (kg/L). Assuming that K_d is independent of TSM, as is usually observed at low TSM values, then $\phi = 0$ in equation 3 and K_d (TSM) is the ratio of forward and reverse rate coefficients (K_f/K_r). K_f/K_r values (Table 1) indicate a relative balance between forward and reverse rate constants in the upper 20 meters. Forward rate constants dominate in the region immediately below the oxic-anoxic interface, and reverse rate constants dominate below 40 meters. Therefore, the 22-32 meter zone appears to be a region of enhanced sorption, whereas thorium distribution in the zone below 40 meters appears to be governed primarily by desorption/remineralization reactions.

Conclusions

(1) ^{234}Th is equally distributed between dissolved (< 0.22 μm) and particulate (> 0.22 μm) phases in the oxic surface waters of Framvaren Fjord. K_d values for ^{234}Th in the surface 40 meters exceed 10^6 L/kg, consistent with other marine environments. Particulate ^{234}Th reaches a maximum at 24 meters, coincident with the TSM maximum and in the region of abundant bacterial biomass. K_d values are consistently low (10^4 - 10^5 L/kg) below 40 meters, with an average of 93% of total ^{234}Th in the dissolved phase in the bottom 100 meters of the water column.

(2) The net flux of ^{234}Th into particulate phases decreases with depth due to a decreased rate of sorption and/or an increased rate of desorption/remineralization in the anoxic bottom waters of Framvaren Fjord.

ACKNOWLEDGEMENTS

We greatly appreciate the efforts of Dr. Jens Skei (Norwegian Institute for Water Research; Oslo, Norway) in coordinating the sampling program in Framvaren Fjord. We also thank Pete Swarzenski and Greg Booth for laboratory assistance. This work was partially supported by NATO International Collaborative Research Grant #08721/87 (to JFT).

References

1. Spencer, D.W. and Brewer, P.G., Vertical advection diffusion and redox potentials as controls on the distribution of manganese and other trace metals dissolved in waters of the Black Sea. Journal of Geophysical Research, 1971, 76, 5877-5892.

2. Emerson, S., Cranston, R.E., and Liss P.S., Redox species in a reducing fjord: equilibrium and kinetic considerations. Deep Sea Research, 1979 26A, 859-878.

3. Jacobs, L. and Emerson, S., and Skei, J., Partitioning and transport of metals across the O_2/H_2S interface in a permanently anoxic basin: Framvaren Fjord, Norway. Geochimica et Cosmochimica Acta, 1985, 49, 1433-1444.

4. Landing, W.M. and Westerlund, S., The solution chemistry of iron (II) in Framvaren Fjord. Marine Chemistry, 1988, 23, 329-343.

5. Lewis, B.L. and Landing W.M., The biogeochemistry of manganese and iron in the Black Sea. Deep Sea Research, 1991, In Press.

6. Langmuir, D. and Herman, J.S., The mobility of thorium in natural waters at low temperatures. Geochimica et Cosmochimica Acta 1980, 44, 1753-1766.

7. Moore. W.S. and Sackett, W.M., Uranium and thorium series inequilibrium in seawater. Journal of Geophysical Research, 1964, 69, 5401-5405.

8. Bhat, S.G., Krishnaswami, S., Lal, D., Rama, and Moore, W.S., ^{234}Th/^{238}U ratios in the ocean. Earth and Planetary Science Letters, 1969, 5, 483-491.

9. Broecker, W.S., Kaufman, A., and Trier, R.M., The residence time of thorium in surface seawater and its implications regarding the fate of reactive pollutants. Earth and Planetary Science Letters, 1973, 20, 35-44.

10. Santschi, P.H., Adler, D., Amdurer, M., Li, Y-H., and Bell, J., Thorium isotopes as analogues for "particle-reactive" pollutants in coastal marine environments. Earth and Planetary Science Letters, 1980, 47, 327-335.

11. Bacon, M.P. and Anderson, R.F., Distribution of thorium isotopes between dissolved and particulate forms in the deep sea. Journal of Geophysical Research, 1982, 87, 2045-2056.

12. Coale, K.H. and Bruland, K.W., $^{234}Th/^{238}U$ disequilibria within the California Current. Limnology and Oceanography, 1985, 30, 22-33.

13. Nozaki, Y., Yang, H-S., and Yamada, M., Scavenging of thorium in the ocean. Journal of Geophysical Research, 1987, 92, 772-778.

14. Skei. J., The biogeochemistry of Framvaren - A permanent anoxic fjord near Farsund, South Norway. Norwegian Institute for Water Research. November 1986.

15. Barth, T.W. and Dons, J.A., Precambrian of southern Norway. In Geology of Norway, ed. O. Holtedahl, Norges Geol. Underskelse #208, 1960, pp. 6-67.

16. Sorensen, K., The distribution and biomass of phytoplankton and phototrophic bacteria in Framvaren, a permanently anoxic fjord in Norway. Marine Chemistry, 1988, 23, 229-241.

17. Anderson, L.G., Dyrssen, D., and Hall, P., On the sulphur chemistry of a super-anoxic fjord, Framvaren, South Norway. Marine Chemistry, 1988, 23, 283-293.

18. Millero, F.J., The oxidation of H_2S in the Framvaren Fjord. Limnology and Oceanography, 1991, In Press.

19. Anderson, L.G., Dyrssen, D., and Skei, J., Formation of chemogenic calcite in super-anoxic seawater - Framvaren, South Norway. Marine Chemistry, 1987, 20, 361-376.

20. Codispoti, L.A., Friederich, G.E., Murray, J.W., and Sakamoto, C.M., Chemical variability in the Black Sea: Implications of continuous vertical profiles that penetrated the oxic/anoxic interface. Deep Sea Research, 1991, In Press.

21. Aller, R.C. and Cochran, J.K., $^{234}Th/^{238}U$ disequilibrium and diagenetic time scales. Earth and Planetary Science Letters, 1976, 29, 37-50.

22. Wei, C-L. and Murray, J.W., $^{234}Th/^{238}U$ disequilibria in the Black Sea. Deep Sea Research, 1991, In Press.

23. LaFlamme, B.D. and Murray, J.W., Solid/solution interaction: the effect of carbonate alkalinity on adsorbed thorium. Geochimica et Cosmochimica Acta, 1987, 51, 243-250.

24. Honeyman, B.D., Balistrieri, L.S., and Murray, J.W., Oceanic trace metal scavenging: the importance of particle concentration. Deep Sea Research, 1988, 35, 227-246.

25. Nyffeler, U.P., Li, Y-H., and Santschi, P.H., A kinetic approach to describe trace-element distribution between particles and solution in natural aquatic systems. Geochimica et Cosmochimica Acta, 1984, 48, 1513-1522.

THE USE OF ^{210}Po/^{210}Pb DISEQUILIBRIA IN THE STUDY OF THE FATE OF MARINE PARTICULATE MATTER*

GEORGE D RITCHIE AND GRAHAM B SHIMMIELD
Department of Geology and Geophysics
University of Edinburgh
West Mains Road, Edinburgh, EH9 3JW, UK

ABSTRACT

The oceanic disequilibria of ^{210}Pb and ^{210}Po has been determined in an attempt to study the fate of euphotic zone particles. Levels of dissolved (<0.45µm) and particulate (>0.45µm) ^{210}Pb and ^{210}Po were determined to a depth of 1000 m in a four station south-north transect of the North Atlantic (at 20°W) in the late summer of 1989. An estimation of parent ^{226}Ra levels was made from determined silicate levels. A surface ratio of dissolved ^{210}Po/^{210}Pb of <1 and particulate ratio of >1 indicate scavenging by planktonic organisms. Two regions of high particulate ^{210}Po$_{xs}$ are observed: (1) in the mixed layer and (2) at about 500 m. Net ^{210}Po release beneath the mixed layer only occurs at the southernmost station. Box model calculations indicate residence times of 0.20 - 0.03 y for ^{210}Po and 2.58-1.54 y for ^{210}Pb both decreasing to the north. Evidence suggests that the area of greatest biological productivity was centred on 55°N. Atmospheric deposition of ^{210}Pb in this area of the NE Atlantic is calculated to be approximately 3.5 mBq cm^{-2} y^{-1}. These radionuclides give an indication of the depth of remineralisation of particulate material produced within the euphotic zone.

INTRODUCTION

The study of the disequilibria of the non-conservative ^{238}U-series radionuclides ^{210}Pb (half life=22.3y) and its daughter ^{210}Po (half life=138d) in the dissolved (<0.45µm) and particulate (>0.45µm) phases of seawater can be used to observe relatively short term oceanic particle flux processes. The high particle reactivities of these nuclides and their convenient half lives facilitate maintenance of radioactive disequilibrium by biogeochemical processes. By knowing the nuclide supply rates (determined from the concentration of the parent nuclide) the oceanic residence times of these radionuclides in the dissolved and particulate phases can be calculated giving an indication of the fate of associated particles in the water column.
The first measurements of ^{210}Pb in seawater were determined in surface samples by Rama et al. (1), with dissolved ^{210}Po first reported by Shannon et al. (2) since when much data has been reported for the world ocean. The ^{210}Pb content of seawater is supplied by in situ decay of its parent ^{226}Ra (via ^{222}Rn) and a smaller amount of surface input by decay of the atmospheric gaseous ^{222}Rn (a product of land-based ^{226}Ra). The airborne supply of ^{210}Po is negligible so that nearly all seawater ^{210}Po is derived from in situ decay of parent ^{210}Pb. It was suggested by Broecker et al. (3) and Fisher et al. (4) that the flux of particulate matter

---

*NERC Biogeochemical Ocean Flux Study (BOFS) Contribution No. 42

determines the distribution of particle reactive species in the water column and accounts for the short residence times of these nuclides in surface waters. In the case of ^{210}Pb and ^{210}Po the uptake of the nuclides by plankton has been shown to support disequilibria in both the dissolved and particulate phases (5,6,7,8,9,10).

The two nuclides are concentrated to different extents by plankton. ^{210}Po is approximately three times enriched with respect to ^{210}Pb in phytoplankton and twelve times in zooplankton (2). The greater particle reactivity of oceanic ^{210}Po has been demonstrated by Bishop et al. (11) by showing that on the small size fraction (<1μm) ^{210}Po is in great excess of ^{210}Pb. This particle reactivity is the means by which the organic cell surface of phytoplankton assimilates both nuclides (12). Surface adsorption of the nuclides by bacteria will only contribute to nuclide transport when associated with larger particles and so need not be considered separately. The higher concentration of ^{210}Po in zooplankton cannot be accounted for solely by surface adsorption. Grazing zooplankton also concentrate ^{210}Po through ingestion of particulates and adsorption of ^{210}Po into digestive organs (13,14), this being the dominant process. Zooplankton can also repackage the nuclides into faecal pellets (13) producing a ^{210}Po/^{210}Pb ratio of 2.2 (15).

It has been shown (11,16) that although faecal material contributes only 5% of the total suspended matter it accounts for as much as 99% of particulate flux reaching the sea floor. This process is obviously important when considering ^{210}Pb and ^{210}Po disequilibria, due to the elevated levels of the nuclides found in faecal pellets. As the faecal material has a high sinking velocity it is unlikely to be sampled by bottle collection which will only collect standing and slowly sinking particulates. At depth this may include an increasing proportion of faecal material disaggregated by bacterial action.

The closest comparative data comes from the equatorial Atlantic (6) and the Labrador Sea (7). This enables both latitudinal and temporal comparisons of data the validity of which is discussed below. The hydrography of the areas sampled changes to the north. The mixed layer at station 11877, is only about 10 m depth probably due to recent calm sea. The maximum mixed layer was found at station 11892 (about 50m), decreasing to the south and to the north. σ_t values (Fig. 1.) change significantly between 11877 and the more northerly stations with a very low surface value at 11877 and an intense gradient below the mixed layer. Hydrographic data are held at the British Oceanographic Data Centre, Bidston.

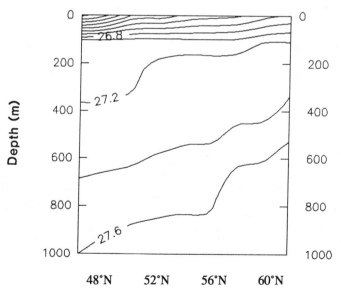

Figure 1. σ_t surfaces from 20°W transect between 48°N and 60°N.

The work presented here forms part of the UK BOFS program which is attempting to develop models of the oceanic carbon cycle. These radiotracers are being used to study euphotic zone particulate cycling thus avoiding the problems associated with traditional methods of particulate study (e.g. swimmer problems with sediment traps).

SAMPLING AND METHODS

The data presented here are from samples collected during RRS *Discovery* Cruise 184, Leg 3 of the UK BOFS 1989 program along a 20°W transect of the North Atlantic from 47°N, to 60°N. Twenty litre samples were obtained from seven depths in the upper water column and processed as in Fleer and Bacon (17).

The samples were gravity filtered through 0.45 μm Asypor^{TM} filters to remove particulates, the filters being stored for later shore based determination of the ^{210}Pb and ^{210}Po content of the particulates. The water was then acidified with HCl (40ml:conc) and spiked with stable ^{206}Pb (0.5ml:2000ppm) and radiogenic ^{208}Po (approx 67mBq). Upon equilibration cobalt (10 mg) and 1 g of APDC (ammoniumpyrrolydine dithiocarbamate) co-precipitant were added, precipitating the intrinsic ^{210}Pb and ^{210}Po and the added spikes. Collection of the precipitate was by glassfibre (GF/D) filters.

Sample analysis is by first plating out the ^{210}Po and ^{208}Po (both intrinsic and spike) and then, after a period of at least six months, measuring ^{210}Po in-growth from the intrinsic ^{210}Pb. The Asypor^{TM} filter containing the particulates is first spiked with ^{206}Pb (0.5ml:2000ppm) and ^{208}Po (approx 17mBq), acetone is added and evaporated off, helping to break up the filter. It is then digested by a mixture of nitric, perchloric and hydrofluoric acids and fumed to near dryness. The glassfibre filter was treated with an acetone (95%)/nitric acid (0.1M:5%) mixture to dissolve the co-precipitate which is then filtered off, the solution heated to dryness and the residue digested with nitric and perchloric acids and fumed to near dryness. The residues were then taken up in 1M HCl in preparation for plating onto a silver disc as described by Flynn (18). The plates were counted on an Ortec^{TM} alpha spectrometer for at least three days. Sample solutions were then stored for about one year, re-spiked, and replated to determine ^{210}Po in-growth. They were then diluted to 100 ml and analysed by flame AAS (bracketing standards) to determine stable lead recovery, this being the tracer of ^{210}Pb recovery.

The method of calculation was as reported in Fleer and Bacon (17). Sample ^{210}Po and in-grown ^{210}Po levels were determined with ^{210}Pb levels calculated from the in-growth. Activities were corrected for tracer recovery, decay and in-growth from the time of collection. Errors include combined 1σ counting error and propagated analytical errors.

RESULTS

Radium-226

The ^{226}Ra values presented here are estimated from silicate levels determined on board ship by photometric autoanalysis (Technicon Method Number 186-72W). The Ra/Si relationship used was that given in Bacon *et al.* (5) for the Eastern North Atlantic (Ra=0.217Si+6.91) giving the radium values in dpm/100kg, then corrected to mBq/10l. A surface radium value of 11.4 - 11.8 mBq/10l is observed increasing to approximately 15.0 mBq/10l at 1000 m. Moving northwards points of equal radium activity are found at progressively greater depths. This is shown in Figure 2 with results of the analyses presented in Table 1. Note that conversion from mBq/10l to dpm/100l is by multiplication by 0.6.

TABLE 1
^{210}Pb and ^{210}Po levels from BOFS Leg 3, 1989.

	Depth (m)	Dissolved Po-210 (mBq/10l)	+/-	Dissolved Pb-210 (mBq/10l)	+/-	Particulate Po-210 (mBq/10l)	+/-	Particulate Pb-210 (mBq/10l)	+/-
STATION 11877	2.0	*	*	*	*	2.74	0.45	0.89	0.12
-------------	25.4	*	*	8.92	1.52	0.96	0.12	2.33	0.13
48.00 N	54.4	13.78	3.50	6.03	1.48	0.36+	0.07	0.90	0.05
19.32 W	74.4	3.50	0.25	19.50	1.02	0.66	0.10	0.82	0.07
	103.3	7.08	0.47	18.57	0.97	0.82	0.12	0.89	0.07
19/7/89	202.1	8.20	0.58	16.97	0.98	1.68	0.60	0.35	0.12
4495m	349.0	4.20	0.32	15.18	0.93	0.49	0.10	0.64	0.07
STATION 11885	2.0	1.92	0.15	17.52	1.03	1.69	0.35	1.09	0.07
-------------	30.1	6.33	0.68	13.77	1.13	4.23	0.93	0.95	0.10
52.20 N	54.3	4.73	0.38	15.73	0.95	3.05	0.65	0.66	0.05
22.25 W	128.4	7.67	0.58	16.13	0.88	1.71	0.42	0.63	0.07
	302.8	9.38	0.80	17.28	1.10	1.09	0.25	0.84	0.05
29/7/89	501.7	6.95	0.68	13.02	1.00	2.90	1.20	0.23	0.08
4305m	994.7	3.47	0.40	11.00	1.02	2.10	0.45	0.97	0.07
STATION 11892	2.0	0.00	0.05	15.27	1.07	3.45	0.72	1.39	0.10
-------------	32.0	1.52	0.15	14.98	1.07	3.41	0.72	1.32	0.08
55.41 N	52.3	4.97	0.47	15.03	1.07	1.68	0.38	1.04	0.07
21.58 W	126.0	12.37	1.17	17.45	1.30	0.96	0.22	1.05	0.07
	300.5	7.17	0.75	13.43	1.15	1.76	0.45	0.86	0.12
2/8/89	501.4	6.08	0.65	10.75	0.95	2.96	0.73	0.49	0.07
2030m	991.7	5.42	0.60	12.13	1.03	2.07	0.53	0.73	0.07
STATION 11899	2.0	3.57+	0.33	14.13	0.97	3.44	0.75	1.68	0.10
-------------	32.4	1.78	0.17	16.07	1.02	3.68	0.82	1.89	0.10
59.05 N	54.1	7.02	0.60	16.03	1.03	2.79	0.67	0.99	0.07
21.58 W	130.3	5.57	0.43	15.83	1.05	2.42	0.55	0.76	0.07
	307.4	5.07	0.40	14.27	0.98	1.94	0.43	0.99	0.08
5/8/89	496.6	2.60	0.23	11.47	0.90	2.90	1.13	0.76	0.25
2853m	993.3	1.50	0.15	10.40	0.87	2.12	0.47	1.00	0.07

* No sample
+ Correction made for poor plating efficiency.

Comparisons with our data are made with equatorial Atlantic data (6), sampling being carried out in December 1973 at stations 32 (33°N, 13°W) and 34 (38°N, 11°W) somewhat to the south of the samples presented here. Also with the data from the Labrador Sea (7), sampling being carried out in June 1975, at a similar latitude to our data.

Lead-210

High surface particulate ^{210}Pb (0.89-1.68 mBq/10l) is observed (Fig. 2.) dropping to between 0.23 and 0.76 mBq/10l at 500 m and increasing again below this. Surface dissolved levels of ^{210}Pb are high (up to 17.52 mBq/10l) due to atmospheric input to the upper water column. Below the surface layer levels drop reaching approximately 11 mBq/10l at about 1000 m. A small increase in surface particulate ^{210}Pb is seen to the north but no significant change in dissolved levels. These values are largely in agreement with previous Atlantic work (6,7).

From the ^{226}Ra profiles a surface dissolved ^{210}Pb excess is seen except at station 11877 where there is a ^{210}Pb deficit above 100 m, possibly due to recent scavenging. A dissolved ^{210}Pb deficit becomes apparent between 300 and 450m depth.

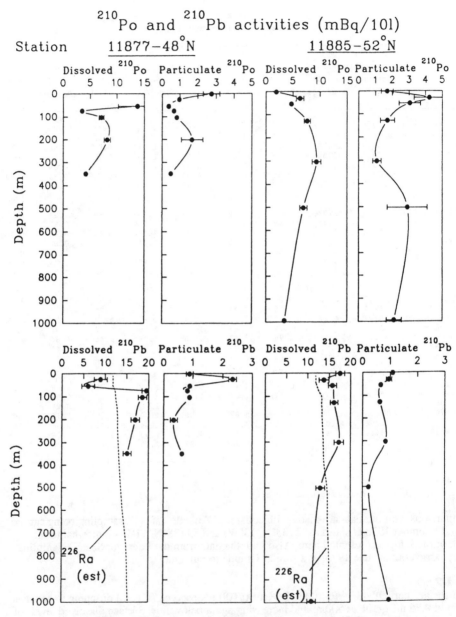

Figure 2(a). ^{226}Ra, ^{210}Pb and ^{210}Po profiles for stations 11877 and 11885 (mBq/10l)

147

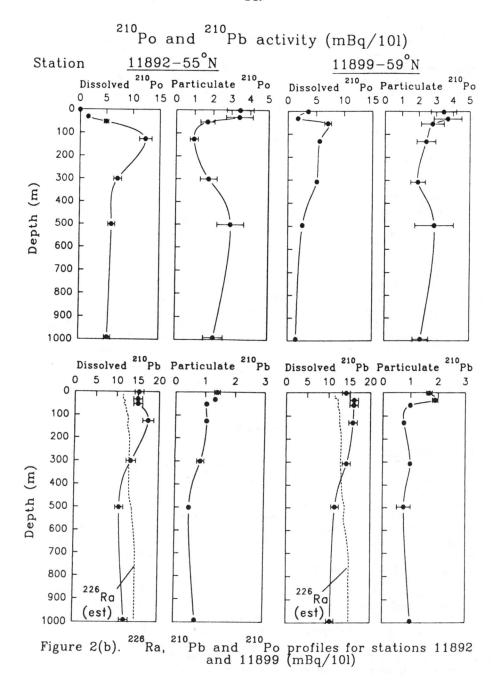

Figure 2(b). ^{226}Ra, ^{210}Pb and ^{210}Po profiles for stations 11892 and 11899 (mBq/10l)

Polonium-210

The general trend of low (about 2 mBq/10l) surface dissolved ^{210}Po increasing to a mid-water maximum and again decreasing to low values at about 1000m, shown in Figure 2, is similar to previously recorded trends (5,6,7,8,9). This follows the pattern of dissolved ^{210}Pb below the mixed layer. The peaks appear either on, or just below, the pycnocline as defined by the σ_t profiles except at station 11885 where a deeper maximum appears. The particulate values are high (up to 3.45 mBq/10l) at the surface decreasing to a minimum coincident with the dissolved maximum. They increase again to a similar high at about 500m and then shadow the decrease in dissolved values to 1000m following a similar pattern to the particulate ^{210}Pb. The total depletion of surface dissolved ^{210}Po at station 11892 is unusual, although this has been observed previously in the Atlantic by Bacon (6) and in the Indian Ocean by Cochran *et al.* (8).

The results presented here show increasing particulate ^{210}Po from south to north for any given depth sampled, but no systematic change in the dissolved levels, although the lowest values are found at the most northerly station. These values are approximately between those of the Labrador Sea (7) and the north equatorial Atlantic (6). The comparison with the equatorial Atlantic data shows the dissolved ^{210}Po activities to be lower at the BOFS stations, closer to the Labrador Sea levels. This anomaly may be accounted for by recent productivity differences in the area of sampling. The Meteor (6) samples were collected during December with probable low recent productivity compared to the early summer sampling of the Labrador Sea and the late summer sampling of the BOFS cruise. The more recent productivity and associated particle flux at the BOFS stations may have stripped dissolved ^{210}Po from the water column leaving depleted values of both particulate and dissolved ^{210}Po. This hypothesis is supported by bathymetric photography carried out during the cruise which showed large amounts of organic phytodetritus lying on the sediment surface. The levels of dissolved ^{210}Pb however are approximately the same as the equatorial Atlantic. This is probably due to the lower particle reactivity of the ^{210}Pb preventing significant depletion of dissolved ^{210}Pb.

DISCUSSION

Excess Polonium-210

Excess ^{210}Po (^{210}Po$_{xs}$) values are simply the subtraction of the activity of ^{210}Pb from that of ^{210}Po in either phase at each depth. The units are therefore the same as those for the discrete values. The profiles (Fig. 3.) show a large dissolved ^{210}Po deficit at the surface. Two processes may account for this; (1) preferential assimilation of ^{210}Po by plankton indicated by high particulate ^{210}Po$_{xs}$, and (2) the atmospheric input of dissolved ^{210}Pb. It is probable that both mechanisms contribute. There is rapid increase in dissolved ^{210}Po$_{xs}$ from the surface to a maximum at the base of the mixed layer. This increase only becomes a net excess at 50m at station 11877, confirmed by a minimum in the ^{210}Po$_{xs}$ particulate profile at the coincident depth, suggesting rapid remineralisation. Station 11877 appears to show different characteristics to the other stations. As previously shown the hydrology, particularly the σ_t profile, at station 11877 is significantly different to that of the other stations. This indicates a significant influence of the water stratification at this station on nuclide disequilibria and hence particulate sinking rates. At the less stratified northern stations a surface deficit and a subsurface peak of dissolved ^{210}Po was observed however there was no subsurface net dissolved ^{210}Po$_{xs}$. This influence of water column stratification was previously suggested by Bacon *et al.* (7) who concluded that only highly stratified waters, for example the north equatorial Atlantic (6), could produce such a net ^{210}Po release.

Below the mixed layer the dissolved ^{210}Po$_{xs}$ gradually increases peaking again at about 100 m between -9 and -5 mBq/10l. The particulate excess is shown to be high between 1 and 3 mBq/10l at the surface or immediate subsurface, drops to a mid-water low and rises to a maximum similar in magnitude to the surface at a depth of 500 m before decreasing to approximately 1 mBq/10l around 1000 m. If simple adsorptive processes were responsible for

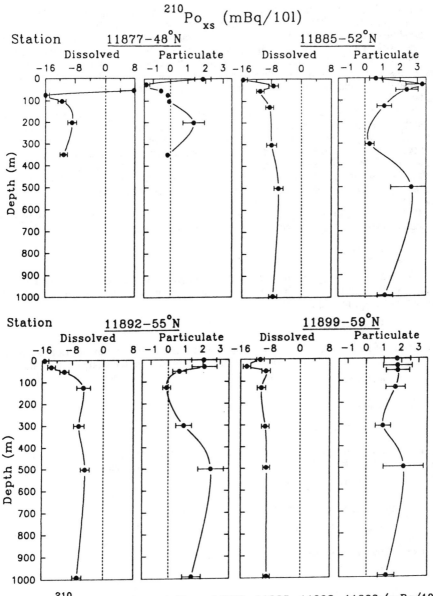

Figure 3. $^{210}Po_{xs}$ profiles, stations 11877, 11885, 11892, 11899 (mBq/10l)

the high particulate $^{210}Po_{xs}$ there would be a coincident drop in the dissolved $^{210}Po_{xs}$. However the dissolved $^{210}Po_{xs}$ either stays constant or increases slightly. An explanation may be the breakup of faecal pellets, leading to some remineralisation and an increase in standing particulate stock. This may give a small increase in the dissolved excess but a large one in particulate excess due to an increased contribution to standing stock from fragmenting faecal pellets. If this increase were due to advective processes then a noticeable change in dissolved nuclide activities would be expected. However it is not seen. The particulate $^{210}Po_{xs}$ in the mixed layer drops from south to north.

Mass balance calculations are used to determine fluxes from specific depth ranges of a box model of the water column. Mass balance equations assume that the only processes affecting the activity of the two nuclides are supply from parent decay, decay of the nuclide, transfer between the dissolved and particulate phases and loss of particulates through sedimentation. From these assumptions the flux of the ^{210}Pb and ^{210}Po in respect of adsorption from solution and sedimentation of particulates can be determined. The corresponding residence time within any depth zone is calculated from the reciprocal of the nuclide flux.

First we state that in the dissolved phase production of ^{210}Po by dissolved ^{210}Pb decay is balanced by decay and transfer of ^{210}Po to the particulate phase:

$$\lambda_{Po} A^d_{Pb} + J_{Po} = \lambda_{Po} A^d_{Po} \qquad (1)$$

Where λ is the decay constant (yr^{-1}), A is the inventory (mBq cm^{-2}), J is the net flux (mBq cm^{-2} yr^{-1}), negative values indicating adsorption. From this J value the dissolved ^{210}Pb residence time (τ^d_{Po}) in years is calculated:

$$\tau^d_{Po} = A^d_{Po}/-J_{Po} \qquad (2)$$

Supply of particulate ^{210}Po by particulate ^{210}Pb decay and by transfer from solution is assumed to be balanced by decay and settling of particulates from the study depth range:

$$\lambda_{Po} A^p_{Pb} - J_{Po} = \lambda_{Po} A^p_{Po} + P_{Po} \qquad (3)$$

The particulate ^{210}Po residence time is given by:

$$\tau^p_{Po} = A^p_{Po}/P_{Po} \qquad (4)$$

By assuming that the ^{210}Pb is removed from the mixed layer by the same particles which remove particulate ^{210}Po we can find its flux from the study range:

$$P_{Pb} = (A^p_{Pb}/A^p_{Po})P_{Po} \qquad (5)$$

The flux of dissolved ^{210}Pb can then be determined:

$$-J_{Pb} = \lambda_{Pb} A^p_{Pb} + P_{Pb} \qquad (6)$$

Its residence time is therefore:

$$\tau^d_{Pb} = A^d_{Pb}/J_{Pb} \qquad (7)$$

The atmospheric ^{210}Pb input is then determined (assuming it is in the dissolved phase):

$$\lambda_{Pb} A_{Ra} + I_a + J_{Pb} = \lambda_{Pb} A^d_{Pb} \qquad (8)$$

Residence times were calculated from inventories of ^{226}Ra, ^{210}Pb and ^{210}Po over depth bands representing the mixed layer and the thermocline.

TABLE 2
Calculated residence times and comparative data.

Station (Refs.)	Latitude	Depth Range (m)	I_a (mBq cm^{-2} y^{-1})	τ^p_{Po} (yr)	τ^d_{Po} (yr)	τ^d_{Pb} (yr)	Part. Po/Pb
11877	48°00N	0-30					1.14
		54-350		0.05	0.43	1.09	1.12
11885	52°20N	0-30	1.83	0.17	0.20	2.58	3.74
		0-50	2.50	0.18	0.21	3.08	3.34
		30-500		0.18	0.47	4.65	3.71
		50-500		0.17	0.58	4.59	2.93
11892	55°41N	0-30	2.83	0.15	0.03	1.69	2.53
		0-50	4.67	0.14	0.09	1.65	2.28
		30-500		0.18	0.64	2.98	2.04
		50-500		0.24	0.87	4.20	2.47
11899	59°05N	0-30	3.00	0.18	0.12	1.54	2.00
		0-50	4.17	0.19	0.20	1.91	2.17
		30-500		0.18	0.30	2.90	2.85
		50-500		0.17	0.26	2.76	2.66
Arctic Ocean (19)	85°N	0-50	0.50	1.4	>3	5	0.94
N. Eq. Atlantic (5) approx.	20°N	0-50	8.50	0.08	0.27	1.4	1.72
Labrador Sea (7)	50°N	0-50	3.33	0.5	0.15	6*	5.25
			6.67	0.6	0.16	7*	
Phytoplankton (2)							3
Zooplankton (2)							12

* Values variable with water sinking rates

The data indicates that the area of most rapid surface depletion of dissolved nuclides is centred on station 11892. At depth, the adsorbed nuclides then appear to be regenerated intensely with high dissolved and particulate residence times. The spatial distribution of biological uptake and assimilation of ^{210}Po, and resulting regeneration, is indicated by Fig. 4.

Comparison with the water column structure (Figure 1) suggests that biological processes within the euphotic zone dominate the dissolved ^{210}Po distribution, whilst particle sinking rates and resulting disaggregation of faecal pellet material is largely affected by the density structure of the water column beneath the seasonal thermocline. At the time of sampling (late summer) the distribution of dissolved ^{210}Po and the calculated residence times suggest that the locus of biological production and recycling was centred around 55°N (station 11892). At 59°N (station 11899) high surface values of particulate ^{210}Po and associated high residence time suggests that a bloom was developing but was in an early stage such that no significant sinking or regeneration of organic material had occurred. Consequently, no subsurface regions of high dissolved ^{210}Po are observed. Similar changes in dissolved nuclide residence times have also been related to productivity variation by Nozaki et al. (20) for the North Pacific gyre.

A surface particulate Po/Pb ratio maximum of 3.74 is found at station 11885. This suggests a greater proportion of zooplankton in the particulate load at this station compared to those to the north. This increase in relative zooplankton population, lagging the phytoplankton bloom, supports the suggestion of bloom progression to the north.

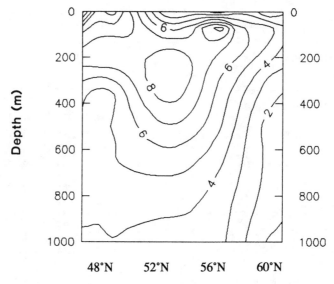

Figure 4. Dissolved ^{210}Po profile from 20°W transect (values in mBq/10l).

The particulate activity ratio at station 11877 of 1.14, with virtually no change at depth, and low particulate nuclide concentrations indicates that this area is probably post-bloom.

Estimates of the atmospheric input of ^{210}Pb increase to the north, and are closer to those for the Labrador Sea (7) than the equatorial Atlantic (6), indicating latitudinal influence. The I_a values for the transect show consistently greater input to the fifty metre than the thirty metre band although the mixed layer was seen to vary within this range. This indicates that penetration of atmospheric ^{210}Pb is not governed by the depth of the mixed layer.

CONCLUSIONS

1. In the eastern North Atlantic ^{210}Pb and ^{210}Po are removed from the surface mixed layer by ingestion, assimilation and adsorption by plankton and faecal products, followed by subsequent vertical flux of the organodetritus.
2. A net subsurface release of dissolved ^{210}Po (exceeding the activity of its parent ^{210}Pb) is only seen in the highly stratified water at station 11877 (48°N).
3. The residence time in the surface mixed layer of dissolved ^{210}Po is approximately 0.2 years. That of particulate ^{210}Po is approximately 0.18 years. Within the thermocline ^{210}Pb residence times are approximately 5-10 times greater than dissolved ^{210}Po.
4. The disequilibria produced between ^{210}Pb and ^{210}Po indicate the depth and intensity of recycling of organic matter.
5. Break-up of faecal pellets appears to contribute significantly to the standing particulate load at about 500m.
6. The atmospheric input of ^{210}Pb is latitude dependent.

ACKNOWLEDGEMENTS

We wish to thank the captain, crew and scientific party of RRS *Discovery* Cruise 184 for help with sample collection. We would also like to thank Mike Saunders for help with the chemical analyses. This work was supported by NERC Grant No. GST/02/300.

REFERENCES

1. Rama, M., Koide, and Goldberg, E.D., Lead-210 in natural waters. Science, 1961, **134**, 98-99.
2. Shannon, L.V., Cherry, R.D. and Orren, M.J., Polonium-210 and lead-210 in the marine environment. Geochim. Cosmochim. Acta, 1970, **34**, 701-711.
3. Broeker, W.S., Kaufman, A. and Trier, T.M., The residence time of thorium in surface seawater and its implications regarding the rate of reactive pollutants. Earth Planet. Sci Lett., 1973, **20**, 35-44.
4. Fisher, N.S., Teyssie, J-L., Krishnaswami, S. and Baskaran, M., Accumulation of Th, Pb, U, and Ra in marine phytoplankton and its geochemical significance. Limnol. Oceanogr., 1987, **32(1)**, 131-142.
5. Bacon, M.P., Spencer, D.W. and Brewer, P.G., $^{210}Pb/^{226}Ra$ and $^{210}Po/^{210}Pb$ Disequilibria in seawater and suspended particulate matter. Earth Planet. Sci. Lett., 1976, **32**, 227-296.
6. Bacon, M.P., ^{210}Pb and ^{210}Po results from F.S."Meteor" Cruise 32 in the North Atlantic, "Meteor" Forschungsergeb. Reihe A, 1977, **19**, 24-36.
7. Bacon, M.P., Spencer, D.W. and Brewer, P.G., Lead-210 and Polonium-210 as Marine Tracers: Review and Discussion of Results from the Labrador Sea.(1978), From: Natural Radiation Environment III, (T.F. Gesell and W.F. Lowder, eds.), Vol. 1, pp. 473-501. U.S. Department of Energy Report CONF-780422. Proceedings of a symposium held at Houston, Texas, April 23-28, 1978.
8. Cochran, J.K., Bacon, M.P., Krishnaswami, S., and Turekian, K.K., ^{210}Po and ^{210}Pb distributions in the central and eastern Indian Ocean. Earth Planet. Sci. Lett., 1983, **65**, 433-452.
9. Thompson, J. and Turekian, K.K., ^{210}Po and ^{210}Pb distributions in ocean water profiles from the Eastern South Pacific. Earth Planet. Sci. Lett., 1976, **32**, 297-303.
10. Nozaki, Y. and Tsunogai, S., $^{226}Ra, ^{210}Pb$ and ^{210}Po Disequilibria in the Western North Pacific. Earth Planet. Sci. Lett., 1976, **32**, 313-321.
11. Bishop, J.K.B., Edmond, J.M., Ketten, D.R., Bacon, M.P. and Silkers, W.B., The chemistry, biology, and vertical flux of particulate matter from the upper 400m of the equatorial Atlantic Ocean. Deep-Sea Res., 1977, **24**, 511-548.
12. Fisher, N.S., Burns, K.A., Cherry, R.D. and Heyraud, M., Accumulation and cellular distribution of $^{214}Am, ^{210}Po$, and ^{210}Pb in two marine algae. Mar. Ecol. Prog. Ser., 1983, **11**, 233-237.
13. Cherry, R.D., Fowler, S.W., Beasley, T.M. and Heyraud, M., Polonium-210: Its vertical oceanic transport by zooplankton metabolic activity. Mar. Chem., 1975, **3**, 105-110.
14. Skwarzec, B. and Falkowski, L., Accumulation of ^{210}Po in Baltic Invertibrates. J. Env. Rad., 1988, **8**, 99-109.
15. Beasley, T.M., Heyraud, M., Higgo, J.J.W., Cherry, R.D. and Fowler, S.W., ^{210}Po and ^{210}Pb in zooplankton faecal pellets. Mar. Biol., 1978, **44**, 325-328.
16. McCave, I.N., Vertical flux of particles in the ocean. Deep-Sea Res., 1975, **22**, 491-502.
17. Fleer, A.P. and Bacon, M.P., Determination of ^{210}Pb and ^{210}Po in seawater and marine particulate matter. Nucl. Instr. and Meth. in Phys. Res., 1984, **223**, 243-249.
18. Flynn, W.W., The determination of low levels of polonium-210 in environmental materials. Analyt. Chim. Acta., 1968, **43**, 221-227.
19. Moore, R.M. and Smith, J.N., Disequilibria between $^{226}Ra, ^{210}Pb$ and ^{210}Po in the Arctic Ocean and the implications for chemical modification of Pacific water inflow. Earth Planet. Sci. Lett., 1986, **77**, 285-292.
20. Nozaki, Y., Thomson, J., and Turekian, K.K., The distribution of ^{210}Pb and ^{210}Po in the surface waters of the Pacific Ocean. Earth Planet. Sci Lett., 1976, **32**, 304-312.

STUDY OF THE SCAVENGING OF TRACE METALS IN MARINE SYSTEMS USING RADIONUCLIDES

Roland WOLLAST and Michèle LOIJENS
Laboratory of Chemical Oceanography
University of Brussels
Campus Plaine, C.P. 208, Bd. du Triomphe, 1050 Brussels, Belgium.

ABSTRACT

The rate of scavenging of Mn, Co, Zn, and Cd by particles and the possible role of living organisms in the uptake of these metals has been investigated using water samples collected in various marine environments. The method consists of measuring the transfer of the corresponding radioactive trace metals from the dissolved to the particulate phase simultaneously with measurements of the uptake of labeled ^{14}C bicarbonate, as classically performed for estimations of primary production.

Relative uptake measurements were conducted in incubators under constant light conditions. Experiments in the dark or using inhibitors of the biological activity have also been carried out. Preliminary experiments have been tested in the laboratory using pure cultures of diatoms.

Results of uptake experiments of radioactive trace elements in waters of various marine environments indicate that the method provides an easy way to quantify the rate of uptake of radioactive tracers. The elements investigated exhibit a different but specific behaviour, similar in the various environments studied. The transfer of radionuclides occurs mainly through passive uptake except for Mn. The uptake of this element is significantly enhanced during photosynthesis.

INTRODUCTION

The transfer of dissolved components to particles is a fundamental step for the removal of elements from the water column. At the scale of the residence time of elements in the marine system, the particles are rapidly removed by sedimentation and the scavenging process constitutes thus the rate limiting step. Several physical, chemical and biological mechanisms have been proposed for this transfer. Many trace elements exhibit vertical concentration profiles of the dissolved species which show significant surface depletion and deep water regeneration. The similarity of these profiles to those for the vertical distribution of nutrients (see e.a. Bruland, 1980) and the known metabolic role of several of these trace elements (e.g. V, Cr, Mn, Fe, Co, Ni, Cu, Zn, Ba,..) strongly suggest that they tend to be incorporated into the biological cycle. In addition, there are several elements that have no apparent biological function but show nutrient like vertical profiles (e.g. As, Sr, Ag, Cd). These elements may be assimilated simultaneously with the essential elements because of the lack of discrimination in the biological uptake mechanism and incorporated for example in the skeleton of the organisms. The elements can also diffuse passively through their membranes.

Futhermore, many trace elements form strong complexes with organic ligands, which favors their scavenging from the water column by adsorption onto the organic surfaces of living or dead organisms. It is therefore not surprising that certain elements are scavenged at higher rates in areas of high productivity (Tanaka et al., 1983; Bacon et al., 1985; Coale and Bruland, 1985 and 1987) and exhibit seasonal trends in relation with the annual cycle of plankton (Tanaka et al. 1983; Jickels et al. 1984; Bacon et al. 1985). Morel and Hudson (1985) have therefore suggested the consideration of metals as essential elements of living organic matter that are removed by organisms from seawater according to a Redfield type ratio.

There have been several in vitro studies of the uptake of trace elements by marine plankton, mainly performed on pure cultures and often conducted in order to evaluate the toxic effect of these elements. There have been however few attempts devoted to the evaluation of the uptake of trace elements under natural conditions. Sorption of radiotracers by natural sediments has been used successfully to estimate the distribution coefficients between the dissolved and particulate phase (e.g. Balistiery and Murray, 1983; Li et al. 1984; Buchtoltz et al. 1985) or to study the uptake rates of metals (e.g. Nyffeler et al. 1984; Santschi et al. 1986; Jannasch et al. 1988). Usually, very fast sorption of the radiotracers is followed by a much slower and extended uptake. This suggests that there is a rapid adsorption at the surface of the particles which is followed by slower physical, chemical or biological transfer processes.

The purpose of the present study is to evaluate the rate of scavenging of trace metals by particles in natural samples and to investigate the possible role of living organisms in the uptake of these metals. The method used consists of measuring the transfer of various radiotracers (Mn, Co, Zn, and Cd) from the dissolved to the particulate phase simultaneously with measurements of the uptake of ^{14}C labeled bicarbonate, as classically performed for primary productivity estimations of marine samples. The biological uptake of these trace elements was first tested on a pure laboratory culture of a diatom population (Fragilaria sp.).

The method has been tested on water samples collected in various contrasting marine environments. These samples spiked with radionuclides ^{59}Mn, ^{60}Co, ^{65}Zn and ^{109}Cd were incubated at sea-water temperature, on board of the ship under natural or constant light conditions. The experiments are carried out with low levels of radioactivity and of concentration of the metal carriers in order to maintain the experimental conditions as close as possible to the natural ones.

We will present here the results of the laboratory experiment carried with a pure culture and of the incubation experiments performed on water samples collected at a few selected marine stations corresponding to coastal and open sea environments.

LABORATORY EXPERIMENT ON PURE CULTURE

Methods

A preliminary experiment was carried out with a pure culture of Fragilaria sp. in sterile artificial sea-water spiked with four radiotracers ^{54}Mn, ^{60}Co, ^{65}Zn, ^{109}Cd). A separate sample of the same culture was spiked with ^{14}C bicarbonate (6 uCi per 100 ml) in order to evaluate the primary productivity. The initial concentration of the diatom biomass was fixed at 20 mg/l dry weight. The activity of each metal added was approximately 50 nCi corresponding to the addition of 1 ug of Mn and Cd, 2 ug of Zn and 2.7 ug of Co in 100 ml samples of the artificial culture. The cultures were subjected to light for a total of 10 hours per day during the 24 hour experimental runs. One culture was kept in the dark for 24 hours and two cultures were inhibited with 100 ul of saturated $HgCl_2$ solution. All experiments were performed in an incubator at 19 °C. At given time intervals, 100 ml bottles spiked with the trace radioelements were withdrawn from the incubator and immediately filtered through 0.45 um Millipore filters. The filters were rinsed with prefiltered sea-water. Both the filters and the solutions were gamma counted at the "Institut des Radioéléments" (detector : HPGE Camberra, relative efficiency : 40%, resolution : 1.8 KeV; spectrometer : IN96B Intertechnique). Counting of both filters and solutions allowed to verify the assumption that no significant amount of the radiotracers was adsorbed on the walls of the polystyrene containers.

At similar time intervals, 100 ml samples spiked with radiocarbon were filtered on GFC filters and the filters analysed by liquid scintillation (Packard Tri-carb 3255). The efficiency was determined with a ^{14}Chexadecane standard solution.

Results

In order to evaluate the relative uptake of the various radiotracers used in these experiments, we have used the activity of the solid phase collected on the filters divided by the total activity (filter plus solution) at a given time, expressed in parts per thousand (ppt).

The evolution of the relative uptake in the pure culture, represented in figure 1 indicates that there is a very significant transfer of the four elements considered from the dissolved to the particulate phase. After a single diurnal cycle, 30 per cent of the initial zinc and manganese, 17 per cent of the initial cadmium and 6 per cent of initial cobalt have been transfered to the particulate phase.

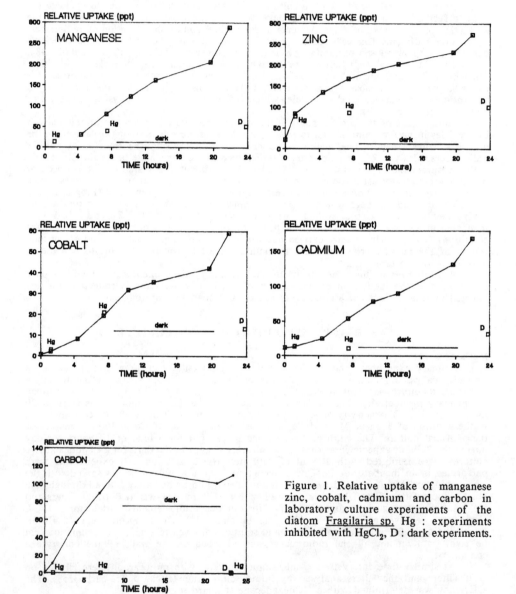

Figure 1. Relative uptake of manganese zinc, cobalt, cadmium and carbon in laboratory culture experiments of the diatom _Fragilaria sp._ Hg : experiments inhibited with $HgCl_2$, D : dark experiments.

Furthermore, there is a systematic increase of transfer under light conditions, as indicated by the much lower values of the activities recorded in the dark experiment after 24 hours. The effect of light is also confirmed by the fact that the rate of scavenging decreases for all the trace elements during the 14h dark period. Nevertheless, the transfer of trace elements to the particulate phase persists at a lower rate under these conditions although there is no primary production during this period, as shown in figure 1.

The experiments carried with samples poisoned by $HgCl_2$ confirm the total inhibition of photosynthesis as shown by the absence of ^{14}C fixation after 8 hours of illumination (fig 1). However, there is a significant transfer of Mn, Zn, Co and to a lesser extend of Cd after the same time. Except for Co, there is nevertheless a higher uptake when the biological population is active.

These results are similar to previously published observations made during laboratory experiments with pure cultures using the radiotracer methodology but with much higher levels of concentration of carriers and radionuclides. Fisher (1981, 1985) for example found that accumulation of metals by eukariotic phytoplankton or blue-green picoplankton was rapid and only slightly affected by the absence of light. He concluded that the uptake is mainly passive and that the partition between the particulate and dissolved phases reached an adsorption equilibrium which can be described by a Freundlich isotherm.

It is thus very likely that the scavenging of the trace elements observed during our experiment may be attributed to the active and passive uptake of these elements by the organisms. It is however important to recognize that other removal mechanisms of these trace elements from the solution are possible. Among them, chemical precipitation of Mn^{++} with coprecipitation of other metals (especially Co) is very likely even at the low concentrations used in these experiments. It should be pointed out that the inhibition of the uptake of Cd in $HgCl_2$ spiked seawater may also reflect the result of a competitive adsorption process between the two metals. In the presence of an excess of Hg^{++}, the adsorption of Cd may be completely hindered. We have therefore used sodium azide in the field experiments.

FIELD MEASUREMENTS

Methods

A slightly modified procedure was used to estimate the uptake of trace elements in natural samples. Usually, the water samples were collected in surface waters at depths corresponding to the maximum of fluorescence or of dissolved oxygen as recorded during the lowering of the CTD rosette. The collected seawater samples were immediately spiked with the radionuclides and incubated under constant light conditions, at the temperature of surface water. A few experiments were also carried out under variable light conditions. In order to reduce the perturbations of the natural conditons during the incubation experiment the concentration of the carrier was reduces and the addition of radionuclides corresponds to a final concentration of 10 nmoles/l for Mn, Co, Zn and 5 nmoles/l for Cd. The activities were measured with a HPGE Canberra detector with a relative efficiency of 20 % and a multichannel spectrometer serie 20 model 2802. The activities recorded on the filters after incubation of open-sea water samples are often of the order 0.1 dpm and requires long counting times. The minimum number of counts was fixed at 1000 to reduce the statistical error of counting.

Results

The simplest experiment consists in the incubation of a spiked sample under constant light. Figure 2 shows typical results obtained with a seawater sample collected at the limit between the Channel and the North Atlantic (49°03'32" N - 05°12'30" W). This area is characterized by the presence of a tidal front (the Ouessan front) which is responsible for the high productivity observed during the cruise (september 1989). The production estimated from the ^{14}C fixation was equal to 0.2 umoles C $l^{-1}h^{-1}$ and correspond to a eutrophic conditions.

During the incubation experiment, the fixation of C remained constant over the 24 hours of constant illumination. The behaviour of the trace elements is very different. There is first a very rapid initial transfer of the radionuclides to the solid phase. The first measurements, taken 10 minutes after the addition of the radionuclide spike, show that the initial uptake is especially important for zinc. It is followed by a slower transfer of this element from the dissolved to the particulate phase during a second step. At stations with a lower productivity this second step is less pronounced and the concentration of the particulate radionuclide reaches rapidly a steady state situation. For the other elements, the evolution of the concentration of the radionuclide in the solid phase follows roughly a parabolic curve. The amount of metals transferred to the particulate phase can be easily calculated from its concentration in the aqueous phase. After 24 hours the amounts transferred were respectively equal to 4 umoles of C, 65 pmoles of Mn, 50 pmoles of Zn, 10 pmoles of Co and 5 pmoles of Cd per liter of sea water.

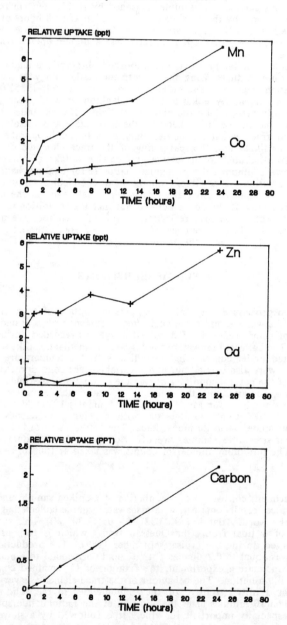

Figure 2. Scavenging of dissolved radionuclides and ^{14}C fixation during an incubation experiment of a water sample collected in the Channel. Constant light conditions.

The results obtained during incubation experiments carried out in the dark or with addition of sodium azide are compared to the experiments under constant light in figure 3. The results are very similar to what was observed in the laboratory experiments. A large fraction of the transfer of the radionuclides occurs in the dark or in samples were the biological activity has been inhibited. There seems however to be an active biological uptake of some trace elements and especially of manganese.

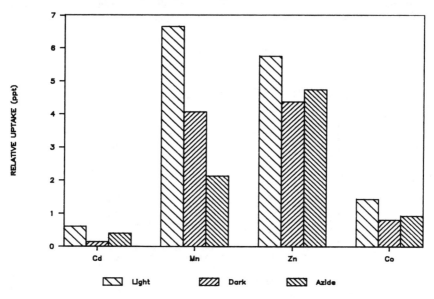

Figure 3. Influence of dark and azide on the transfer of radionuclides after an incubation period of 24 hours on the Channel water samples.

The uptake experiments were repeated during this cruise at other stations in the Gulf of Biscaye where much less productive waters are encountered. The central part of the Gulf is dominated by oligotrophic surface waters of the North Atlantic. The results, shown in table 1, indicate that the transfer of metals were much less important in the area of low productivity. The effect is more important for manganese which seems to be, from the other experiments, the element which is the most significantly influenced by the biological activity.

Station	Primary Production nmC $l^{-1}h^{-1}$	Chlor.a ug/l	S.M. ug/l	P.O.C. ugC/l	Relative uptake % after 8 hours			
					Mn	Co	Zn	Cd
2	200	2.9	806	276	3.67	0.76	3.83	0.76
6	45	0.46	135	44	0.57	0.56	0.94	0.56
5	20	0.28	81	22	0.15	0.28	1.03	0.28

Table 1. Uptake of radionuclides at 3 stations of the Channel and Gulf of Biscaye, compared to the primary production and the concentrations of chlorophyll a, suspended matter and particulate organic carbon. Station 2 (49°03'32"N-05°12'30"W) Station 5 (47°19'50"N-08°31'30"W) Station 6 (46°00'09"N-08°30'06"W)

It is however difficult to identify clearly the parameter controlling the uptake of the radionuclides. As shown in table 1, there is a high correlation between the primary production and other master variables like the concentration of suspended matter and that of particulate organic carbon.

Similar results have been obtained in the Mediterranean sea. Figure 4 compares the transfer of the radionuclides from the dissolved to the particulate phase observed at two stations in the Gulf of Lion. Station 8 is near the mouth of the river Rhone (43°14'16"N 04°55'33"E) and station 25 (42°27'15"N 05°18'31"E) corresponds more to an open ocean station in the central part of the gulf. Primary production observed during the summer cruise were extremely low at both stations and of the order of 36 nmoles $l^{-1}.h^{-1}$ and 6 nmoles $l^{-1}.h^{-1}$ in the central part and in the coastal area respectively.

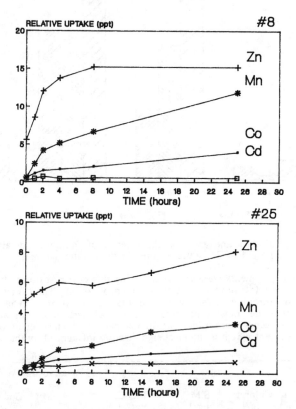

Figure 4. Scavenging of radionuclides during an incubation experiment under constant light of water samples collected in the Gulf of Lion, Mediterranean sea, July 1989.

The shapes of the uptake curves are very similar to the kinetics observed in figure 2 for the Channel. The fast initial transfer of ^{65}Zn is especially well demonstrated in both samples. The relative uptake of all the metals after 24 hours is significantly lower than in the previous experiments.

Numerous uptake experiments were performed during two cruise in the Northwestern Mediterranean sea during 1988 and 1989. The results were always very similar to those presented in figure 4. During these incubations experiments we have observed a striking correlation between the cobalt and the manganese transfer between the dissolved and the particulate phases (figure 5). It is well established that oxy-hydroxydes of Mn adsorb many trace metals and that redox cycling of Mn in natural waters can greatly affect the fate of many trace metals (Balistieri and Murray, 1986). The correlation here is remarkable.

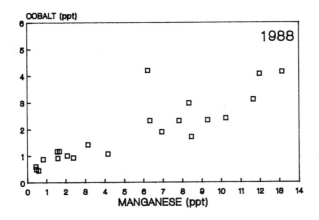

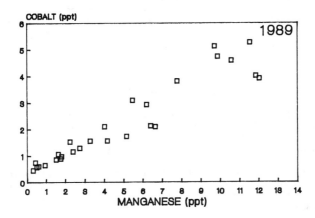

Figure 5. Comparison between the uptake of cobalt and of manganese of water samples collected during the Discovery Cruise (December 1988) and the Bannock Cruise (July 1989) in the Mediterranean sea.

CONCLUSIONS

Results of the uptake experiments of radioactive trace elements in water samples collected in various marine environments indicate that the method provides an easy way to quantify the rate of scavenging of trace elements.

The transfer of the radionuclides from the dissolved to the particulate phase starts first with a very rapid uptake (in less than 10 minutes) of the dissolved radionuclide by the particulate matter. This fast step may be due to the adsorption and exchange of the radionuclides with the surface of the particles. This phenomenon is especially important for zinc. During a second step there is a slower uptake of the dissolved radionuclides which follows roughly a parabolic law. In most cases the rate of transfer after 24 hours becomes very low and the concentration of the particulate radionuclide tend to reach a steady-state. These results are very similar to those obtained during laboratory studies of the kinetics of trace elements uptake by marine particles (Li et al, 1984 ; Nyffeler et al, 1984 ; Santchi et al, 1987 ; Jannach et al, 1988).

From the concentration of the particulate matter in the water samples it is also possible to calculate a distribution coefficient (Kd) of the radionuclides between the solid and the liquid phase after 24 hours of reaction. This "fast" distribution coefficient is of the order of 10^3 ml/mg for Mn and Zn, and 10^2 ml/mg for Cd and Co. These values are also similar to those found in the laboratory experiments cited except for Co which exhibits usually a higher for K_d.

One of the aims of this study was also to evaluate the influence of the organisms and especially of the photosynthesis on the scavenging process of trace elements. The comparison of various stations characterized by different productivities in the North-Atlantic indicates that the transfer of radionuclides is proportional to the primary production and to the concentration of chlorophyle a. However, the concentrations of particulate matter and of particulate organic carbon are also well correlated with the primary production at these stations. Light/dark incubations or experiments performed with inhibition of the biological activity indicate that most of the transfer of radionuclides occurs through a passive uptake. There is obviously a contribution of the biological activity for manganese and to a smaller extend for cobalt, which seems to be related to an active process.

The use of radiotracers applied to natural samples appears to be very fruitful and may allow a better understanding of the mechanism of transfer of trace metals from the dissolved to the particulate phase in the marine environment. Furthermore if the rate of transfer of radionuclides is measured simultaneously with the concentration of the dissolved trace metals in the water column, it is possible to estimate the rate of scavenging of these elements in the natural environment, provided that the spikes used the manipulation during the incubations does not disturb significantly the samples.

ACKNOWLEDGMENTS

This work was supported by the EEC contract EROS-2000 Nr. EV4V0111F and Impulse programme "Global Change" of the Belgian State-Prime Minister's Service. Science Policy Office Contract Nr GC/11/009.

The cruises were performed on board of the R.R.V. Discovery (1988), R.V. Bannock (1989) and R.V. Belgica (1988-1989).

We thank very much the crew of these vessels for their help during the cruises. The measurements of chlorophyl reported here were performed by Ch Lejaer (Liège University) and the concentration of particulate organic carbon by L. Goyens and F. Dehairs (Vrije Universiteit Brussel).

BIBLIOGRAPHY

Bacon, M. P., C. A. Huh, A. P. Fleer and W. G. Deuser. 1985. Seasonality in the flux of natural radionuclides and plutonium in the deep Sargasso Sea. Deep-Sea Research, 32, 273-286.

Balistrieri, L. S. and J. W. Murray. 1983. Metal-solid interactions in the marine environment : Estimating the apparent equilibrium binding constants. Geochim. Coscmochim. Acta 47, 1091-1098.

Balistrieri, L.S. and Murray, J.W., 1986. The surface chemistry of sediments from the Panama Basin: the influence of Mn oxides on metal adsorption. Geochim. Cosmochim. Acta, 50, 2235-2243.

Bruland, K. W. 1980. Oceanographic distribution of cadmium, zinc, nickel and copper in the North-Pacific. Earth Plan. Sc. Letters, 47, 176-198.

Buchholtz, M. R., P. H. Santschi and W. S. Broecker. 1985. Comparison of radiotracer K_d values from batch equilibration experiments with in-situ determinations in the deep sea using the MANOP lander : The importance of geochemical mechanisms in controlling ion uptake and migration. In : Proc. Scientific Seminar on the Application of Distribution Coefficients to Radiological Assessement Models. CEC, Brussels.

Coale, K. H. and K. W. Bruland. 1985. ^{234}Th-^{238}U disequilibria within the California Current. Limnol. and Oceanogr., 30, 22-33.

Coale, K. H. and K. W. Bruland. 1987. Oceanic stratified euphotic zone as elucidated by ^{234}Th-^{238}U disequilibria. Limnol. and Oceanogr., 32, 189-200.

Fisher, N.S. 1981. On the selection for heavy metal tolerance in diatoms from the Derwent Estuary, Tasmania. Austr. J. Mar. Freshwater Res. **32**, 555-561.

Fisher, N.S. 1985. Accumulation of metals by marine picoplankton. Marine Biology, **87**, 137-142.

Jannasch, H. W., B. D. Honeyman, L. S. Balistrieri and J. W. Murray. 1988. Kinetics of trace element uptake by marine particles. Geochim. et Cosmochim. Acta, **52**, 567-577.

Jickells, T. D., W. G. Deuser and A.H. Knap. 1984. The sedimentation rate of trace elements in the Sargasso Sea measured by sediment trap. Deep-Sea Res., **31**, 1169-1178

Li, Y. H., L. Burkhardt, M. Buchholtz, P. O'Hara and P. H. Santschi. 1984. Partition of radiotracers between suspended particles and seawater. Geochim. et Cosmochim. Acta, **48**, 2011-2019.

Morel, F. M. and R. J. Hudson. 1985. The geobiological cycle of trace elements in aqueous systems : Redfield revisited. **In** : Chemical Processes in Lakes (ed. W. Stumm). J. Wiley & Sons, 389-426.

Nyffeler U. P., Y. H. Li and P. H. Santschi. 1984. A kinetic approach to describe trace element distribution between particles and solution in natural systems. Geochim. et Cosmochim. Acta, **48**, 1513-1522.

Santschi, P. H., U. P. Nyffeler, R. F. Anderson, S. L. Schiff and P. O'Hara. 1986. Response of radioactive trace metals to acid-base titration in controlled experimental ecosystems : Evaluation of transport parameters for application to whole-lake radiotracer experiments. Can. J. Fish Aquat. Sci. **43**, 60-77.

Tanaka, N., Y. Takeda and S. Tsunogao. 1983. Biological effect on removal of Th-234, Po-210 and Pb-210 from surface water in Funka Bay, Japan. Geochim. et Cosmochim. Acta, **47**, 1783-1790.

LEAD-210 BALANCE AND IMPLICATIONS FOR PARTICLE TRANSPORT ON THE CONTINENTAL SHELF, MID-ATLANTIC BIGHT, USA

MICHAEL P. BACON, REBECCA A. BELASTOCK
Woods Hole Oceanographic Institution
Woods Hole, MA 02543
USA

MICHAEL H. BOTHNER
U. S. Geological Survey
Woods Hole, MA 02543
USA

ABSTRACT

Investigations carried out as part of the Shelf-Edge Exchange Processes (SEEP) program show that a consideration of the geochemical balance of the natural radionuclide Pb-210 (half-life 22 years) can provide important insights concerning the fate of fine particulate matter on the continental shelf. This is because (1) the supply of Pb-210 is dominated by the well-known input from the atmosphere and (2) the removal is dominated by scavenging and particle transport. An important and somewhat counter-intuitive finding is that there are significant inventories of Pb-210 (similar in magnitude to the inventory predicted from the atmospheric supply) in the shelf sediments, even though the sediments are predominantly coarse-grained sands that are not believed to be accumulating at present. Analyses of size-fractionated sediment samples show that the Pb-210 accumulations are due mostly to the presence of small amounts (a few percent) of silt- and clay-sized material mixed into the sands. The hypothesis is advanced that particulate matter introduced to the shelf is trapped efficiently on the shelf because deposition is followed by downward mixing into the sediment column by benthic organisms. Eventually there is completion of the cycle by mixing back up to the sediment surface followed by resuspension, but completion of the cycle and the eventual export of the particles take many decades to complete (time for most of the Pb-210 to decay). The resulting long detention of particles on the shelf has important implications for the fate of particulate organic matter and the possible entry of particle-associated contaminants into food chains on the shelf.

165

SURFACE COMPLEXATION MODELLING OF PLUTONIUM ADSORPTION ON SEDIMENTS OF THE ESK ESTUARY, CUMBRIA

DAVID R. TURNER, SUSAN KNOX, FRANCISCO PENEDO[1]
Plymouth Marine Laboratory, Prospect Place, West Hoe, Plymouth PL1 3DH, UK

JOHN G. TITLEY[2], JOHN HAMILTON–TAYLOR, MICHAEL KELLY
Institute of Environmental and Biological Sciences, Lancaster University,
Lancaster LA1 4YQ, UK

GERAINT L. WILLIAMS[3]
Department of Chemistry, Imperial College, London SW7 2AY, UK

1. Present address: Departamento de Química Física, Universidad de Santiago de Compostela, E–15706 Santiago de Compostela, Spain
2. Present address: National Radiological Protection Board, Chilton, Didcot, Oxon, OX11 0RQ
3. Present address: Admiralty Research Establishment, Holton Heath, Poole, Dorset BH16 6JU

ABSTRACT

Previous work has shown that plutonium is remobilised from Esk sediments at low salinities of overlying water. A constant capacitance surface complexation model has been developed in order to understand and model the chemical processes occurring. The model is based on detailed chemical characterisation of sediment samples from the estuary. The following measurements were carried out to provide input parameters for the model: (i) specific surface area (by EGME adsorption and BET); (ii) total surface sites (by tritium exchange); (iii) proton and major ion exchange (by potentiometric titration); and (iv) plutonium partition coefficient (K_d) as a function of pH and salinity (by α spectrometry). A chemical model for the plutonium adsorption process has been developed.

INTRODUCTION

The Esk estuary has been the focus of much interest with regard to artificial radionuclides because of its proximity (10 km to the south) to the nuclear fuel reprocessing plant at Sellafield. Artificial radionuclides, originating in the authorised low–level discharges from Sellafield, have been transported into the estuaries of the area by tidal inflows. Large gross fluxes of Pu into and out of the estuary occur over a tidal cycle, associated mainly with suspended sediment movements [1,2]. That transport of Pu is dominated by sediment movement is in agreement with the high observed distribution coefficients (K_d) for this element. Pentreath [3] reported K_d values for Irish Sea waters of around 10^5 l.kg^{-1}, implying that at equilibrium the majority of the Pu remains on the sediments until they are dispersed in the water column to a concentraton of less than 10 mg.l^{-1}.

Field observations and laboratory experiments indicate that Pu behaves non–conservatively in the estuary with release from sediment to solution (<0.2μm) occurring at salinities less than ≈5ppt [4–8]. It has been observed that (i) both Pu(III/IV) and Pu(V/VI) display non–conservative behaviour, (ii) the specific activity of Pu in solution is related to sediment concentration, (iii) the desorption process appears to be a reversible exchange reaction reaching quasi–equilibrium within 30 minutes, (iv) H$^+$ (and probably major seawater cations) is exchanged concurrently with Pu, and (v) only ≈5% of the total sediment bound Pu is available in this way [6,7].

In order to improve our understanding of these processes, a project was initiated to develop detailed chemical models for the reactions of Pu at Esk sediment surfaces. In this paper we report measurements and modelling of Pu(IV) adsorption on Esk sediments. Further details, and information on other actinides, can be found in a recent contract report [9].

THEORY

This work uses a surface complexation model, which treats the process of adsorption at particle surfaces as entirely analogous to complexation in the dissolved phase, with the key difference that surface complexing groups are immobilised on the surface, leading to the buildup of a charge. A range of models have been used for surface complexation reactions, differing in their assumptions about the surface reactions occurring and the charge/potential relationships at the solid/solution interface. We have chosen to use the constant capacitance model which is conceptually simple, and where the double layer structure is described by a single parameter, the capacitance [10]. In order to model the adsorption process, the following information is required:

* complexation reactions in solution (major components)
* complexation reactions in solution (trace adsorbing component)
* surface characteristics (surface area, site density, capacitance)
* surface reactions (major components)
* surface reactions (trace adsorbing component)

Experimental measurements were carried out for the surface processes, while the solution processes were modelled on the basis of information available from the literature. The major ionic components were modelled using the ion–pairing model of Dickson and Whitfield [11], while the plutonium dissolved phase model was based on stability constants derived from the literature, fitted to an extended Debye–Hückel equation in the manner used previously for other trace components [12,13].

EXPERIMENTAL METHODS

Sampling

Sediment samples were collected by surface scrape from a number of sites along the Esk estuary (Fig. 1). Following physicochemical characterisation, one sample was selected for detailed examination.

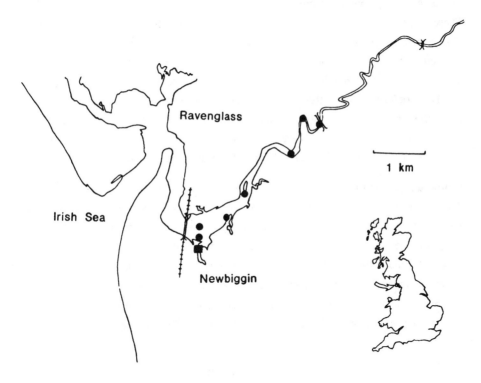

Figure 1. Site of sediment sample selected for this work (■) and other sample sites (●) in the Esk estuary.

Surface site density

Surface site density was estimated from separate determinations of specific surface area (by BET N_2 adsorption and by ethylene glycol monoethylether (EGME) adsorption), and of the concentration of surface sites (by tritium exchange). It was found that the results obtained by the tritium exchange technique were critically dependent on the outgassing temperature used. Thermogravimetric analysis, differential scanning calorimetry, and X–ray diffractometry were used to identify outgassing conditions which resulted in loss of adsorbed water without damaging the solid phase structure [14]. The optimum outgassing conditions were found to be 200°C for 15 hrs.

Proton and major ion exchange

These processes were characterised by acid/base potentiometric titrations of suspensions of Esk sediments. Following Balistrieri and Murray [15], who carried out a similar characterisation for a pure solid phase (goethite), titrations were carried out in each of the four major salt solutions NaCl, Na_2SO_4, $MgCl_2$, $CaCl_2$, each at three ionic strengths (0.01, 0.1, 0.7M). In order to obtain information relevant to the Esk estuarine system, the sediment concentration was kept low (of the order 1 g.l^{-1}), and the maximum time allowed for equilibration after additions of base was of the order of 30 minutes, corresponding to the timescale of the quasi–equilibrium adsorption process noted above.

Plutonium adsorption

As with the potentiometric titrations, the experimental conditions were designed to approach those of the estuary as closely as possible, *viz.* low sediment concentration (50–100 mg.l^{-1}), low Pu concentration ($\approx 10^{-16}$M), and reaction times of the order of 30 minutes. ^{236}Pu was used as the tracer for these experiments, and all adsorption experiments were carried out at constant temperature (25°C) and pH, through use of a pH–stat system. Pu was recovered and determined by α spectrometry from both the solid and dissolved phases in order to avoid the uncertainties associated with estimating the Pu content of one phase by difference.

RESULTS

Surface site density

Tritium exchange measurements yielded a surface site concentration of 0.16 mol.kg^{-1}. The two surface area measurements gave 4.5 m^2.g^{-1} (BET), and 40 m^2.g^{-1} (EGME), corresponding to surface site densities of 21 and 2.4 sites.nm^{-2} respectively when combined with the surface site concentration value. Davis and Kent [16] recommend using a density of 2.31 sites.nm^{-2} for all minerals where direct measurements are unavailable: the BET–based value, which is unrealistically high in any case, is therefore rejected in favour of the EGME–based value.

Proton and major ion exchange

The results of the potentiometric titrations were converted to proton surface charge as a function of pH. The point of zero charge was found to be poorly defined for this material, and a value of 7.0 was estimated from separate salt titration experiments [17]. Non–linear regression techniques were then used to fit values of intrinsic constants for surface complexation reactions and of capacitance (assumed constant in all solutions): the results are summarised in Table 1. The capacitance obtained is lower than that commonly obtained for oxide phases with the constant capacitance model (e.g. 1.28 $F.m^{-2}$ for alumina [18]). This may, however, reflect the differences between pure oxide phases and natural estuarine particles.

TABLE 1

Constants derived from potentiometric titration data

Reaction[a]	log K
$SOH = SO^- + H^+$	–7.29
$SOH + H^+ = SOH_2^+$	6.77
$SOH + Na^+ = SO^-Na^+ + H^+$	–8.46
$SOH + Cl^- + H^+ = SOH_2^+Cl^-$	3.52
$SOH + Mg^{2+} + H_2O = SO^-MgOH^+ + 2H^+$	–16.97
$SOH + Ca^{2+} + H_2O = SO^-CaOH^+ + 2H^+$	–24.7
$SOH + SO_4^{2-} + 2H^+ = SOH_2^+HSO_4^-$	7.96
Capacitance/$F.m^{-2}$	0.89

[a] SOH represents surface binding site

Plutonium adsorption

Three sets of measurements were carried out: in 0.7M NaCl and artificial seawater at a range of different pH values, and in artificial seawater at a range of different salinities, all at pH 7.2. The results were then expressed as log K_d, and non–linear regression techniques were used to identify the Pu surface complexes consistent with these data, and to estimate their values. The results are shown in Figs. 2–4, and the values of the Pu surface constants are summarised in Table 2. The constants are similar to those obtained by Sanchez *et al.* [19] for Pu adsorption on goethite, although direct comparison may be misleading since these authors were using a triple layer model for surface complexation, and different constants for other surface reactions and Pu dissolved phase reactions. We were unable to define the constants for less hydroxylated surface complexes since we did not carry out measurements at pH<4.

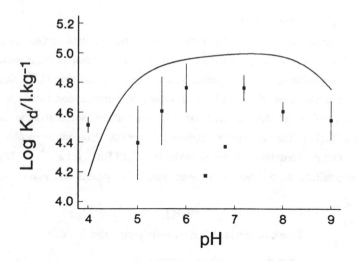

Figure 2. Pu partition coefficient (K_d) in 0.7 M NaCl as a function of pH. Individual points show measured values ± 1 s.e. (based on replicate experiments); the continuous line shows the fitted curve.

TABLE 2

Constants derived from Pu adsorption data

Reaction[a]	log K	
	this work	Sanchez et al. [19][b]
$SOH + Pu^{4+} + H_2O = SO\text{-}PuOH^{3+} + 2H^+$	–	2.5
$SOH + Pu^{4+} + 2H_2O = SO\text{-}Pu(OH)_2^{2+} + 3H^+$	–	–2.0
$SOH + Pu^{4+} + 3H_2O = SO\text{-}Pu(OH)_3^+ + 4H^+$	–4.7	–5.9
$SOH + Pu^{4+} + 4H_2O = SO\text{-}Pu(OH)_4^0 + 5H^+$	–12.2	–12.0

[a] SOH represents surface binding site
[b] adsorption on goethite measured over pH range 2–9; triple layer surface complexation model used

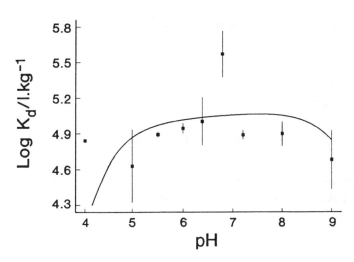

Figure 3. Pu partition coefficient (K_d) in 35‰ artificial seawater as a function of pH. Individual points show measured values ± 1 s.e. (based on replicate experiments); the continuous line shows the fitted curve.

DISCUSSION

Surface complexation modelling developed from studies of acid–base and adsorption reactions at the surfaces of pure oxide surfaces such as goethite, alumina, and silica. Application to natural, composite solid phases is attractive since it offers a means of replacing the wholly empirical cataloguing of K_d values with a modelling approach which can deal with changes in salinity and pH of the aqueous phase in a coherent manner. Such an approach is best described as semi–empirical, since the assumptions which lie behind the surface complexation model are not strictly justified for composite materials with a significant organic content: the organic content of estuarine particles is known to have a major influence on their surface properties [20]. Previous applications of surface complexation models to soils, groundwaters, and fresh waters have demonstrated the value of this approach in dealing with natural materials [16]. The system studied here is particularly challenging because (i) Pu has a complicated solution chemistry, with some key complexes still poorly defined, and (ii) the salinity range in estuaries encompasses a wide variation in major ion composition. It can be seen that certain features of the experimental data are not consistent with the current adsorption model:

* K_d values at pH 4, which are predicted by the model to be lower
* the model cannot reproduce the full extent of the adsorption peak at intermediate salinities

* the model understates the difference in K_d between 0.7NaCl and seawater

Despite these problems, the progress achieved in modelling the observed Pu adsorption behaviour in terms of just two surface complexes is extremely encouraging.

A key conclusion which can be drawn from Figs. 3 and 4 concerns the mechanism of desorption of Pu on the ebb tide which has been observed in the Esk, and which provided the main impetus for this project. During the ebb tide, both the salinity and pH of the water overlying the intertidal mudflats fall. It can be seen from Fig. 3 that a fall in pH produces little change in K_d, while in contrast a reduction in salinity below the 10–15‰ range results in a sharp drop in K_d, and it is therefore most likely to be the salinity change which triggers the desorption of Pu.

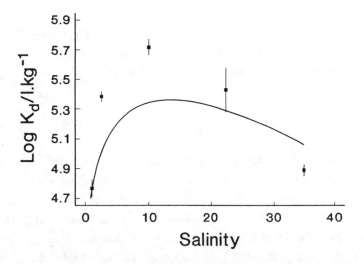

Figure 4. Pu partition coefficient (K_d) in artificial seawater as a function of salinity at pH 7.2. Individual points show measured values ± 1 s.e. (based on replicate experiments); the continuous line shows the fitted curve.

ACKNOWLEDGEMENTS

This work was supported by the Department of the Environment (UK) under contract number PECD 7/9/429. FP thanks the University of Santiago de Compostela for the award of a grant enabling him to work in Plymouth.

REFERENCES

1. Burton P.J. and Yarnold L.P., Remobilisation of actinides from intertidal sediments of the Ravenglass Estuary. Sci.Tot.Environ., 1988, **69**, 239– 260.
2. Kelly M., Emptage M., Mudge S., Bradshaw K., and Hamilton–Taylor J., The relationship between sediment and plutonium budgets in a small macrotidal estuary: Esk Estuary, Cumbria, UK. J.Environ.Radioactivity, 1991, in press.
3. Pentreath R.J., Radioactive discharges from Sellafield (UK). In: Behaviour of Radionuclides Released into Coastal Waters, IAEA–TECDOC–329, IAEA, 1985, Vienna, pp 67–110.
4. Assinder D.J., Kelly M., and Aston S.R., Conservative and non–conservative behaviour of radionuclides in an estuarine environment with particular respect to the behaviour of plutonium. Environ.Technol.Lett.,, 1984, **5**, 23–30.
5. Burton P.J., Laboratory studies on the remobilisation of actinides from Ravenglass Estuary sediment. Sci.Tot.Environ., 1986, **52**, 123–145.
6. Hamilton–Taylor J., Kelly M., Mudge S., and Bradshaw K., Rapid remobilisation of plutonium from estuarine sediments. J.Environ.Radioactivity, 1987, **5**, 409–423.
7. Mudge S., Hamilton–Taylor J., Kelly M., and Bradshaw K., Laboratory studies of the chemical behaviour of plutonium associated with contaminated estuarine sediments. J.Environ.Radioactivity, 1988, **8**, 217.
8. Kelly M., Mudge S., Hamilton–Taylor J., and Bradshaw K., The behaviour of dissolved plutonium in the Esk Estuary, UK. In: Radionuclides: a Tool for Oceanography, (Guary J.C., Guegueniat P., and Pentreath R.J. Eds.), Elsevier, 1988, pp 321–330.
9. Turner D.R., Knox S., Titley J.G., Hamilton–Taylor J., Kelly M., and Williams G.L., Application of a Surface Complexation Model to the Interactions of Pu and Am with Esk Estuary Sediments. Report on contract PECD/7/9/429, submitted to DOE, Plymouth Marine Laboratory, 1991, 156pp.
10. Dzombak D.A. and Morel F.M.M., Surface Complexation Modelling: Hydrous Ferric Oxide, John Wiley, New York, 1990, 393pp.
11. Dickson A.G. and Whitfield M., An ion–association model for estimating acidity constants (at 25°C and 1 atm. pressure) in electrolyte mixtures related to seawater (ionic strength < 1 mol.kg^{-1}H$_2$O). Mar.Chem., 1981, **10**, 315–333.
12. Turner D.R., Whitfield M., and Dickson A.G., The equilibrium speciation of dissolved components in freshwater and seawater at 25°C and 1 atmosphere pressure. Geochim.Cosmochim.Acta, 1981, **45**, 855–881.
13. Turner D.R. and Whitfield M., An equilibrium speciation model for copper in sea and estuarine waters at 25°C including complexation with glycine, EDTA and NTA. Geochim.Cosmochim.Acta, 1987, **51**, 3231–3239.

14. Williams G.L., Characterisation of Structural Hydroxyl Groups of Estuarine Sediments, Ph.D. Thesis, Imperial College, London, 1991,327pp.
15. Balistrieri L.S. and Murray J.W., Surface of goethite (α–FeOOH) in seawater. In: Chemical Modelling in Aqueous Systems, (Jenne E.A. Ed.), American Chemical Society, Washington DC, 1979, pp 275–298.
16. Davis J.A. and Kent D.B., Surface complexation modelling in aqueous geochemistry. In: Mineral–Water Interface Geochemistry, (Hochella M.F. and White A.F. Eds.), Reviews in Mineralogy, 1990, 23, 177–260.
17. Davis J.A. and Leckie J.O., Surface ionisation and complexation at the oxide/water interface. II. Surface properties of amorphous iron oxyhydroxide and adsorption of metal ions. J.Coll.Interfac.Sci., 1978, 67, 90–107.
18. Lövgren L., Sjöberg S., and Schindler P., Acid/base reactions and Al(III) complexation at the surface of goethite. Geochim.Cosmochim.Acta, 1990, 54, 1301–1306.
19. Sanchez A.L., Murray J.W., and Sibley T.H., The adsorption of plutonium(IV) and (V) on goethite. Geochim.Cosmochim.Acta, 1985, 49, 2297–2307.
20. Hunter K.A. and Liss P.S., Organic matter and the surface charge of suspended particles. Limnol.Oceanogr., 1982, 27, 322–335.

Processes in the Seabed

The Determination of ^{240}Pu/^{239}Pu Atomic Ratios and ^{237}Np Concentrations Within Marine Sediments

K.E. Sampson, R.D. Scott, *M.S. Baxter and **R.C. Hutton.
SURRC, East Kilbride, Glasgow, G75 OQU.
*International Laboratory of Marine Radioactivity, IAEA, Monaco, MC98000.
**VG Elemental, Ion path, Road Three, Winsford, CW7 3BX.

ABSTRACT

Inductively coupled plasma mass spectrometry is a sensitive multielemental detection technique capable of determining isotopic ratios. An account is given of its use in the simultaneous determination of neptunium and plutonium isotopes in marine sediments. Matrix suppression of the ion signal and the low concentration of these radionuclides in the environment imply that some chemical separation and preconcentration is required prior to analysis. A procedure has been developed to extract plutonium isotopes and ^{237}Np simultaneously from the bulk matrix.

Using electrothermal vaporisation to introduce samples into the plasma, ^{237}Np, ^{239}Pu and ^{240}Pu concentrations have been determined from 1g aliquots of a bulk sediment sample taken from the Esk estuary, Cumbria. The results were in good agreement with those obtained using liquid nebulisation ICP-MS and alpha-spectrometry.

An assessment of the accuracy and precision of ETV-ICP-MS is given from data obtained using certified standard solutions. As an application of the technique, the analysis of a marine sediment core will be outlined.

INTRODUCTION

In the early 1980's a new analytical technique was developed by the successful combination of an inductively coupled plasma ion source (ICP) with a mass spectrometer (MS). The development and applications of ICP-MS have been reviewed by Date and Gray, and Houk [1, 2].

From the outset, it was recognised that ICP-MS offered potential advantages over radiometric techniques for the determination of long-lived radionuclides [3]. The detection of decay products is often inappropriate for radionuclides with low specific activity as long counting times are required. ICP-MS however is an atom-based technique with the sensitivity to detect a few picograms of material from solution, in a matter of minutes. In addition, ICP-MS does not require such a high standard of chemical separation as techniques such as α-spectrometry where thick sources and overlapping α-energies cause interference. The main disadvantage of ICP-MS apart from isobaric interferences, is the restriction in the dissolved solid content of the sample which must be kept to below 0.5% to prevent clogging of the sampling orifices [4]. This limitation, combined with the low concentrations of some radionuclides in the environment, implies that separation of trace analytes from the major matrix constituents is still required.

The latest improvement in the sensitivity of ICP-MS has been through the development of electrothermal vaporisation (ETV) as a sample introduction technique [5]. ETV employs 50 µl sample volumes and therefore has the advantage of an increased potential for sample preconcentration and enables the analysis of small samples.

For nearly 40 years BNFL Sellafield, Cumbria have discharged a cocktail of radionuclides into the marine environment via the Irish Sea. Of the long lived radionuclides discharged, ^{237}Np, ^{239}Pu and ^{240}Pu are free from isobaric interferences and so are suitable for determination by ICP-MS. Although ^{237}Np is of long-term radiological significance through ingrowth from ^{241}Pu and ^{241}Am, there is little environmental data available [6]. Techniques previously used to determine ^{237}Np, such as α-spectrometry and neutron activation, require lengthy chemical separation procedures to remove uranium isotopes which cause interferences. ICP-MS has been shown to be as sensitive as these techniques [7]. ICP-MS also offers the advantage of determination of the ^{240}Pu/^{239}Pu ratio, which provides information on the source term [8], but which can be determined radiometrically only by a laborious combination of α-spectrometry and fission track analysis after neutron irradiation [9]. The ^{240}Pu/^{239}Pu atom ratio is dependent on the degree of burn up of the reactor fuel, and for Magnox fuel varies from 0.07, for low burn up material, to 0.30 [10]. The average fallout value is 0.18. The ^{240}Pu/^{239}Pu atom ratio has been used to determine the degree of mixing in Irish Sea sediments by relating the degree of burn up to the time of discharge [11].

METHODS

A bulk intertidal sediment sample was taken from the Esk estuary, Cumbria. This was then dried and homogenised. 1g aliquots of this sample were taken for analysis by ETV-ICP-MS and 20g for both liquid nebulisation ICP-MS and α-spectrometry. The chemical separation procedure used, shown in figure 1, was developed at SURRC by Hursthouse [12] from a method by Byrne [13]. ^{239}Np, milked from a ^{243}Am solution (Amersham International Ltd.), was used as a chemical yield tracer for ^{237}Np whilst ^{242}Pu was used both as yield tracer and internal standard for determination of plutonium isotopes. The yield tracers were added to the ashed sample which was then leached with aqua regia for at least 48 hours to allow the tracers time to equilibrate with the sample. ^{236}U was used as an internal standard for neptunium analysis to correct for matrix effects. All reagents used were of AnalaR quality.

The ICP-MS instrument used was a VG PlasmaQuad (PQ1) with a Microtherm ETV (VG Elemental Limited, Cheshire, UK). The instrument parameters are shown in table 1. To assess the capabilities of ETV-ICP-MS compared to liquid nebulisation a plutonium isotopic standard reference material with certified isotope ratios (AEA Technology, Harwell Laboratory, Oxforshire, UK) was analysed.

Table 1: ICP-MS Operating Parameters

Operating Parameters for ETV and Liquid Nebulisation

Instrument	VG PlasmaQuad (PQ1)
Forward r.f.power	1400W
Sampling cone	Platinum 1mm aperture
Skimmer cone	Nickel 0.75mm aperture
Load coil-sampling distance	10mm
Coolant gas flow	14 l/min Ar
Auxiliary gas flow	0.4 l/min Ar
Nebuliser gas flow	0.75 l/min
Data system	IBM PC-AT (version 3.02)

Scanning Parameters

Scanning mass range : 235 to 246 a.m.u.

	Liquid Nebulisation	ETV
No. scan sweeps	1600	100
Dwell time (μs)	80	80
Approx. run time (s)	66	4

Figure 1
Analytical procedure for Neptunium and Plutonium Analysis.

SAMPLE
1 - 20g
│   + yield monitors, Pu-242 and Np-239

DISSOLUTION by HCl and HNO_3
│   + 50ml 9M HCl

REMOVE IRON
Di-isopropylether extraction
│

ADJUST OXIDATION STATES
1g NH_3OHCl
│

ADD TO ANIONIC EXCHANGE COLUMN
Dowex 1X8, 100-200 mesh
│

REMOVE URANIUM
200ml 8M HNO_3
│

ELUTE Pu AND Np
200ml 1.2M HCl
│

TTA Extraction from 0.5M HCl
Back extract with 8M HNO_3
│

TAKE UP IN 2% HNO_3
1ml ETV, 5ml liquid nebulisation.
│

COUNT ON Ge(Li) (^{237}Np recovery)
│
ADD INTERNAL STANDARD (^{236}U)
│
ASPIRATE INTO ICP

RESULTS AND DISCUSSION

ETV-ICP-MS

In ETV-ICP-MS the sample enters the plasma as a concentrated plug so there is little time in which to collect data. Data can be collected in two ways, single ion monitoring where the quadrupole is left resting at the mass unit of interest, or multielement scanning where the quadrupole is rapidly scanned over a discrete number of mass units. Detection limits (based on 3σ background counts) obtained by both these methods and by liquid nebulisation, are shown in table 2. Single ion monitoring is the most sensitive technique but does not enable isotope ratios or internal standard corrections to be determined. Multielement scanning using the ETV offers improved sensitivity over liquid nebulisation when scanning over a narrow mass range. It should be noted that these detection limits are subject to severe variation depending on the optimisation of the instrument parameters.

Table 2. Detection Limits for ICP-MS

Isotope	Single Ion ETV Bq/ml	Scanning ETV Bq/ml	Scanning liquid neb. Bq/ml
U-236	4.8×10^{-8}	6.8×10^{-7}	2.4×10^{-6}
Np-237	5.3×10^{-7}	7.3×10^{-6}	2.6×10^{-5}
U-238	2.5×10^{-10}	3.6×10^{-9}	1.3×10^{-8}
Pu-239	4.5×10^{-5}	6.8×10^{-4}	2.3×10^{-3}
Pu-240	1.7×10^{-4}	2.3×10^{-3}	8.4×10^{-3}
Pu-241	0.077	1.1	3.85
Pu-242	2.9×10^{-6}	4.1×10^{-5}	1.4×10^{-4}

Table 3 shows the results of 10 analyses of the plutonium isotope standard (at approximately 300 pg/g) by liquid nebulisation and by ETV. The results by both techniques agree with the expected values although the precision is slightly better by liquid nebulisation (3% compared to 4%).

Table 3. Plutonium Standard Isotope Ratios

Isotope Ratio	$^{240}Pu/^{239}Pu$	$^{242}Pu/^{239}Pu$	$^{244}Pu/^{239}Pu$
ETV	0.99 ± 0.04	1.07 ± 0.06	0.36 ± 0.02
Liquid Nebulisation	0.98 ± 0.03	1.04 ± 0.02	0.346 ± 0.008
Certified Value	0.966 ± 0.001	1.025 ± 0.002	0.3358 ± 0.0008

Figure 2 shows a linear calibration for ^{237}Np in the range 1-100 pg by ETV with a correlation coefficient of 0.996. This implies that quantitative analysis is possible even at such low concentrations.

Figure 2. Calibration for ^{237}Np by ETV-ICP-MS

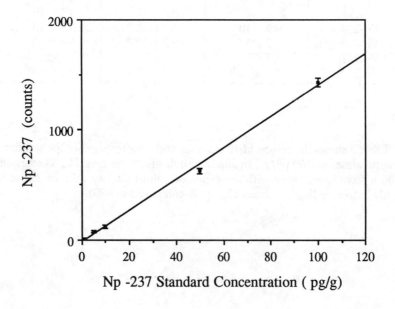

Sediment Samples

A typical spectrum obtained by ETV-ICP-MS for a 1g Ravenglass sediment sample is shown in figure 3. A comparison of results obtained by ETV-ICP-MS, liquid nebulisation and alpha spectrometry shown in table 4 indicates satisfactory agreement within the precision of each technique. ETV resulted in a range of concentrations for ^{237}Np, ^{239}Pu and ^{240}Pu indicating that 1g samples were not representative of the bulk sample.

Table 4 Results for Intertidal Ravenglass Sediment

Comparison of ETV with Liquid nebulisation

Method	Np-237 (pg/g)	Pu-239 (pg/g)	Pu-240 (pg/g)	^{240}Pu/^{239}Pu
ETV	77 ± 27	250 ± 36	65 ± 17	0.26 ± 0.03
Liquid Nebulisation	104 ± 15	210 ± 6	46 ± 4	0.219 ± 0.009

Comparison of ICP-MS with α-spectrometry

Method	$^{238+239}$Pu (Bq/g)
ETV ICP-MS	1.11 ± 0.33
Liquid Nebulisation	0.86 ± 0.04
α-Spectrometry	0.860 ± 0.004

Figure 3. Spectrum Obtained from 1g Ravenglass Sediment Sample

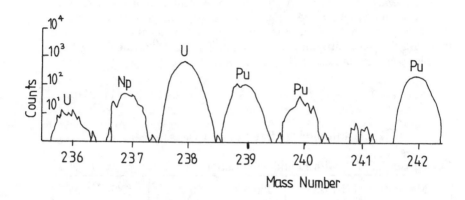

CONCLUSIONS

ETV-ICP-MS has been shown to be a sensitive technique for the simultaneous determination of ^{237}Np, ^{239}Pu and ^{240}Pu, with the advantage of increased potential for sample preconcentration. In the future the technique will be applied to the determination of these isotopes in a sediment core taken from Maryport harbour, Cumbria. The detection limits achieved here with ETV-ICP-MS are adequate to analyse 2g samples from the core. The data from analysis of several radionuclides down the core show some correlation with discharge data from BNFL Sellafield, Cumbria [14]. It is hoped the data made available using this technique will enable the reconstruction of available ^{237}Np data and give an indication of the amount discharged previous to 1978. Also it is hoped that the ^{239}Pu/^{240}Pu ratio in the core will yield information on the source term.

ACKNOWLEDGEMENTS

I would like to thank Mr. Keith McKay (I.A.E.A., International Laboratory of Marine Radioactivity, Principality of Monaco) for his technical assistance, Dr. Kershaw (Ministry of Agriculture, Fisheries and Foods, Marine Radioactivity Laboratory, Lowestoft) for the donation of the Maryport core samples, and the Natural Environment Research Council for the funding of the project.

REFERENCES

1. Date, A.R. and Gray, A.L., Applications of Inductively Coupled Plasma Mass Spectrometry, Blackie, Glasgow, 1989.

2. Houk, R.S. and Thompson, J.J., Inductively Coupled Plasma Mass Spectrometry. Mass Spectrom. Reviews, 1988, **7**, 425-461.

3. Hislop, J.S., Long, S.E., Brown, R.M., Morrison, R. and Pickford, C.J., The Use of Argon Plasma Sources for Measurement of Long Lived Radionuclides. ICRM - Low level techniques group meeting, Warenlingen, 1987.

4. Beauchemin, D., McLaren, J.W., Mykytiuk, A.P. and Berman, S.S., Determination of Trace Metals in an Open Ocean Water Reference Material by Inductively Coupled Plasma Mass Spectrometry. J. Anal. Atomic Spectrom., 1988, **3**, 305-308.

5. Hulmston, P. and Hutton, R.C., Analytical Capabilities of Electrothermal Vaporization-Inductively Coupled Plasma-Mass Spectrometry. Spectroscopy, 1991, **6**, 35-38.

6. Harvey, B.R., Potential for Post-depositional Migration of Neptunium in Irish Sea Sediments. In Impact of Radionuclide Release in the Marine environment, Vienna, IAEA, IAEA-SM-248, 1981, 93-103.

7. Kim, C.K., Takaku, Y., Yamamoto, M., Kawamura, H., Shiraiski, K., Igarashi, Y., Igarashi, S., Takayama, H. and Ikeda, N., Determination of ^{237}Np in Environmental Samples using Inductively Coupled Plasma Mass Spectrometry. J. Radioanal. Nucl. Chem. Art., 1989, **132**, 131-139.

8. Buesseler, K.O. and Sholkovitz, E.R., The Geochemistry of Fallout Plutonium in the North Atlantic: Part 2 ^{240}Pu/^{239}Pu Ratios and their Significance. Geochim. Cosmochim. Acta, 1987, **51**, 2623-2637.

9. Kim, C.K., Oura, Y., Takaku, Y., Nitta, H., Igarashi, Y. and Ikeda, N., Measurement of ^{240}Pu/^{239}Pu Ratio by Fission Track Method and Inductively Coupled Plasma Mass Spectrometry. J. Radioanal. Nucl. Chem. Lett., 1989, **136**, 353-362.

10. Tyror, J.G., A Series of Lectures on Reactor Physics of Burn Up, 1971, AEEW-M999.

11. McCarthy, W. and Nicholls, T.M., Mass Spectrometry Analysis of Plutonium in Soils near Sellafield. J. Environ. Radioactivity, 1990, **12**, 1-12.

12. Hursthouse, A.S., Ph.D. Thesis, Unpublished, 1990.

13. Byrne, A.R., Determination of ^{237}Np in Cumbrian Sediments by NAA: Preliminary Results. J. Environ. Radioactivity, 1986, **4**, 133-144.

14. Kershaw, P.J., Woodhead, D.S., Malcom, S.J., Allington, D.J. and Lovett M.B., A Sediment History of Sellafield Discharges. J. Environ. Radioactivity, 1990, **12**, 201-241.

A NEW TRACER TECHNIQUE FOR *IN SITU* EXPERIMENTAL STUDY OF BIOTURBATION PROCESSES.

G. GONTIER[1], M. GERINO[2], G. STORA[2], JP. MELQUIOND[1]

[1]Service d'Etudes et de recherches sur l'environnement du CEA,

Station Marine de Toulon

Base IFREMER, BP330, 83507 La Seyne sur mer - France.

[2]Centre d'Océanologie de Marseille

Rue de la batterie des Lions, 13007 Marseille - France.

ABSTRACT

An experimental method has been developed to study material and radioactivity fluxes at the sediment-water interface and in the sedimentary column. This method has been applied in the Gulf of Fos, an area that is affected by deposits from the river Rhone, and where biodeposition products, resulting from the presence of intensive mussel cultures, can induce a concentration of trace elements at the sediment-water interface.

Sediment surface materials were labelled with a mixture of radionuclides (Cerium-144, Cobalt-60 and Cesium-137), in experimental cores filled either with sediment containing *in situ* fauna or with defaunated sediment.

The coupling of this mixture with luminophores - inert colored sediment particles - enabled us to measure radionuclide flux in both solute fraction and solid fraction. At the sediment-water interface, the tracer balance indicates that migrations into deeper sediment are estimated to be until 25 times greater in presence of macrofauna, depending on the tracer examined. Bioturbation may equally enhance exportation to the water column, to a factor ranging from 1.5 to 2.0. During a period of 14 days, in presence of macrofauna, we observed a migration of radionuclides to a maximum depth of 11 cm. A similar distribution pattern of luminophores at the same sediment depths indicates the preponderance of particle reworking in migration.

INTRODUCTION

The first links of the trophic chain have considerable opportunity to accumulate heavy metals or radioactive elements [1]. These elements, in environments of intensive primary productivity such as the Carteau mussel park in the Northwestern Mediterranean basin, are quickly concentrated and the biological debris is accumulated at the water-sediment interface [2], [3], [4], [5], [6], [7].

Burial or resuspension of these elements depends on both physical factors such as diffusion [8] and hydrodynamic action, and biological factors via bioturbation [9], [10], [11], [12], [13], [14], [15], [16], [3],[17], [18].

In order to quantify the incidence of these various processes on material transfers in the benthic boundary layer, an *in situ* experiment was carried out using several types of tracers. The tracers used were 3 radionuclides and one particle specific type of tracer: the luminophores. ^{137}Cs and ^{60}Co were chosen because they are more frequently found in *in situ* cores; ^{144}Ce was added because of its greater ability to adsorb on particles. The use of a mixture of tracers means that the solute and particulate fractions of sediment can be labelled differentially.

METHODS

1/ Experimental site. The Gulf of Fos is located near the mouth of the river Rhone and is a semi-enclosed environment (figure 1).

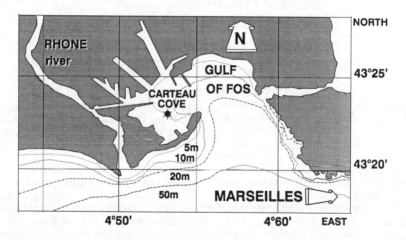

Figure 1. The experimental station in the Gulf of Fos (Carteau bay) in the Northwestern Mediterranean basin.

The experimental site is situated in a mussel park at a depth of 5 m. Three cores were collected on the experimental site to analyse the sediment parameters.

2/ Experimental procedure. A mixture of tracers was deposited simultaneously in the cores. These tracers are various radionuclides (^{144}Ce, ^{60}Co and ^{137}Cs), with about 50000 Bq of each element, to label both solute and particulate fractions of sediment, and 2 g of luminophores [18], [19], which are inert colored particles with a diameter range of between 40 and 200µm. The *in situ* experiment involved implanting 4 PVC cores with a diameter of 10.3 cm to a depth of 30cm. Two cores isolated a portion of the *in situ* ecosystem, and the other 2 cores were filled with defaunated sediment (figure 2). The defaunated sediment was previously collected on the experimental site and frozen to kill all the fauna without homogenizing the sediment. The control cores were maintained without macrofauna by sediment contamination with tetraethyl lead [20]. The tracers were deposited simultaneously in each core at the beginning of the experiment.

Core 2 contained sediment with *in situ* macrofauna, and tracers were deposited at the sediment-water interface. Core 1 was a control core with defaunated sediment and it too was labelled at the sediment-water interface. In core 4, containing *in situ* macrofauna, the labelling was placed at a depth of 6.3 cm in the sediment. Control core 3, without macrofauna, was labelled identically to core 4.

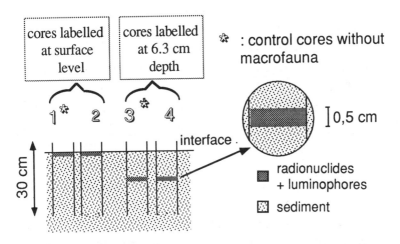

Figure 2. Diagram showing the experimental cores implanted in the sediment.

The experiment was carried out over a period of 14 days (from 21/12/89 to 04/01/90). At the end of this period, all the cores were collected and immediately frozen. In the laboratory, they were cut into horizontal sections of 0.9 cm. After homogenisation of each section, subsamples were taken to determine:

- macrofauna abundance (by sorting, identification and counting)
- the amounts of the 3 radionuclides (by gamma spectrometry on GeLi cristal)
- the number of luminophores (under ultraviolet light).

RESULTS

The sediment and macrofauna of the area are those typically found in muddy-sand sheltered ecosystems. The main features of the sediment on the site, at different levels in the sedimentary column, are summarised in Table 1.

TABLE 1
Sediment caracteristics at the experimental station in Carteau cove.

Level (cm)	Fraction <200 μm (%)	Organic matter (%)	137 Cs (Bq/kg (dry))	134 Cs (Bq/kg (dry))	60 Co (Bq/kg (dry))
0-2	90.7	5.0 ± 0.6	20.6 ± 4.2	2.6 ± 0.8	0.7 ± 0.3
2-4	93.1	6.3 ± 1.3	20.7 ± 1.1	< 0.2	< 0.2
4-6	96.4	6.1 ± 0.9	13.6 ± 0.7	< 0.2	< 0.2
6-8	96.0	5.6 ± 1.0	7.8 ± 0.8	< 0.2	< 0.2
8-10	98.6	6.2 ± 1.8	4.3 ± 0.4	< 0.2	< 0.2
10-14	98.9	6.7 ± 1.0	1.9 ± 0.3	< 0.2	< 0.2
14-18	97.6	6.2 ± 1.6	1.5 ± 0.3	< 0.2	< 0.2
18-22	97.3	6.6 ± 0.6	< 0.3	< 0.2	< 0.2

1/ Surface labelled sediment. Figure 3 shows macrobenthos abundance in the different sections of cores 1 and 2. Core 2 analyses revealed the presence of 98 individuals distributed over the 10 first centimeters of sediment, with 85% of the individuals in the 5 first cm. The polychaetes are heavily dominant. In the defaunated control core, only 10 individuals were found.

Figures 4 and 5 show the distribution of radionuclides and luminophores in the sedimentary column of cores 1 and 2. Quantities of radionuclides and luminophores are expressed in each section as a percentage of the total quantity present in the whole sediment sample at the end of the experiment. In the core with macrofauna, a deeper migration of radionuclides and luminophores in comparison with the control core is apparent. Proportions and penetration depth vary according to the tracer. Maximum penetration depth is 11 cm with ^{137}Cs. In the control core, the tracers stayed in the surface levels of sediment.

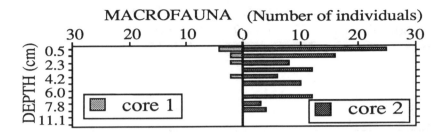

Figure 3. Macrobenthos abundance as a function of sediment depth in cores 1 and 2.

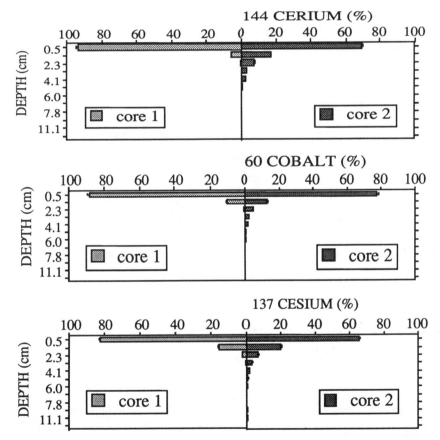

Figure 4. ^{144}Ce, ^{60}Co, and ^{137}Cs distribution as a function of sediment depth in cores 1 and 2.

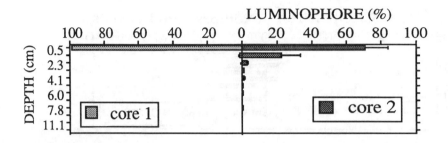

Figure 5. Luminophore distribution as a function of sediment depth in cores 1 and 2.

2/ Labelling of sediment at 6.3 cm. In core 4, 177 individuals were counted, whereas only 21 individuals were found in control core 3 (Figure 6).

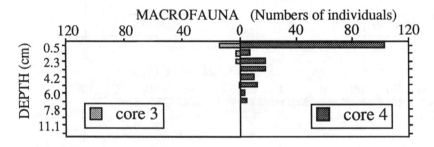

Figure 6. Macrobenthos abundance as a function of sediment depth in cores 3 and 4.

At the labelled depth (6.3 cm), a lower abundance of macrofauna is apparent. The radionuclides and luminophores have the same distribution pattern in cores 3 and 4 and are restricted to the initially labelled sections.

DISCUSSION

Quantitative analyses of *in situ* radionuclides were carried out on site sediment at the beginning of the experiment (Table 1). The maximum values are at least 10000 times lower than labelled sediment samples, which confirms that the experimental results are not influenced by the *in situ* radionuclide concentrations.

By comparing the amounts of tracer introduced at the beginning of the experiment with concentrations in the core sediment at the end, we have calculated 3 tracer percentages corresponding to -1- exportation from the labelled sediment to the overlying water, -2- migration from the labelled sediment into the surrounding sediment, and -3- stationary tracers.

1- Surface labelled sediment

a) Control core 1. Table 2 shows the percentages of these 3 fractions in core 1. The migratory capability of a tracer depends on the degree to which it adsorbs on particulate material or remains in solution. The results reflect this diversity.

For all the radionuclides, migration within the sedimentary column is low. ^{144}Ce is distinguished by a high capacity for adsorption on particles and its behaviour is similar to that of the luminophores, which are tracers specific for sediment particles. The zero migration of luminophores indicates the absence of particle reworking. It may therefore be assumed that radionuclide flux in cores results simply from their diffusion in the interstitial water. Exportation processes to the water column are more intensive than those within the sediment, and are the consequence of both resuspension by hydrodynamic action and diffusion.

TABLE 2

Exportations (%) to the overlying water and into the deeper sediment in core 2 with macrofauna and control core 1 labelled at the sediment surface (0-0.5 cm), after 14 days, and the factors F_{water} and $F_{sediment}$ by which exportation increased in presence of macrofauna.

traceurs	Exportation into overlying water			Exportation into deeper sediment		
	Core 1 control	Core 2 with macrofaune	F water	Core 1 control	Core 2 with macrofaune	F sediment
144 Cerium	13.1 %	26.6 %	2.0	5.5 %	22.3 %	4.1
60 Cobalt	27.8 %	42.3 %	1.5	8.6 %	13.0 %	1.5
137 Cesium	42.6 %	73.9 %	1.7	10.4 %	9.3 %	0.9
Luminophores	20.0 %	30.0 %	1.5	0.8 %	20.0 %	25.0

^{60}Co exportation (27,8%) and ^{137}Cs exportation (42,6%) to the water column are more intensive than that of ^{144}Ce (13,1%). This may be explained by the tendency of these 2 elements, in marine environment, to be particularly present in solution [21] and [22].

b) Core 2 (with macrofauna). In this core with macrofauna, the results (Table 2) are very different from those observed in control core 1. The effects of bioturbation are quantified by the factors $F_{sediment}$ and F_{water} which represent the rate of increase of fluxes within the surrounding sediment and into the overlying water in presence of macrofauna, and are calculated from the ratio exportation in presence of macrofauna and exportation in the control cores.

For all 4 tracers, exportation processes to the overlying water increased by a factor F_{water} of between 1.5 and 2. On the other hand, migrations within the sedimentary column, in presence of macrofauna, are more intensive for the luminophores (Fsed.=25), ^{144}Ce (Fsed. = 4,1), and ^{60}Co (F sed. = 1,5), whereas ^{137}Cs migrations have the same intensity. The fact that F is higher in the case of exportation of the particle specific tracers 144 Ce and luminophores within the sediment, indicates that biological reworking has both a greater impact on migration within the sediment than on exportation to the overlying water, and a greater impact on sediment particle migration than on interstitial water fluxes.

2/ Labelling of sediment at 6.3 cm. Results shown in table 3 show that a small proportion of radionuclides and luminophores migrated within the sediment from the initially labelled section, both in cores with macrofauna and in control cores.

Comparison between these two cores shows that, over a period of 14 days, the abundance of macrofauna present in the core at the labelled level is insufficient to induce any significant sediment mixing.

TABLE 3

Percentage of tracers remaining in the initially labelled section at 6.3 cm in control core 3 and core 4 (with macrofauna), after 14 days.

%	core 4 (with macrofauna)	core 3 (control)
Cerium-144	99.5	97.6
Cobalt-60	95.5	93.7
Cesium-137	91.6	90.4
Luminophore	99.8	99.4

CONCLUSIONS

The use of a mixture of tracers to label simultaneously the solute and particulate fractions of the sediment is a means of obtaining accurate information on the specific incidence of physical processes such as hydrodynamic action or diffusion and biological processes such as bioturbation. The results show that bioturbation quantification may vary according to the type of tracer used. Of those used here, it is the particle specific tracers 144 Ce and the luminophores which are the most affected by bioturbation. Although the duration of the experiment was relatively short, the results indicate that benthic activity in a littoral environment results in the rapid burial of material initially deposited at the interface; the migrations are estimated to be until 25 times greater in presence of macrofauna, depending on the tracer examined. Bioturbation may equally enhance exportation to the water column, to a factor ranging from 1.5 to 2.0. This type of experiment, involving the simultaneous use of a variety of tracers, provides data which could form the basis of comparisons with results already obtained with monospecific tracers.

REFERENCES

1. Ancellin, J., Guegueniat, P., Germain, P., 1979. Radioécologie marine : Etude du devenir des radionucléides rejetés en milieu marin et application à la radioprotection. Eyrolles Ed., 256p.

2. Cadee, G.C., 1984. Biological activity and sediments. In : sediments and pollutions in waterways. General considerations.IAEF Technical docmentation: IAEA-TECDOC-302, IAEA VIENNA 1984 : 111-126.

3. Kershaw, P., J., 1985. ^{14}C and ^{210}Pb in NE Atlantic sediments : Evidence of biological reworking in the context of radioactive waste disposal. J. Environ. Radioactivity, vol2 : 115-134.

4. Fowler, S.W., Buat-Ménard, P., Yokoyama, Y., 1987. Rapid removal of Chernobyl fallout from Mediterranean surface waters by biological activity. Nature 329 : 56-58.

5. Whitehead, N.E., Ballestra, S., Holm, E., Huynh-Ngoc, L., 1988. Chernobyl Radionucleides in Shelfish. J. Environ. Radioactivity, 7 : 107-121.

6. Grenz, C. 1989. Quantification et destinée de la biodéposition en zones de production conchylicoles intensives en Méditerranée. Thèse Doct. Univ. Aix-Marseille 2, 145pp.

7. Gontier, G., Grenz, C., Calmet, D., Sacher, M., 1991. Contribution of Mytilus sp in radionuclide transfert between water column and sediments in the estuarine and delta system of the Rhône river (N.W. Mediterranean sea). (submitted to referee).

8. Duursma, E.., K., 1984. Some typical examples of the importance of the role of sediments in the propagation and accumulation of polluants. In : Sediment and pollution in waterways. Vienna, AIEA-TECDOC-302, 127-137.

9. Rhoads, D.C., 1974. Organism-sediment relations on the muddy sea floor. Oceanogr. Mar. Biol. Ann. Rev., 12 : 263-300.

10. Aller, R.C., Cochran, J.K., 1976. $^{234}Th / ^{238}U$ disequilibrium in near-shore sediment : particles reworking and diagenetic time scale. Earth and Planetary Sciences letters, 29 : 37-50

11. Benninger, L.K., Aller, R.C., Cochran, J.K., Turekien, K.K., 1979. Effects of biological sediment mixing on the ^{210}Pb chronology and trace metal distribution in a Long island sound sediment core. Earth and Planetary Science letters, 43 : 241-259.

12. Cochran, J.K., Aller, R.C., 1979. Particle reworking in sediments from the New York bight apex. Evidence from the ^{234}Th/^{238}U disequilibrium. Estuarine and Coastal marine Science, 9 : 739-747.

13. Livingston, H.D., Bowen, V.T., 1979. Pu and ^{137}Cs in coastal sediments. Earth and Planetary Science lettrers, 43 : 29-45.

14. Koide, M., Golberg, E.D., 1980. ^{241}Pu and ^{241}Am in sediments from coastal basins off California and Mexico. Earth and Planetary Science letters, 48 : 250-256.

15. Cochran, J.K., 1985. Particle mixing rates in sediment of the eastern equatorial Pacific : Evidence form ^{210}Pb, $^{239-240}$Pu and ^{137}Cs distributions at MANOP sites. Geochimica et Cosmichimica Acta, 49 : 11-95-1210.

16. DeMaster, D.J., McKee, B.A., Nittrouer, C.A., Breswter, D.C., Biscaye, P.E., 1985. Rates of sediment reworking at the Hebble site based on measurement of ^{234}Th, ^{137}Cs and ^{210}Pb. Marine Geology, 66 : 133-148.

17. Anderson, R.F., Bopp, R.F., Buesseler, K.O., Biscaye, P.E., 1988. Mixing of particles and organic constituents in sediments from the continental shelfs and slope off Cape Cod : SEEP-1 results. Continental Shelf Research, 8(5-7) : 925-9046.

18. Gérino, M., 1990. The effects of bioturbation on particle redistribution in Mediterranean coastal sediment. Preliminary results. In : Flux between trophic levels and through the water-sediment interface. Eds Bonin Golterman : 251-258.

19. Mahaut, M.L., Graf, G., 1987. A luminophore tracer technique for bioturbation studies. Oceanol. Acta, 10 : 323-328.

20. Arnoux, A., Stora, G., Vacelet, E., Vitiello, P., 1988. Etude expérimentale dans le milieu naturel, de sédiments artificiellement contaminés par différentes formes chimiques d'un métal (Plomb) : Evolution chimique et biologique du sédiment. Rapport PIREN-ATP Ecotoxicologie. Contrat 1294.

21. Fukai, R., Ballestra, S., Thein, M., 1981. Vertical distribution of transuranic nuclides in the Mediterranean Sea, in : Techniques for identifiyng transuranic speciation in aquatic environments. International Atomic Energy Agency, Vienna : 79-87.

22. Mahler, P., 1985. Comportement du cesium-137, chrome-51, cobalt-60, manganèse-54, sodium-22 et zinc-65 en milieux d'embouchures simulés ; Inflluence des particules minérales en suspension et des matières organiques dissoutes. Thèse de 3ème cycle de Géochimie, Université de Nice, 190pp.

LE ^{137}Cs: TRACEUR DE LA DYNAMIQUE SEDIMENTAIRE SUR LE PRODELTA DU RHONE (Méditerranée nord-occidentale).

*FERNANDEZ JM, **BADIE C, *ZHEN.Z et *ARNAL I.
*Station Marine de Toulon, SERE/DPEI/IPSN/CEA
Base IFREMER BP330, 83500 LA SEYNE SUR MER, FRANCE.
**Département de Protection de l'Environnement et des Installations, IPSN/CEA
CEN-Cadarache, 13108 SAINT PAUL LEZ DURANCE

RESUME

Le Rhône est le principal fleuve du bassin méditerranéen nord-occidental avec un débit moyen de 1800 m^3/s. Sa charge particulaire provient essentiellement du lessivage d'un important bassin versant (98.000 km2). A l'embouchure, une fraction des matières en suspension véhiculées flocule à l'interface eau douce/eau salée et se dépose rapidement pour donner naissance à une construction sédimentaire en forme de prisme: le prodelta. Une autre fraction de la charge particulaire est transportée vers le large dans une couche turbide (néphéloïde benthique) qui est chenalisée vers le bassin algéro-provençal.
Les retombées atmosphériques (explosions d'engins nucléaires, Tchernobyl) et l'industrie nucléaire installée le long du fleuve, en particulier le centre de retraitement des combustibles irradiés de Marcoule, sont les deux principales sources de radionucléides artificiels des eaux du Rhône. Présents sous forme dissoute mais aussi associés à la matière en suspension ils permettent le traçage des apports rhodaniens en mer.
Compte tenu de certains phénomènes (diffusion, bioturbation...) l'étude du dépôt marqué par le ^{137}Cs permet l'élaboration d'un bilan sédimentaire dans le prodelta du Rhône. Le calcul des volumes déposés est réalisé par deux techniques mathématiques distinctes:
 (i) - L'intégration sectorielle de fonctions continues décroissantes exprimant l'importance de la couche marquée avec la distance à l'embouchure.
 (ii) - La numérisation de la couverture sédimentaire marquée grâce à une technique d'interpolation et d'approximation (*krigeage*), à partir du réseau constitué par les stations d'observation.
Le volume estimé des apports du Rhône est de l'ordre de 4,7.10^6 m^3/an. A cette estimation s'ajoute une comparaison avec des résultats obtenus antérieurement par d'autres études et des valeurs moyennes pour d'autres fleuves français.

INTRODUCTION

Le Rhône est le principal fleuve de Méditerranée nord-occidentale avec un bassin versant de 98.000 km^2 et un débit de moyen de 1800 m^3/s. Il draine l'ensemble des effluents rejetés par les industries installées le long de sa vallée, et notamment il transporte un certain nombre de radionucléides artificiels dont la présence s'explique:

(i) - soit par le lessivage de son bassin versant marqué par les retombées radioactives dues aux essais atmosphériques d'engins nucléaires et plus recemment par celles issues de l'explosion du réacteur N°4 de Tchernobyl.

(ii) - soit par le rejet d'effluents faiblement radioactifs déversés par l'industrie nucléaire civile. Près de 90% des radionucléides artificiels, excepté le ^3H, proviennent du centre de retraitement des combustibles irradiés de Marcoule (1).

Au niveau du delta rhodanien, la confrontation des eaux continentales avec les eaux marines donne naissance à des phénomènes physico-chimiques de floculation. Il en résulte la formation en mer d'une structure hydrologique complexe permanente (néphéloïde), constituée principalement d'une couche turbide de surface, soumise aux vents et aux courants dominants et d'une couche turbide de fond qui s'étale plus largement sur le plateau continental (2)(3)(4)(5). La couche profonde (néphéloïde benthique) se caractérise par une forte charge particulaire (30 à 50 mg/l), dotée d'une grande capacité d'adsorption.

Sur le plateau continental, ce phénomène se traduit par la présence d'un envasement précoce (prodelta), dès 20 mètres de profondeur qui s'étend sur plusieurs centaines de km^2. L'ordre de grandeur des accumulations est de quelques mètres près de l'embouchure et de quelques centimètres sur le bord extérieur du plateau continental. De ce fait, le prodelta constitue un vaste piège où se concentre une fraction des composés chimiques transportés par les eaux fluviales.

Le ^{137}Cs associé à la matière particulaire rhodanienne qui sédimente en mer ne (FIGURE 1) semble représenter que 1% du césium issus des retombées atmosphériques du bassin versant et 10% de celui provenant des rejets du centre de Marcoule (6)(1)(7). Cependant en tenant compte notamment de sa diffusion dans le sédiment (8)(9), il est possible d'utiliser ce marqueur pour évaluer les taux de sédimentation actuels aux environs immédiats de l'estuaire rhodanien (10)(11)(12).

Les carottages réalisés sur la construction prodeltaïque ont permis:

(i) - de montrer que la sédimentation des particules n'est pas isotrope mais qu'elle s'organise, globalement, suivant 5 axes dont 1 est dominant en direction du S-SW (13)(14)(15);

(ii) - d'étudier l'enfouissement du ^{137}Cs dans les couches sédimentaires en fonction de la distance à l'embouchure (10).

L'ensemble de ces connaissances conduit à tenter une estimation du volume de sédiments marqués, apportés depuis l'origine de l'injection du radiotraceur dans le fleuve, et une évaluation de la charge solide annuellement déposée par le Rhône.

Deux méthodes de calculs sont proposées: l'une fait appel à des fonctions décroissantes continues et intégrables par secteur sur un domaine défini, l'autre est basée sur les techniques numérisées de krigeage.

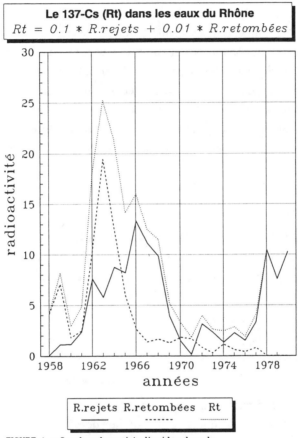

FIGURE 1 — Courbes des rejets liquides dans le Rhône et des retombées atmosphériques. Courbe (Rt) résultante. (modifié Fernandez, 1984).

METHODES ET TECHNIQUES

A - LES ECHANTILLONS DE SEDIMENT

La stratégie d'échantillonnage s'appuie sur l'existence de nombreux levés sismiques réalisés au cours de travaux sédimentologiques antérieurs. Toutefois, l'exploration de certains secteurs a été complétée par une reconnaissance préalable avec un sondeur de 3.5 kHz (mud-penetrator).

Au total 29 carottes ont été prélevées à des profondeurs comprises entre -20m et -100m à l'aide de carottiers de type **box-corer** ou **Kullenberg**.

(i) - Les carottes obtenues avec les box-corer ont une section de 625 à 700cm^2 pour une hauteur prélevée comprise entre 20 et 70cm. Pour étudier la distribution verticale du ^{137}Cs, un découpage centimétrique a été réalisé sur chaque carotte.

(ii) - Les carottes obtenues grâce au Kullenberg ont une section plus réduite (50 à 80 cm^2) mais la pénétration dépasse 3m. Pour obtenir la quantité nécessaire aux analyses le découpage des horizons s'est effectué tous les 5cm.

Les échantillons, congelés à bord, sont lyophilisés ou séchés à 40° Celsius, une fois à terre. Le sédiment est broyé et tamisé à 1 mm avant d'être conditionné dans des boîtes de 350 cm^3 pour subir une mesure des radionucléides présents par spectrométrie gamma directe.

B- LES OUTILS DE CALCUL

1- INTEGRATION DE FONCTIONS CONTINUES

Entre 4 et 5 carottes ont été prélevées le long de chacun de ces 5 axes pour permettre d'ajuster des fonctions mathématiques exprimant la décroissance de l'épaisseur marquée en fonction de la distance à l'embouchure. Les expressions choisies sont:

$H(r) = H_0 e^{-ar}$ (1)   - $H(r)$, l'épaisseur de sédiment marqué;

$H(r) = H_0 r^{-a}$ (2)   - H_0, la constante à l'origine;

$H(r) = -ar + H_0$ (3)   - a, la constante de décroissance;

     - r, la distance à l'embouchure (km).

Chacun des 5 axes ainsi déterminés permet de délimiter 5 secteurs de dépôt, d'aires inégales, où le domaine de validité de chaque fonction est défini par (FIGURE 2 et TABLE.1):

- α, l'angle d'ouverture du secteur;
- r_1, la limite de rupture de pente prodeltaïque (km), au deçà des phénomènes sédimentaires différents interviennent (10);
- r_2, la distance à l'embouchure (km) maximale définie par la station de carottage la plus éloignée dans chaque secteur.

L'évaluation du volume de sédiments marqués V, s'obtient alors en effectuant la somme des 5 volumes partiels V_i correspondant à chacun des 5 secteurs.

$$V = \sum_{i=1}^{5} V_i = \sum_{i=1}^{5} \left(\frac{2\pi\alpha_i}{360} \int_{r1}^{r2} H(r) r dr \right)$$

On notera que dans le dépôt en place, la hauteur de sédiments contenant du ^{137}Cs est le résultat de:

(i) - la sédimentation naturelle des particules marquées;

(ii) - la contamination, par diffusion, des horizons inférieurs, déposés antérieurement à l'apparition de ce radionucléide dans le milieu marin.

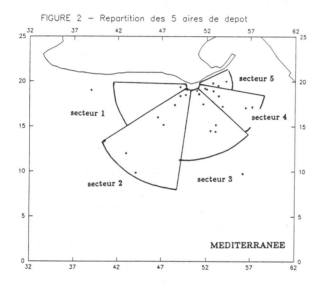

FIGURE 2 — Repartition des 5 aires de depot

2- LA METHODE PAR TRAITEMENT NUMERIQUE

Deux fichiers nécessaires aux calculs (RHONE et DFU) ont été constitués à partir des résultats de mesure de 29 carottes et contiennent:

(i) - les valeurs des hauteurs totales de sédiment marqué par le ^{137}Cs;

(ii) - les hauteurs de sédiment marqué par le ^{137}Cs qui a diffusé (diffusion physico-chimique + bioturbation).

Un logiciel de calcul (SURFER), outre les représentations cartographiques et tri-dimensionnelles des isopaques, permet d'obtenir des évaluations de volumes selon 3 méthodes différentes: Trapézoïdale, Simpson et Simpson 3/8 (16).

A partir de chacun des deux fichiers, une grille de maille régulière est créée par krigeage afin d'obtenir les corrélations entre les points, par secteur de $2\pi/4$. En théorie, cette technique offre une très bonne précision de calcul tout en respectant les données d'origine. Toutefois certaines anomalies apparaissent dans la représentation des données lorsque celles-ci donnent naissance à la co-existence de reliefs plats et de reliefs accidentés. De ce fait, les aires de relief accidenté ont été lissées par une fonction *Spline* classique (puissance 3) qui ne change pas les valeurs initiales. Le lissage s'effectue par ajout de point intermédiaires.

Pour les aires de faible relief, le lissage a été effectué à partir de la création d'une matrice (n*m) où chaque élément est pondéré par un facteur $1/(d^2)$, d étant la distance entre 2 éléments de la matrice: La fonction possède un effet de lissage important et elle convient lorsqu'il s'agit d'atténuer les artéfacts créés par le krigeage dans les zones plates.

RESULTATS

A - LE CALCUL PAR INTEGRATION

Les études géotechniques dans le prodelta du Rhône (17) montrent que les caractéristiques sédimentologiques varient avec l'éloignement à l'embouchure. De ce fait la valeur de la diffusion ne peut être considérée comme constante en tout point du prodelta, le coefficient D dépend notamment de la densité du sédiment (18).

En première approximation, sur le bord extérieur du plateau continental, une valeur du coefficient D couramment admise est de $5.10^{-8} cm^2/s$ (19)(20)(21). Près des côtes la valeur déduite, après l'injection du ^{134}Cs issus de Tchernobyl, peut être considérée comme voisine de 3 à $5.10^{-7} cm^2/s$ (12). Dans la suite des calculs on supposera que le cocfficient D évolue de façon continue entre ces deux valeurs extrêmes. Dans ce cas, à la hauteur en sédiment marqué par le ^{137}Cs, il faut soustraire, pour chaque profil de radioactivité, entre 12 cm (bord extérieur du plateau continental) et 40 cm (talus prodeltaïque).

Les coefficients de corrélation des ajustements se situent entre 0,94 et 0,99 avec un seuil de signification (Fisher-Snedecor) compris entre 0,01 et 0,05.

L'évaluation du volume V déposé sur le prodelta du Rhône, entre 1958 et 1980, fournit des valeurs comprises entre $1,04.10^8$ et $1,27.10^8$ m^3 (TABLE 1).

Si on admet en première approximation, pour l'ensemble de l'unité prodeltaïque rhodanienne, des valeurs moyennes de la densité de 1,7 et de la teneur en eau de 25%, la masse déposée annuellement pourrait alors être estimée entre $6,03.10^6$ et $7,39.10^6$ tonnes/an pour une surface d'environ 260 km^2, selon les expressions mathématiques choisies.

TABLE 1
Principaux résultats du calcul par intégration

RADIALE		1	2	3	4	5
r_1 en km		2	2	2	2	2
r_2 en km		11,1	20,0	11,0	9,8	4,6
type de fonction		(2)	(1) (2)	(1) (2)	(1) (2)	(3)
H_0		$4,6.10^5$	3,24 $2,1.10^5$	2,75 $4,3.10^6$	6,59 $4,5.10^8$	2,86
a		1,62	$1,8.10^{-4}$ 1,45	$2,2.10^{-4}$ 1,85	$3,6.10^{-4}$ 2,35	$5,5.10^{-4}$
$\dfrac{2\pi\alpha}{360}$		0,60	0,75	0,75	0,65	0,50
V en 10^6 m^3	MINIMUN	12,2	56,7	19,4	23,4	2,5
	MAXIMUM	-	61,7	25,9	25,0	-
Total V.10^8 m^3			1,04 à 1,27			
V/an.10^6 m^3			4,73 à 5,80			

B - LE CALCUL PAR KRIGEAGE

L'évaluation de l'importance de l'ensemble des phénomènes de diffusion est obtenue a partir de l'étude des courbes décrivant l'évolution historique des retombées atmosphériques et des rejets du centre de retraitement depuis 1958 (FIGURE 1). Dans les deux cas on observe des valeurs maximales en ^{137}Cs, enregistrées respectivement en 1963 et 1966. La composante de ces deux courbes (R.totale = 0,1*R.rejets + 0,01*R.retombées d'après (6)), permet d'obtenir un profil *théorique* de radioactivité en ^{137}Cs des sédiments si l'on suppose que la sédimentation est régulière, homogène et continue en tout point du prodelta. Le **pic de radioactivité résultant** est alors exploitable comme un repère chronologique dans les profils de concentration du ^{137}Cs expérimentaux. La superposition des profils "théoriques" et "expérimentaux", donne pour chaque carotte, la hauteur de la couche concernée par les phénomènes de diffusion et de bioturbation.

Le traitement de l'ensemble des données par une technique de numérisation présente l'avantage de visualiser la totalité de la construction prodeltaïque édifiée au droit du Rhône et de vérifier la vraisemblance des résultats obtenus:

(i) - La cartographie des lignes d'iso-dépôt semble indiquer l'existence de deux axes de sédimentation dominants orientés l'un vers le S-SW et l'autre vers le S-SE (FIGURE 3).

(ii) - L'accumulation maximale de sédiments est localisée, face au Rhône, à environ 1 mille de l'embouchure.

(iii) - Plus de 50% de la masse sédimentaire constituant le prodelta du Rhône est stockée dans un rayon approximatif de 7km centré sur l'embouchure. Au-delà l'importance de la couverture marquée diminue rapidement et devient millimétrique en direction des principaux chenaux (canyons du Grand et du Petit Rhône).

L'estimation du volume est calculé pour une aire limitée à environ 610 km^2 (TABLE 2).

TABLE 2
Principaux résultats des calcul par krigeage

METHODE DE CALCUL	V.10^8 m^3	V/an.10^6 m^3
Trapèzoïdale	1,045	4,75
Règle de Simpson	1,041	4,73
Règle de Simpson 3/8	1,045	4,75

La charge annuellement déposée peut être approchée en prenant les mêmes valeurs que précédemment pour la densité et la teneur en eau du sédiment. Le résultat est alors compris entre 6,03.10^6 et 6,06.10^6 tonnes/an.

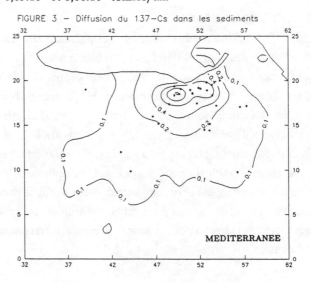

FIGURE 3 - Diffusion du 137-Cs dans les sediments

DISCUSSION ET CONCLUSION

Bien que les résultats obtenus par les méthodes de calcul par ajustement et par krigeage soient du même ordre de grandeur, ils laissent apparaître des divergences sensibles dans l'évaluation des volumes de sédiments apportés par le Rhône.

En effet, pour des surfaces comparables, la méthode par ajustement de fonction donne systématiquement des estimations plus élevées qui peuvent s'expliquer essentiellement par une appréciation différente des phénomènes globaux de diffusion du ^{137}Cs dans les couches sédimentaires antérieures: les coefficients de diffusion D fournis dans la bibliographie sont issus, le plus souvent, d'études expérimentales en laboratoire et ne reflètent que partiellement la réalité et notamment les effets de la bioturbation.

Il semble que la méthode numérique qui tient compte de l'ensemble des données soit plus représentative car:

(i) - Malgré l'insuffisance en stations de prélèvement, le traitement numérique fourni une description vraisemblable de l'importance et de l'extension géographique de la couverture sédimentaire déposée par le fleuve;

(ii) - L'ensemble des phénomènes de diffusion du ^{137}Cs dans le prodelta est pris en compte et intègre de façon plus complète les variations spatiales du coefficient D, qui dépendent notamment des caractéristiques sédimentologiques;

(iii) - L'image de la couverture sédimentaire disposée autour de l'embouchure, élaborée par le calcul, coïncide avec un phénomène de développement d'une barre silto-sableuse, fortement brassée, qui rend difficile la sédimentation des particules fines (10).

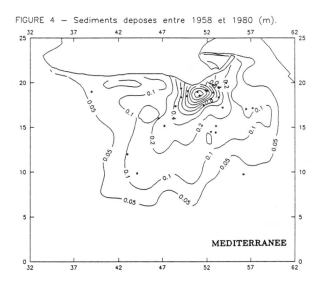

FIGURE 4 — Sediments deposes entre 1958 et 1980 (m).

Celle-ci ne reprend qu'en direction du large, à 1 mille de l'embouchure (Bouées de Roustans), et correspond à une zone de dépôt maximal (FIGURE 4).

Ces observations permettent de considérer que la valeur du volume la plus probable est sans doute obtenue par la méthode numérique. Celle-ci estime l'apport du Rhône à $4,7.10^6$ m^3/an, ce qui correspond environ à 6.10^6 t/an de matière en suspension sédimentée et transportée par charriage près du fond. Des calculs antérieurs semblent montrer que les apports du Rhône ont diminué en importance au cours de ce siècle: ceux-ci ont été évalués entre 30.10^6 t/an en 1847 (22) et $5,5.10^6$ t/an en 1954 (23). Une des dernières estimations (24) fournie une valeur encore plus faible: $2,5.10^6$ t/an.

Pour d'autres fleuves français, des calculs relativement récents (25)(26) ont permis d'évaluer les apports en sédiments fins (sédimentation des matières en suspension et charriage particulaire) dans les estuaires de la Gironde à $2,9.10^6$ t/an et de la Loire à $1,2.10^6$ t/an.

L'appréciation de la charge massique est incertaine car les valeurs de la densité et de la teneur en eau prises en compte dans les calculs ne sont que des moyennes alors que celles-ci varient sensiblement, notamment avec l'éloignement à l'embouchure et avec l'enfouissement: à 1 mille de l'embouchure, les valeurs sont proches de d=1,2 et w=65%, dans le premier demi-mètre (17).

Une nouvelle campagne de prélèvement couvrant toute l'aire prodeltaïque vient de s'achever (novembre 1990). Cette campagne permettra de combler certaines lacunes et de ce fait améliorer les calculs, notamment près de l'embouchure où la couverture sédimentaire, très épaisse, est à échantillonner dans sa totalité.

REFERENCES

1. MARTIN J.M., THOMAS A.J., 1990
   Origines, concentrations and Distributions of artificial radionuclides discharged by the Rhône river to the Mediterranean sea.
   J. Environ. Radioactivity 11, p.105-139

2. ALOISI J.C., DUBOUL-RAZAVET C.A., 1974
   Deux exemples de sédimentation deltaïque actuelle en Méditerranée: Les deltas du Rhône et de l'Ebre.
   Bull. Centre Rech. PAU-SNPA, 8, 1, p.227-240

3. ALOISI J.C., MONACO A., PAUC H., 1975
   Mécanismes de la formation des prodeltas dans le golfe du Lion. Exemple de l'embouchure de l'Aude (Languedoc).
   Bull. Inst. Géol. Bassin Aquitain, n° 18, p.3-12

4. ALOISI J.C., MILLOT C., MONACO A., PAUC H., 1979
Dynamique des suspensions et mécanismes sédimento-génétiques sur le plateau continental du golfe du Lion.
C. R. Acad. Sc. Paris, t.289, p.879-882

5. ALOISI J.C., 1986
Sur un modèle de sédimentation deltaïque. Contribution à la connaissance des marges passives.
Thèse de Doctorat d'Etat, Université de Perpignan, 160 p.

6. MARTIN J.M., 1970
Variations saisonnières de la radioactivité de la matière en suspension dans les fleuves.
C. R. Acad. Sc. Paris, 271, p.1934-1937

7. LIVINSTON H.D., BOWEN V.T., 1979
$239\text{-}240$Pu and 137Cs in coastal sediments.
Earth and Planetary Science letters, 43, p.29-45.

8. STANNER D.A., ASTON S.R., 1981
Factors controlling the interaction of 137Cs with suspended and deposited sediments in estuarine and coastal envoronnements.
In Impacts of radionuclides releases in the marine environnements. IAEA Symposium, IAEA-SM-248-141, p.131-142

9. STANNER D.A., ASTON S.R., 1982
Desorption of 106Ru, 134Cs, 137Cs, 144Ce and 241Am from intertidal sediment contaminated by nuclear fuel reprocessing effluents.
Estuarine, Coastal and Shelf Science, 14, p.687-691

10. FERNANDEZ J.M., 1984
Utilisation de quelques éléments métalliques pour la reconstitution des mécanismes sédimentaires en Méditerranée occidentale: apport du traitement statistique.
Thèse de 3ème cycle, Université de Perpignan, 233 p.

11. BADIE C., BARON Y, FERNANDEZ J.M., FERNEX F., 1984
Une technique d'estimation du taux de sédimentation dans le delta du Rhône.
XXIX Congrès CIESM, Comité Géologie et Géophysique Marines, Lucerne, 15-20 Octobre 1984, p.189-192

12. CALMET D., FERNANDEZ J.M., 1990
Caesium distribution in northwest Mediterranean seawater, suspended particles and sediments.
Continental Shelf Research, Vol.10, Nos 9-11, p.895-913

13. BLANC F., CHAMLEY A., LEVEAU M., 1969
Les minéraux en suspension, témoins du mélange des eaux fluviatiles en milieu marin. Exemple du Rhône.
C. R. Acad. Sc. Paris, t.269, p.2509-2512

14. GOT H., PAUC H., 1970
Etude de l'évolution dynamique récente au large de l'embouchure du Grand Rhône par l'utilisation des rejets du centre nucléaire de Marcoule.
C. R. Acad. Sc. Paris, t.271, p.281-285

15. RINGOT J.L., 1981
L'utilisation de l'analyse factorielle des correspondances et de la classification automatique dans l'étude de la répartition des polluants métalliques devant l'embouchure du Rhône.
C.I.E.S.M., Vème journées, Etude Poll. Mer. Médit., p.109-116

16. SQUIRE W., 1970
    Integration for Engineers and Scientists. New-York: American Elsevier Pub.

17. CHASSEFIERE B., 1990
    Mass-physical properties of surficial sediments on the Rhône continental margin: implications for the nepheloïd benthic layer.
    Continental Shelf Research. Vol. 10, Nos 9-10, p.857-867

18. CRANK J., 1975
    The mathematics of diffusion, 2nd edn, Oxford University Press, 414 p.

19. DUURSMA E.K., EISMA D., 1977
    Theoritical, experimental and field studies concerning diffusing of radioisotopes in sediments and suspended particles of the sea. Part C. Application to field studies
    Netherlands Journal of Sea Research, 6, 3, p.365-324

20. ERICKSON K.L., 1980
    Radionuclide sorption and diffusion studies.
    Sand-81-1095, p.305-321

21. ULLMAN W.J., ALLER R.C., 1982
    Diffusion coefficient in nearshore marine sediments.
    Limnol. Oceanogr., vol.27, N°3, p.552-556

22. SURELL M., 1847
    Mémoire sur l'amélioration des embouchures du Rhône.
    1 vol., in 8°, Imp.Générale, Nîmes

23. VAN STRAATEN L.M.J.U., 1957
    Dépôts sableux récents du littoral des Pays-Bas et du Rhône.
    Geologische Minjlouw, 19ème Jaargang, p.196-213

24. LEVEAU M., COSTE B., 1987
    Impacts des apports rhodaniens sur le milieu pélagique du golfe du Lion.
    Bulletin d'Ecologie, 18, p.119-122

25. CASTAING P., 1981
    Le transfert à l'océan des suspensions estuariennes. Cas de la Gironde.
    Thèse Doct. Sciences, Université de Bordeaux I, N°701, 530 p.

26. JOUANNEAU J.M., 1982
    Matières en suspension et oligo-éléments métalliques dans le système estuarien girondin. Comportement et flux.
    Thèse Doct. Sciences, Université de Bordeaux I, N°732, 150 p.

^{10}BE AND ^{9}BE IN SUBMARINE HYDROTHERMAL SYSTEMS

D.L. BOURLÉS [1], G.M. RAISBECK [1], F. YIOU [1] and J.M. EDMOND [2]

1 Centre de Spectrométrie de Masse et de Spectrométrie Nucléaire,
Bât. 108, 91405 Orsay Campus, France

2 E34-201, Department of Earth, Atmospheric and Planetary Sciences,
Massachusetts Institute of Technology, Cambridge, MA, 02139, U.S.A.

ABSTRACT

The solution chemistry of the hydrothermal fluids emanating from sediment covered ridges is distinct from that observed on the open ridges where the solutions exit directly from basalt. Differences in composition are believed to result from interaction of a "primary" hydrothermal fluid, similar in composition to that characteristic of sediment-starved open ocean ridge axis, with the overlying sediments.

Because most (>99%) of the cosmic ray produced ^{10}Be accumulates in sedimentary deposits and because the relative short half-life (1.5×10^6 y) of this cosmogenic radioisotope precludes its presence in magmas due to long period recycling, ^{10}Be appears to be an ideal tracer of the postulated reactions between "primary" hydrothermal solutions and the sediments overlying the intrusion zone. Both ^{9}Be and ^{10}Be isotopes have therefore been measured in fluids emanating from sediment-starved hydrothermal systems as well as in hydrothermal solutions and associated sediments from sediment hosted environments. The use of these isotopes to constrain both the chemical composition of the precursor fluid and the magnitude and evolution of the interaction between hydrothermal fluids and overlying sediments, will be discussed.

Extended to hydrothermal systems from back-arc basins, this study allows us to investigate the characteristics of the sedimentary component associated with the subducting oceanic plate.

THE ROLE OF RADIOGENIC HEAVY MINERALS IN COASTAL SEDIMENTOLOGY

R.J. de Meijer, H.M.E. Lesscher, M.B. Greenfield and L.W. Put
Kernfysisch Versneller Instituut, Rijksuniversiteit Groningen,
9747AA Groningen, the Netherlands

ABSTRACT

Changes in natural radiation level on selected sites on beaches of the island of Texel, The Netherlands, were measured simultaneously with changes in beach elevation. It was found that radiometric mapping provides a sensitive method for studying sand transport phenomena. Results show a selective transport for wind and water driven transport of sand which can at least qualitatively be understood from rather simple microscopic models of sedimentation and removal along with a simplified model of radiation absorption. The method demonstrated for a linear grid spacing of 2 m should, according to the data, still yield valuable information for grid spacings of about 25 m.

INTRODUCTION

Erosion is a serious problem for many sandy coasts. If the coasts form a natural protection for a highly industrialized and densely populated country like the Netherlands, governments are especially willing to invest large sums of money to reinforce the coast by e.g. beach renourishment programmes. The effect of such programmes is limited, and renourishments often have to be repeated after a number of years.

Presently no methods have proven to be effective in stopping the long term regression of the shore line. Whereas our knowledge on water transport shows large progress compared to about a decade ago, knowledge on sand transport remains insufficient. One of the difficulties encountered is the paucity of data since almost all active detection systems used in the coastal waters only work under calm conditions when hardly any sand transport occurs.

Sand is composed of an ensemble of minerals which have survived the decomposition of rock. In addition to the dominant mineral quartz, minerals with higher density (heavy minerals) are present e.g. ilmenite, garnet, epidote, rutile, zircon and monazite. In sediments one often finds these minerals to have smaller grain sizes than quartz, due to hydraulic equivalence during the sedimentation process. Although these differences are important for both sedimentation and erosion, for this study the main relevant difference between various minerals is that heavy minerals contain several orders of magnitude higher concentrations of U, Th and their decay products than found in quartz. This property does not only allow for the location of concentrations of heavy minerals in an interactive way but also the study of the selectivity in sand transport phenomena.

In recent papers [1-4] the application of radiogenic heavy minerals has been demonstrated in different ways. In [1,2] it has been shown that monitoring elevation and γ-ray intensities of beach sites reveals information on the wind driven transport. It was shown that under conditions of moderate to strong winds, predominantly light minerals were displaced.

A comparison of the radiometric and gravimetric determination of the heavy mineral content in beach sand was made for samples along the Dutch coast[3]. From this study it was found that a remarkably sharp border exists between the radiometric properties of beach sand north and south of the tidal channel south of the island Texel. This difference is attributed to a difference in provenance.

In this paper we apply an existing simple physical model for the effect of forces acting on sand grains to demonstrate a possible origin of selective sand transport. The results of measurements shortly before and after a storm are used to demonstrate the selectivity in water transport. Moreover, the effect of the grid size on the observable features is described. For a more extensive description of the geological setting, the experimental techniques and a simplified absorption model we refer to [1,2].

SEDIMENTATION AND EROSION

This section does not intend to give a comprehensive description of sedimentation and erosion processes but reviews some simplified models to describe the underlying physical processes[6-8].

If two kinds of grains with different radii, R, and densities, ρ, are found in a sediment, they have reacted similarly on the prevailing hydraulic circumstances and are designated as hydraulically equivalent. A criterion for this similar behaviour is a ratio of their settling velocities near unity. This follows from considering a particular grain in a stationary fluid; the settling velocity v is determined by the force balance between the gravitational force and the drag force:

$$v^2 = \frac{8}{3} \frac{R \cdot g \cdot (\rho_m - \rho)}{C \cdot \rho_m} \qquad (1).$$

In this equation g is the gravitational acceleration, ρ_m the density of the medium and C the drag coefficient, depending on the Reynolds number.

For the removal of spherical grains with radius R_1 from a bed of spherical particles with radius R_2, the gravity force, F_g, and the drag force F_d acting on a single grain are schematically presented in figure 1. The drag force F_d is assumed to be proportional to the average shear stress τ times the area

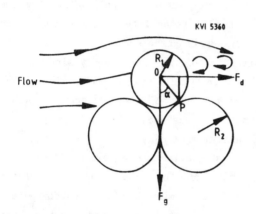

Figure 1. Schematic representation of forces acting on a spherical particle on a bed, taken from [8].

of the grain:

$$F_d = C_d \cdot \tau \cdot R_1^2 \qquad (2).$$

In this equation the drag factor C_d takes into account the geometry and packing (exposed area) of the grain. The gravitational force can be expressed as:

$$F_g = \frac{4}{3} g\pi (\rho - \rho_m) R_1^3 \qquad (3).$$

The angle α which the gravitational force makes with the pivoting axis OP is generally assumed to be equal to the angle of repose. The threshold velocity is the velocity of the medium causing a drag force which obeys the equation

$$F_g \sin\alpha = F_d \cos\alpha \qquad (4).$$

The relation between the velocity and the shear stress is given by:

$$v^2 = \tau/\rho_m \qquad (5).$$

Using eqs. 2, 3, and 4 in eq. 5 results into the expression for the threshold velocity v_t:

$$v_t^2 = A \cdot g \cdot R_1 (\rho - \rho_m)/\rho_m \qquad (6).$$

The coefficient A in eq. 6 depends on the medium, the drag coefficient and on the shape and size of the particle. From eq. 6 it becomes clear that the density of the medium has great influence in the threshold velocity. For e.g. quartz the factor $(\rho-\rho_m)/\rho_m$ is in air in the order of 2000 and in water 2, which means that the threshold velocities in water are one to two orders of magnitude smaller than in air.

In this work a simplified absorption model is used, wherein the measured intensity of the radiation is exponentially reduced by an absorber of known thickness. As derived in Appendix I of ref.[1] the relation between the count rates I_0 above an infinitely extending homogeneous layer of sand, and I after covering this layer with a homogeneous layer of sand with thickness d and a different specific activity, is given by

$$(I - I_1)(I_0 - I_1) = e^{-\mu d} \qquad (7).$$

In this equation I_1 is a constant related to the specific activity of the covering sand and μ is the effective linear absorption coefficient for γ-radiation in sand. From measurements on beach sand [1] it was found that the experimental value of μ ranged between 7 and 12 m^{-1} for uncompacted and compacted sand, respectively.

EXPERIMENTAL TECHNIQUE AND ANALYSIS

On the island of Texel three small (about 30 x 30 m^2) sites and one larger scale site (200 x 120 m^2) were selected. The description of the three smaller sites and their location on the island is given in ref.[1]. The larger scale site (IV) was situated adjacent to the south side of site II in ref.[1]. The measurements on the three smaller sites were carried out between June 1988 and the beginning of January 1990. The measurements at site IV lasted from July 1989 to the beginning of January 1990. The measurements were stopped due to lack of funding prior to the so called superstorms of January and February 1990 each lasting several days, after which large portions of the original sites were removed and considerable amounts of heavy minerals were deposited.

Data were taken at each site on a given day at roughly monthly intervals. The sites I-III were designated as grids with points spaced at 2 m intervals. Both the elevation, measured using a transit and stadia rod and the gamma intensity, integrated over 10 s using a hand-held, γ-monitor Scintrex) at hip height were determined for each grid point. Moreover at approximately half way along each side of the perimeter of each grid samples of about 1 kg of sand were taken concurrent with the above field measurements. An additional sample was taken on June 8, 1988 at site II at a depth of about 80 cm below the surface.

The data points were portrayed in three-dimensional grid plots. "Difference" grid plots are the result of the subtraction of each grid point from the data obtained from the previous measurement. In this way the changes in elevation and radiometry were obtained. These changes were also presented as two-dimensional figures in which the change in height is given as function of $\Delta \ln I = \ln((I_i - C)/(I_{i-1} - C)$, the "effective" change in count rate. In this expression the quantity C represents the background count rate (30 cps) and i and $i-1$ indicate the run number of the present and the previous measurements, respectively.

RESULTS AND DISCUSSION

Hydraulic properties of sand

In a previous study [5] the grain size distribution of various minerals in a sand sample from Ameland was determined. For light and heavy minerals average grain sizes of 180 and 120 µm with average densities of 2.65 and 4.2 10^3 kg.m^{-3} were found, respectively. Assuming similar properties for the sand on Texel the hydraulic properties of the light and heavy fractions can be calculated from eq.1 using graphically interpolated drag coefficients from ref.[9]. The calculated settling velocities for the two fractions are 1.3 and 1.0 m.s^{-1}, respectively, in air and 0.019 and 0.020 m.s^{-1} in water. These values were calculated for spherical grains; for non-spherical grains the values are lower.

From the densities and grain sizes the threshold velocities can be calculated with eq.6. Transforming the shear velocities to wind velocities at 10 m height[6], being the height at which wind velocities are registered, the threshold velocities for dry materials are exceeded at 5.4 and 7.2 m.s^{-1}, for light and heavy minerals, respectively. For wet materials these values may be higher due to the bonding by pore water. Regarding the assumptions on density, size and sphericity, it therefore seems likely that the sediments were deposited in water, but that further sorting on the beach is likely to happen under the influence of wind.

Sand transport by wind

Figure 2 shows in three and two dimensions the change in elevation and the change in radiometry for site I between June 29, 1988 and July 27, 1988. The data have partly been discussed in ref.[1,2]. The part of the grid with

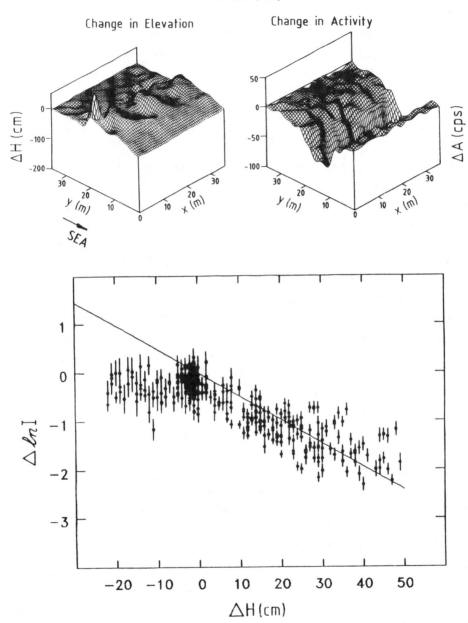

Figure 2. Two- and three-dimensional representations of the change of elevation and the change in radiometry for site I between June 29 and July 27, 1988. The line represents a linear fit to the data with $\Delta H>0$.

4<y<23 m accretion of sand has taken place, for y>23 m the elevation hardly changed, whereas for y<4 m the beach is lowered, probably by water. From the change in activity one notices that the accretion resulted in a reduced γ-ray intensity. Moreover one notices a wave like structure in the sand deposit with a wavelength of about 12 m and 0.4 m amplitude. In the lower part of the figure where the change in the logarithm of the effective count rate has been presented as function of the change in height one notices that for ΔH>0 (corresponding to y>4 m) the data points are well represented by a straight line. This is in agreement with eq.7 and the derived value of μ = 4.8 ± 0.1 m^{-1} is in good agreement with the value for loosely compacted sand.

The wave like structures in fig. 2 are probably the result of southern wind with a velocity of about 13 $m.s^{-1}$, just two days before the measurement in July. The height-length ratio of 1/30 is consistent with empirical relations derived by Bagnold[6]. The composition of the sand corresponds to that of the bulk material and is not enhanced in radioactivity.

Sand transport by water

On December 5, 1988, shortly after the November 30th measurements a west to northwesterly storm with excess water levels of 220 cm occurred. We returned on December 7 for new measurements. The results for site II are presented in fig. 3 as two and three-dimensional plots of change in elevation and change in γ-ray intensity. The data may roughly be divided into two regions: x>8 m (I) and x<8 m (II). The regions may be characterized by lowering of elevation and increase of radiometry, and lowering of both elevation and radiometry, respectively. These regions are apparent in the two-dimensional representation of the data, which more or less form a triangle.

For region I the lowering of the elevation is concurrent with an increased γ-ray intensity indicating selective removal of mainly light minerals. At this site in June 1988 a sample was taken at 80 cm below the surface. From this sample it was found that the heavy minerals were mainly concentrated in the upper 15 cm. Except for the upper part of the site, region I corresponds to the swash-backwash zone where the flow of water becomes less turbulent and the velocity of the flow decreases with increasing elevation. The fact that the radiometry first increases with increasing x-value and then decreases for 22<x<25 m is consistent with a decreasing water velocity with increasing selectivity followed by velocities which are too low for transport.

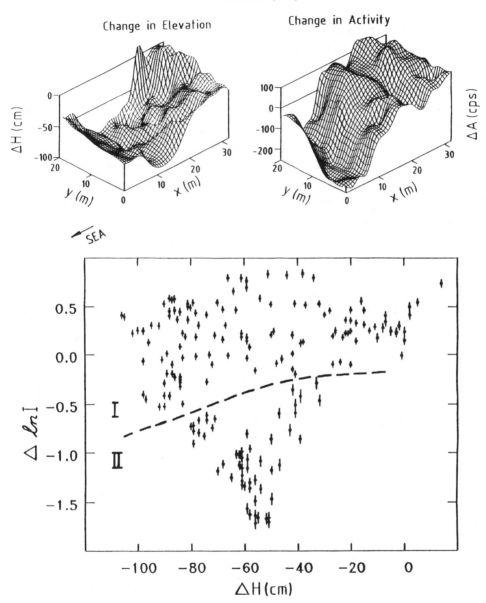

Figure 3. Two- and three-dimensional representations of the change of elevation and the change in radiometry for site II between November 30 and December 7, 1988.

In region II the lowering of the elevation resulted in a decrease of radiometry. This is consistent with a non-selective removal of sand in which all the concentrated heavy minerals were removed. This is not too surprising since the region corresponds to the breaker zone of the waves. Under these circumstances no selectivity is to be expected.

In this explanation the increase in radiometry is due to the selective removal of light minerals. It should however be pointed out that the increase could just as well be due to transport of heavy minerals from region II to region I. This might at first seem unrealistic, but after the superstorms of early 1990, heavy mineral concentrations were observed in quantities at several locations along the Dutch coast which could not be explained by local wash out of beach and dunes. This type of selective transport has also been reported from ore spilled from two sunken carriers near the north west coast of Spain: the light ore (coal) moved seawards, whereas the heavy ore (iron) moved shorewards. Similarly it was reported that coarse sediment may move onshore while finer material may simultaneously move offshore[11].

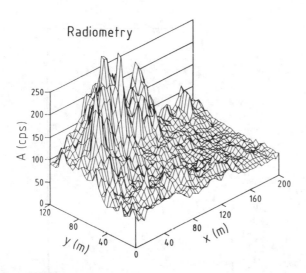

Figure 4. Three-dimensional representation of radiometry for site IV with a linear grid spacing of 4 m.

Optimizing grid size

So far we have made measurements on the beach; in the future we plan to investigate these phenomena on the sea bottom. Under these conditions the grid spacing of 2 m is not feasible. In addition smaller grids limit the amount of information which may be obtained in a reasonable amount of time. To address the question of optimum grid sizes needed to maintain sufficient resolution for observing the features indicative of large scale transport phenomena, we measured the radiometry of a 200 x 120 m^2 site with a 4 m linear grid spacing. The results are presented in figure 4. Next we selectively omitted data points to obtain grid spacings of 8, 16 and 40 m.

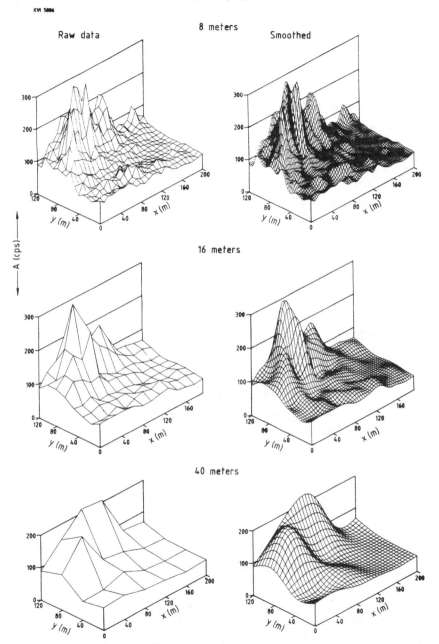

Figure 5. Same data as in fig. 4 but presented for grid spacings of 8, 16 and 40 m. At the right hand side the results of smoothing the data with a spline function are shown.

The data are presented in figure 5 as raw data and smoothed with a spline function. From figure 5 we note that essential information starts to get lost for spacings of 40 meter or more; smoothing of the data reduces the information only slightly and is still quite acceptable for observing large scale features at a spacing of 16 m. From these results we conclude that a linear grid spacing of about 25 m is optimal for measurements on the beaches of the Dutch Wadden islands. We realize that the optimal value on the sea bottom may be different.

CONCLUSIONS

The present investigation has shown that the radiometric properties of heavy minerals in beach sand provide a tool to study selective transport of sand. Simplified microscopic models to describe sedimentation and removal of sand yield results which are at least qualitatively consistent with the data. The negative correlation between changes in elevation and radiometry are well understood with a simplified radiation absorption model.

Studying selectivity in sand transport provides a more critical test for sand transport models and may provide insight into coastal erosion phenomena which are observed but have yet to be explained. In this context we refer to changes in local concentrations of heavy minerals present on the beaches that cannot be explained from local wash out, the observation [10] that light and heavy ores spilled from sunken vessels moved in opposite directions on the sea bottom and the opposite movement of coarse and fine material near beaches[11].

In the present investigation we focussed on small size grids. An attempt to determine the optimal linear grid spacing for observing relevant transport features led to a value of about 25 m. Such a spacing is feasible for remote sensing techniques such as towed seabed detectors and enables the observation of the evolution of such features over areas of 0.5 km^2. It is our intention to investigate transport phenomena using such grid sizes as further preparation for sea bottom experiments.

This investigation is part of the research programme "Environmental Radioactivity Research" of the Kernfysisch Versneller Instituut and the programme "Kustgenese" (Coastal genesis) of Rijkswaterstaat. This paper is a contribution to Project 274 of the International Geologic Correlation Programme (IGCP).

The authors would like to thank the personnel of the Dienstkring Texel of Rijkswaterstaat for their assistance and especially for the elevation measurements. The stimulating discussions with Drs Donoghue (Dept. of Geol., FSU), Wiersma (Rijkswaterstaat) and de Vriend (Delft Hydraulic) are gratefully acknowledged.

REFERENCES

1. Greenfield, M.B., de Meijer, R.J., Put, L.W., Wiersma, J., and Donoghue, J.F., Monitoring beach sand transport by use of radiogenic heavy minerals, Nucl. Geophys., 1989, 3, 231-244.

2. Lesscher, H.M.E., Blown with the wind; radiometry of evolving beaches, KVI internal report 157i, 1989.

3. de Meijer, R.J.,Lesscher H.M.E., Schuiling, R.D., and Elburg, M., Estimate of the heavy mineral content in sand and its provenance by radiometric methods, Nucl. Geophys., 1990, 4, 455-460.

4. Donoghue, J.F. and Greenfield, M.B., Mineral sands as an indicator of coastal sand transport processes, J. Coast. Res., 1991, 7, 189-201.

5. Schuiling, R.D., de Meijer, R.J., Riezebos, H.J., and Scholten, M.J., Grain size distribution of different minerals in a sediment as function of their specific density, Geol. Mijnbouw 1985, 64, 199-203.

6. Bagnold, R.A., The physics of blown sand and desert dunes, Methuen & Co, London, 1941.

7. Slingerland, R.L., The effects of entrainment on the hydraulic equivalence relation ship of light and heavy minerals in sand, J. Sed. Petr., 1977, 47, 753-770.

8. Allen, J.R.L., Physical processes of sedimentation, George Allen & Unwin, Ltd, London, 1970.

9. Janssen, L.P.B.M., and Warmoeskerken, M.M.C.G., Transport phenomena data compendium, Delftse Uitgevers Mij, Delft, 1987.

10. Dalac,J.J., and Lechuga Alvaro, A., Distribution on the sea bed of solid bulk cargoes from sunken vessels, in: J.R. Houston: General report p.3, section SII-2.2. PIANC, 27th Intern. Navigation Congress 1990.

11. Richmond B.M. and Sallinger, A.H., Cross shore transport of bimodal sands, 19th Int. Conf. on Coastal Eng., 1984.

THE USE OF ^{210}Pb AS AN INDICATOR OF BIOLOGICAL PROCESSES AFFECTING THE FLUX AND SEDIMENT GEOCHEMISTRY OF ORGANIC CARBON IN THE NE ATLANTIC

TIM BRAND AND GRAHAM SHIMMIELD
Department of Geology and Geophysics
University of Edinburgh
West Mains Road, Edinburgh, EH9 3JW, UK

ABSTRACT

Measurement of the fine-scale distribution of ^{210}Pb, organic carbon and dissolved pore water oxygen has been conducted to determine the organic carbon degradation and benthic activity rates in surface sediments of the NE Atlantic Ocean along the 20°W transect. Use is made of the simple diffusion analogue model to calculate bioturbation rates and its suitability is reviewed in the light of recent thoughts on the mechanisms of bioturbation. Calculated bioturbation rates show a general increase towards the northerly stations. Organic carbon degradation rates show a less well defined increase towards the north. Water column ^{210}Pb data indicate a progressive increase in seasonal biological scavenging efficiency to the north of the transect. Predicted organic carbon flux rates show comparable values throughout the whole transect. The response of the benthic infauna and associated organic carbon degradation is considered to operate on a similar temporal scale to the seasonal biological activity of the water column. Sediments which are highly bioturbated appear to disrupt the dissolved O_2 gradient across the sediment-water interface.

INTRODUCTION

The U.K. Biogeochemical Ocean Flux Study (BOFS), of which this work forms part, is at present devoted to identifying and quantifying the overall transfer of carbon from atmospheric input to burial within sediments in a region of the northeast Atlantic. Such a study seeks to examine the rate and pathway of carbon transfer at various sites in the water column and sediments ultimately to discover the ocean's capacity to absorb and regulate atmospheric CO_2 levels.

Phytoplankton utilise dissolved CO_2 within the photic zone of the water column during photosynthetic carbon fixation. The phytoplankton eventually either die and decay to form CO_2 and a mixture of dissolved and particulate degradation products, or are grazed by zooplankton. Various stages of decay and remineralisation occur throughout the descent of the material resulting in a mixture of phytoplankton particulate decay products and zooplankton faecal material reaching the seabed. This material is collectively known as marine phytodetritus and it is the aim this study to identify and quantify some of the chemical and biological processes operating on this material as it reaches, and is incorporated into the ocean sediment.

Many of these processes occur at, or just below, the sediment-water interface and consequently a strategy of high resolution sampling of the surficial sediment has been employed. Measurement of the activity of the naturally-occurring radionuclide ^{210}Pb (half-life 22.3 yr) in the marine environment has wide applications in the mapping of carbon movement both within the water column and sediment. Fine-scale determination of this radionuclide in the sediment column can provide a rate model of the surficial sediment mixing processes, and when compared to water column inventories, give an indication of phytodetritus flux rates.

TABLE 1
Core Station Geographic and Bathymetric Data

RRS Discovery Station No.	Station Code No.	Latitude (N)	Longitude (W)	Water Depth (m)	Date of collect.
11878/4	1M	47°09.6'	22°30.2'	3945	20/07/89
11882/5	5M	50°40.4'	21°51.6'	3560	27/07/89
11891/2	11M	55°11.3'	20°21.0'	2080	01/08/89
11898/2	15M	59°04.9'	20°05.8'	2790	04/08/89

STUDY AREA

This study was carried out in the northeast Atlantic along a transect coinciding with the 20°W meridian. Core station geographic and bathymetric data is given above in Table 1. Core sites are described using RRS *Discovery* station numbers and by core code names, the latter being used in this paper. Core Station 1M was within the region of the JGOFS Biotrans site and was taken on the lower flanks of a northeast-southwest trending ridge. Core Station 5M was situated on the lower southwest flanks of the East Thulian rise. Core Station 11M was situated on the southern flanks of the Fangorn Bank, located on the Hatton Rockall Plateau. Core Station 15M was taken from a levee to the northwest of the Maurey Channel between the South Iceland Basin and the Hatton Bank.

METHODS

All cores were collected using a Multicore sampler (Duncan and Assoc.) which retrieves up to 12 cores of approximately 5.5cm in diameter, each 20cm apart, from an area of 1m² of the sediment surface. The cores were normally retrieved with between 10 and 20cm of bottom water, depending on the length of the sediment core, and were tightly sealed at both ends. The overlying bottom water was invariably clear and many fine scale features such as worm tubes, settled planktonic faecal and degradation matter and epifaunal organisms (brittle stars) could be readily seen sitting on the surface, indicating little or no disturbance of the sediment-water interface upon core recovery. Four core sites along this 20 W° transect were chosen and three cores from each site were studied. Upon recovery all cores were placed in cold storage at 3°C for temperature equilibration prior to pore water oxygen analysis. Pore water oxygen was determined using a Clark-type microelectrode (Microsense™) with a tip diameter of 100-200μm inserted into the top of one of the cores above which there was approximately 10cm of bottom water. The current produced (pA) was amplified and measured and calibrated against nitrogen flushed bottom water and oxygenated bottom water electrode readings to give a percentage saturation. These readings were converted to μmol kg⁻¹ using Winkler titration-derived oxygen concentrations of the bottom water at the core site. Between two and three profiles were measured on one core from each site.

Microscale subsampling was performed on a further two cores using a piston and screw jack mechanism fitted to the core barrel. Millimetre slices were taken from the top 1cm, 2mm slices were taken between 1 and 2cm, 5mm slices were taken between 2 and 10cm and 1cm slices were taken thereafter.

On return to Edinburgh the sediment samples were dried at 50°C and porosity and salt corrected dry bulk densities calculated. Samples were then ground using a tungsten carbide Tema mill. Organic carbon (C_{org}) determinations were carried out on a Carlo Erba CNS 1500 Analyser™ after initial carbonate removal with sulphurous acid using a technique similar to Verado et al. (1). Repeat samples gave a standard error of 10%. ^{210}Pb assay was carried out by complete sample digestion with nitric, hydrofluoric and perchloric acids using ^{208}Po as a yield tracer followed by pH adjustment and auto-plating at 30°C on silver discs. ^{210}Pb activity was determined via measurement of its granddaughter ^{210}Po on surface barrier alpha detectors with which it was assumed to be in secular equilibrium. One sigma (σ) errors were calculated from propagation of counting statistics and analytical error. All carbon content and ^{210}Pb activity determinations were salt corrected.

RESULTS

Porosity data are given in Figure 1 and all cores exhibit rapid changes in porosity over the top 1cm except cores 15M1 and 15M2 which both show rather variable surficial profiles. All pore water oxygen profiles shown in Figure 2 exhibit steep gradients near the sediment-water interface indicating high levels of aerobic carbon degradation, and generally reach levels below detection limits within the range of the microelectrode penetration. The one exception to this is core 15 which appears to adopt an asymptotic value between 9 and 10 cm.

The oxygen consumption of the sediment can be estimated by applying Fick's first law of diffusion and calculating the total flux of oxygen across the sediment water interface (2,3).

$$F = -\phi° D_s (\delta C/\delta z)_{z=0} \qquad\qquad 1$$

where $(\delta C/\delta z)_{z=0}$ is the oxygen concentration gradient in units of μmol cm^{-3} cm^{-1} at the sediment water interface and arbitrarily taken over the upper 3mm where the gradient is linear, $\phi°$ is the porosity of the surface sediment layer and D_s is the sediment diffusion coefficient in cm^2 d^{-1}. If pore water irrigation by biological activity can be ignored over this region (see discussion below), the values of the sediment diffusion coefficient can be approximated by $D_o \phi^2$ where D_o is the free solution diffusion coefficient and equal to 1.07 cm^2 d^{-1} at bottom sea water conditions (4). The oxygen flux calculation results shown in Table 2 indicate higher values of O_2 consumption between 13 and 19 μmol m^{-2} d^{-1} in the two southerly stations 1M and 5M compared to values of between 4 and 6 μmol m^{-2} d^{-1} seen in the northerly stations 11M and 15M. The carbon respiration rates calculated from the oxygen flux measurements using a stoichiometric ratio of 0.768 (106 moles of CO_2 produced for every 138 moles of O_2 consumed) (5) are shown in Table 2. These respiration rates decrease towards the north in accordance with the trend of oxygen consumption rates.

Microscale sampling of the sediment indicates the very heterogeneous nature of the organic carbon (C_{org}) content of within each core (Figure 3). Profiles from stations 1M, 5M and 11M indicate limited decrease of C_{org} concentration over the depth sampled, and furthermore cores 1M2 and 11M2 even show a general increase. In order to determine the rate of organic carbon degradation (K_{orgC}) for each core it is necessary evaluate the rate of biomixing in the sediment. To achieve this ^{210}Pb is used as a bioturbation tracer.

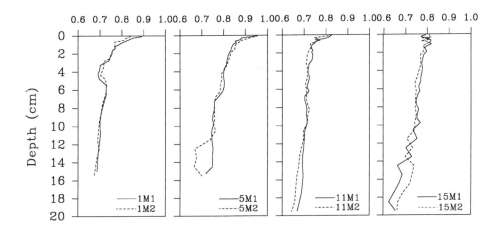

Figure 1. Porosity profiles

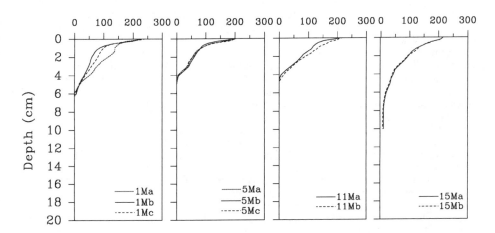

Figure 2. Pore water dissolved oxygen concentration (um kg^{-1}) profiles

TABLE 2
Oxygen Flux Parameters

Core Station	Surface Porosity	Oxygen Gradient (mmol cm^{-3}cm^{-1})	Oxygen Flux (mmol m^{-2}d^{-1})	Mean Carbon Oxidation Rates (mmol m^{-2}d^{-1})
1M	0.889	0.00185	13.91	
		0.00238	17.89	11.68
		0.00184	13.83	
5M	0.948	0.00212	19.33	
		0.00152	13.86	12.63
		0.00175	15.95	
11M	0.823	0.00115	6.86	4.70
		0.00090	5.37	
15M	0.792	0.00086	4.57	3.51
		0.00086	4.57	

^{210}Pb is a particle-reactive radionuclide and in this environment is readily scavenged from the water column by marine particulates to form an excess activity at the sediment surface relative to the *in situ* production. Because of this it can be used in the evaluation of particulate mass transfer rates and as a tracer of particle mixing in pelagic sediments. Owing to its half life of 22.3 years its existence as an excess activity in pelagic marine sediments to any depth greater than a few millimetres is due to bioturbation by benthic organisms. ^{210}Pb$_{xs}$ activity was calculated by subtraction of the *in situ* ^{226}Ra activity, (assumed from the non-zero constant value of ^{210}Pb at the base of each core), from the ^{210}Pb profile. A major difficulty with this approach is that the intermediate radionuclide, ^{222}Rn, diffuses from the uppermost sediment into the overlying water leading to disequilibrium between the ^{226}Ra and ^{210}Pb and an underestimation of the value of ^{210}Pb$_{xs}$ (6). This may have an appreciable effect upon calculated biodiffusion rates. Profiles of ^{210}Pb$_{xs}$ with best-fit curves are shown in Figures 4. One σ propagated error bars are included within the data point symbol.

The profiles shown highlight the transfer of ^{210}Pb$_{xs}$ into the sediment due to bioturbation and the style and shape of the profile is dependent upon the nature and intensity of the bioturbation mechanics. Bioturbation will have the effect of decreasing the steepness of the gradient and lowering the surface specific activity. The surface values of cores 5M1 and 5M2 show significantly higher activities (500-600 Bq kg^{-1}) compared to the other cores (200-400 Bq kg^{-1}). It is interesting to note that there is a significant difference between the shapes of the profiles in core station 5. The presence of a subsurface peak and significantly steeper gradient in core 5M1 contrasts to the shallower, smoother profile seen in Core 5M2. The northerly stations, 11M and 15M, exhibit more shallow ^{210}Pb$_{xs}$ gradients. The presence of the subsurface ^{210}Pb$_{xs}$ peak in core 5M1 coincides with a similar peak in C$_{org}$ indicating the close association of ^{210}Pb with organic carbon.

The distribution of ^{210}Pb$_{xs}$ in marine sediments subject to bioturbation has received much study (6-13) and has traditionally been described in terms of an analogue to random molecular diffusion originally proposed by Goldberg and Koide (14). By its very nature this approximation implies that mixing mechanisms distribute the radionuclide in manner that produces a profile of exponentially decreasing concentration. This can be quantified using Fick's second law of diffusion at steady state:

$$\rho(\delta A/\delta t) = D_b \rho(\delta^2 A/\delta z^2) - \rho\omega(\delta A/\delta z) - \lambda A + \rho P = 0 \qquad 2$$

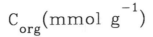
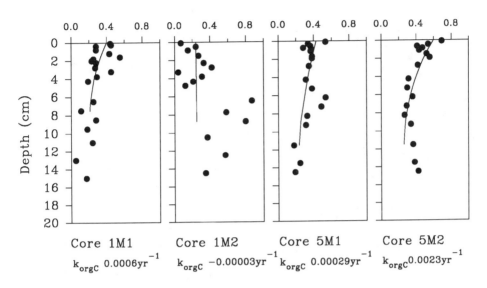
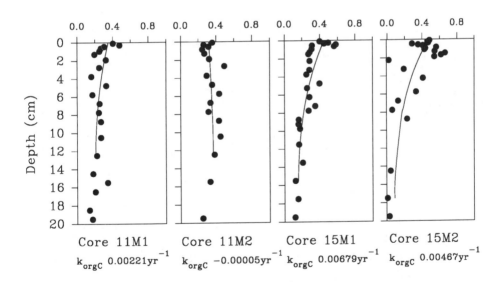

Figure 3. Organic carbon (mmol g^{-1}) profiles showing model best fit lines and calculated degradation constants

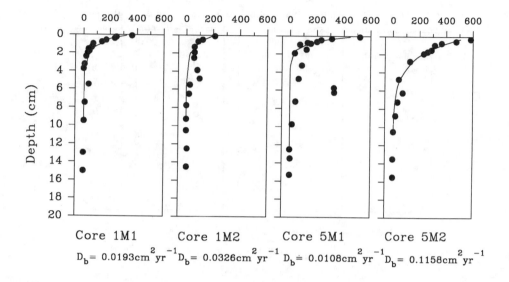

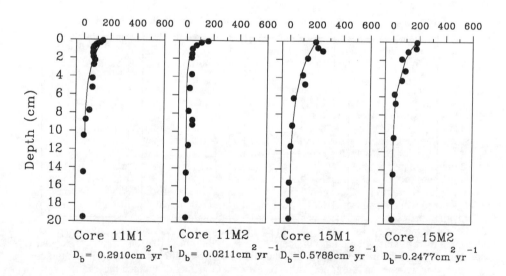

Figure 4. $^{210}Pb_{xs}$ activity (Bq kg^{-1}) profiles showing model best fit lines and calculated biodiffusion coefficients

where A is the activity of the radionuclide (Bq kg^{-1} sediment), D_b is the bioturbation or biodiffusion coefficient (cm^2 yr^{-1}), ω is the bulk sediment accumulation rate (cm yr^{-1}), ρ is the dry bulk density (g dry sed.cm^{-3} wet sed.), P is the *in situ* supported level of activity (Bq kg^{-1}), z is depth (cm) and λ is the radioactive decay constant (yr^{-1}). The above formula assumes constant porosity throughout the sediment column, which as shown in Figure 1, generally decreases rapidly over the top 1 cm. A porosity related bioturbation parameter has been formulated by Officer (11) from which depth-related values of the biodiffusion coefficient can be obtained. However, in order to use the calculated D_B values in the organic carbon degradation rate determinations the constant porosity model has been used here.

Equation 2 can be applied using the general solution:

$$A(z) = M\exp(\alpha x) + N\exp(Bx) \qquad 3$$

where A is the unsupported or excess activity of the radionuclide (Bq kg^{-1}) at depth z (cm) and with boundary conditions of

$$A = A_o \text{ at } z = 0$$

and $D_b(\delta A/\delta z) = 0$ at $z = L$

where L is depth of bioturbation (cm), the constants are calculated from the following:

$$M = A_o - N \qquad 4$$

$$N = -A_o\alpha\exp(\alpha L)/\{\beta\exp(\beta L) - \alpha\exp(\alpha L)\} \qquad 5$$

$$\alpha = \{\omega + \omega^2 + 4\lambda D_b\}/2D_b \qquad 6$$

$$\beta = \{\omega - \omega^2 + 4\lambda D_b\}/2D_b \qquad 7$$

Using sediment accumulation rates for the NE Atlantic given by Balsam and McCoy (15) and depths of bioturbation indicated by the maximum depth of penetration of $^{210}Pb_{xs}$ values of D_b for each core were calculated using an iterative best-fit method to the $^{210}Pb_{xs}$ data. These are shown in Table 3. There is a progressive increase in D_b towards the north of the transect, although considerable variation is seen at the same station (cf. 5M). At the most northerly station (15M) which exhibits the highest D_b the associated porosity profile (Figure 1) displays evidence of more intense sediment reworking.

Organic carbon degradation rates, k_{orgC}, have been calculated over the mixed zone using equations 3 to 7, substituting the calculated values of D_b for each core and applying the iterative best-fit procedure. The results are shown in Figure 2 and Table 3. Negative values have been calculated for cores 1M2 and 11M2 because C_{org} does not decrease with depth and so the above model is not applicable in these circumstances. The degradation rates calculated span an order of magnitude between 0.0003 yr^{-1} seen in core 5M1 to 0.0068 yr^{-1} in core 15M1. There is an overall trend in C_{org} degradation rates calculated by this method increasing towards the north. However, there is also an order of magnitude difference between the two cores of the same station (ie 5M). Core 5M1 which exhibits the large $^{210}Pb_{xs}$ subsurface peak having the lower value, which may indicate that the style of bioturbation directly affects the appropriateness of the model. This model calculates a single value of k_{orgC} within the mixed zone and so represents an integrated value for several C_{org} fractions which may have widely different decay constants.

TABLE 3
Bioturbation and Organic Carbon Degradation Parameters

Core	Sed. rate (cm ky^{-1})	Depth of Bioturb. (cm)	Biodiff. Coeffic. (cm^2 yr^{-1})	Carbon deg. rate (yr^{-1})	Mean Carbon Res. Time (yr)
1M1	2	8	0.019	0.0006	1667
1M2	2	8	0.033	-0.00003	
5M1	4	12	0.011	0.0003	769
5M2	4	11	0.116	0.0023	
11M1	2	12	0.291	0.0022	455
11M2	2	12	0.021	-0.00005	
15M1	2	16	0.579	0.0068	174
15M2	2	17	0.248	0.0047	

DISCUSSION

The use of Fickian diffusion as a model analogue to bioturbation processes has received considerable review of its appropriateness and has undergone several modifications to increase its suitability for describing the mechanisms of bioturbation (13,17,18,19,20,21,22). Of particular relevance to studies of pelagic sediment bioturbation is the common occurrence of subsurface peaks in the tracer profile which are generally considered to represent advective movement of tracer-rich particles from surface layers to some depth below. The existence of such peaks has been recognised for some time (12). Smith *et al.* (13) identified such peaks in NE Atlantic sediments as the result of a species of sipunculide worm living approximately 3cm within the sediment column and transferring tracer-rich surface material to some depth below during feeding. This is in a manner opposite to the "conveyor belt" feeding strategies of head-down deposit feeders identified and modelled by Robbins (20). It is important to consider that such activity is probably temporally- as well as spatially-dependent. Smith *et al.* (13) have shown using ^{210}Pb and ^{239,240}Pu tracers, that the peaks are more likely to be single events which are subject to post-formation smoothing on a time scale of between 3 and 6 years using rates of background biodiffusive bioturbation typically found in deep sea pelagic sediments.

The ^{210}Pb$_{xs}$ profiles shown in this study, although exhibiting in some instances advective particle transfer, generally agree with 'difusive-style' particle movement. The calculated biodiffusion rates show a general increase towards the north of the transect. The organic carbon degradation rate results also show higher levels of degradation probably associated with a more labile fraction towards the north of the transect. It is possible that the bioturbation intensity is reacting towards the larger fraction of labile organic carbon to the north. Bathymetric photography during the cruise (McCave pers. com.) showed evidence of increased quantities of phytodetritus resting on the seafloor during the northward passage, and since it has been shown that such material is quickly degraded and incorporated into the sediment (23,24) it seems likely that a progressively shorter time had elapsed between phytodetritus deposition and sediment coring. As shown in Ritchie and Shimmield (16), in this area away from continental margin scavenging effects, the dominant process of ^{210}Pb removal from the water column is by phytodetrital flux. Consideration of Table 4 indicates that the removal of available ^{210}Pb from the water column ("Scavenging Efficiency") at the time of sampling increases to the north. Since the water column ^{210}Pb dynamics are highly influenced by the timing and intensity of the phytoplankton bloom the data presented represents a 'time snapshot' of water column ^{210}Pb inventories. The increase in scavenging efficiency to north suggests there was increased particulate flux of material from the water column to the sediment in association with the northward drift of the phytoplankton

TABLE 4
Water Column and Sediment ^{210}Pb Inventory and Carbon Flux Data

Stat.	Water Column ^{210}Pb Supply[1] (Bq cm^{-2})	Water Column ^{210}Pb Invent.[2] (Bq cm^{-2})	Water Column ^{210}Pb Defic.[3] (Bq cm^{-2})	Water Column Scavenging Effic.[4] (%)	Mean Sediment Invent. (Bq cm^{-2})	Predicted Corg Flux (μg cm^{-2} yr^{-1})
1M	0.91	0.72	0.19	21	11.8±7.5	11.4
5M	0.77	0.55	0.22	29	24.3±0.9	15.4
11M	0.48	0.27	0.21	43	16.9±2.0	16.3
15M	0.60	0.32	0.29	48	22.6±5.2	16.9

[1] Calculated from water column silica concentrations (16) and atmospheric radon input.
[2] From Ritchie and Shimmield (16) and extrapolating from 1000 m depth.
[3] Calculated from the total supply of ^{210}Pb – water column inventory of ^{210}Pb
[4] Water column deficiency/total supply.

bloom at the time of sampling.

In order to establish whether the increased benthic activity seen towards the north of the transect is a spatial effect, or a temporal property related to a seasonal particulate pulse of phytodetritus, it is necessary to determine the annual C_{org} flux rates over the transect. Moore and Dymond (25), using sediment trap data from the Pacific have shown a linear relationship between particulate ^{210}Pb trap material and organic carbon flux:

$$\text{Flux } C_{org} = 8.5 + 430(\text{Flux } ^{210}\text{Pb/Prod. } ^{210}\text{Pb})$$

where Flux C_{org} is the flux of organic carbon reaching the sediment surface, Flux ^{210}Pb is the flux of ^{210}Pb reaching the sediment surface and Prod. ^{210}Pb is the combined water column production and atmospheric supply of ^{210}Pb. The flux of ^{210}Pb to the sediment surface may be calculated from multiplying the sediment ^{210}Pb$_{xs}$ inventory by the decay constant. The annual C_{org} fluxes calculated are therefore time averaged over the existence period of ^{210}Pb$_{xs}$ in the sediment (ca.100 years). By using such a relationship C_{org} fluxes have been predicted for each station and are shown in Table 4. It evident from the results that averaged organic carbon flux rates are comparable over the whole transect with perhaps a slight increase to the north.

A situation exists, therefore, where bioturbation intensity and C_{org} degradation both increase northwards but time averaged C_{org} flux rates do not show the same clear rise. Water column studies of ^{210}Pb highlight a seasonal pulse of particulate material to the seafloor at the time of sampling which was also identified by bathymetric photography. The rise in D_B and K_{orgC} seen to the north and the associated decrease in residence time of carbon, an indication of an increase in the labile fraction to the north may, therefore, correlate with the higher levels of water column ^{210}Pb scavenging and phytodetritus resting on the seafloor. Thus the benthic activity may be responding in a seasonal fashion to the water column particle production dynamics. Seasonality in the benthic response to a particulate flux has been reported be Graf (26) who demonstrated bioturbation intensity reaction to a seasonal pulse of phytodetritus in shallow northern Atlantic waters. Furthermore, seasonality in sediment oxygen demand has been presented by Smith and Baldwin (27) who identified a four-fold increase in an annual study of benthic oxygen respiration in Pacific pelagic sediments during the main pulse of particulate material. Martin and Bender (28) calculate that in order to achieve any form of observable variability in benthic flux at least 50% of the particulate organic carbon must have degradation constants between 0.2 and 0.02 yr^{-1}.

The oxygen flux measurements shown here do not support the hypothesis of increased benthic activity occurring in stations 11M and 15M, but instead higher respiration rates are observed to the south. However, the assumption that biological mixing has little effect of incorporating oxygenated bottom water into the sediment may not hold if bioturbation is rapid enough. Instead a mechanism of increased bioturbation promoting sediment irrigation and disruption of normal diffusion oxygen gradient may make the use of diffusive flux calculations inappropriate for the measurement of C_{org} degradation rates in highly bioturbated sediments.

CONCLUSIONS

1. Considerable variation is seen in the C_{org} both along the 20°W transect and at individual core stations.
2. Fine-scale $^{210}Pb_{xs}$ determinations reveal the detailed nature of sediment mixing. Diffusive and advective transfer mechanisms can be observed.
3. Application of the simple advection diffusion model to $^{210}Pb_{xs}$ and C_{org} profiles provides information on the relative timescales of both bioturbation and C_{org} degradation. Both processes are seen to increase in intensity to the north of the transect.
4. Consideration of the ^{210}Pb activity balance between water column and sediment indicates efficient particulate scavenging of ^{210}Pb which increased northwards during the time of the sampling (late summer). C_{org} flux rates calculated from sediment ^{210}Pb inventories normalised to the water column ^{210}Pb supply show comparable values over transect.
5. Both bioturbation intensity and C_{org} degradation are considered to respond to the seasonal pulse of phytodetritus to the sea floor.
6. Under conditions of more intense bioturbation, resulting modification of the dissolved oxygen profiles due to pore water irrigation result in inaccuracies of oxygen demand measurements.

ACKNOWLEDGEMENTS

We would like to thank Eliazer Traube for performing the micro-oxygen electrode measurements which we have presented here. The officers and crew of the RRS *Charles Darwin* provided able sea-going support for this project. Two anonymous referees provided helpful and constructive criticism. NERC provided financial assistance through grant number GST/02/427

REFERENCES

1. Verado, D., Froelich, P.N. and McIntyre, A., The determination of organic carbon and nitrogen in marine sediments using the Carlo Erba NA-1500 Analyser. Deep Sea Res., 1990
2. Reimers, C.E., Kalhorn, S., Emerson, S.R., and Nealson, K.E., Oxygen consumption rates in pelagic sediments from the Central Pacific: First estimates from microelectrode profiles. Geochim. Cosmochim. Acta, 1984, **48**, 903-910.
3. Reimers, C.E. and Smith, K.L., Reconciling measured and predicted fluxes of oxygen across the deep sea sediment-water interface. Limnol. Oceanogr., 1986, **31**(2), 305-318.
4. Broecker, W.S.. and Peng, T.H., Gas exchange rates between sea and air. Tellus, 1974, **26**, 21-35
5. Jahnke R., Heggie, D., Emerson, S. and Grundmanis, V., Pore waters of the central Pacific Ocean: nutrient results. Earth Planet Sci. Lett., 1982, **61**, 233-256.
6. Cochran, J.K., Particle mixing rates in sediments of the eastern equatorial Pacific: Evidence from ^{210}Pb, $^{239,240}Pu$ and ^{137}Cs distribution at MANOP sites. Geochim. Cosmochim. Acta, 1985, **49**, 1195-1210.

7. Kershaw, P.J., ^{14}C and ^{210}Pb in NE Atlantic Sediments: Evidence of Biological Reworking in the Context of Radioactive Waste Disposal. J. Environ. Radioactivity, 1985, **2**, 115-134.
8. DeMaster, D.J. and Cochran, J.K., Particle mixing rates in deep-sea sediments determined from excess ^{210}Pb and ^{32}Si profiles. Earth Planet Sci. Lett., 1982, **61**, 257-271.
9. Nozaki, Y., Cochran, J.K. and Turekian, K.K., Radiocarbon and ^{210}Pb distribution in submersible-taken deep-sea cores from Project Famous. Earth Planet Sci. Lett., 1977, **34**, 167-173.
10. Benninger, L.K., Aller, R.C., Cochran, J.K. and Turekian, K.K., Effects of biological sediment mixing on the ^{210}Pb chronology and trace metal distribution in a Long Island Sound sediment core Earth Planet Sci. Lett., 1979, **43**, 241-259.
11. Officer, C.B., Mixing, sedimentation rates and age dating for sediment cores. Mar. Geol., 1982, **46**, 261-278.
12. Jahnke, R.A., Emerson, S.R., Cochran, J.K. and Hirschberg, D.J., Fine scale distributions of porosity and particulate excess ^{210}Pb, organic carbon and $CaCO_3$ in surface sediments of the deep equatorial Pacific. Earth Planet Sci. Lett., 1986, **77**, 59-69.
13. Smith, J.N., Boudreau, B.P. and Noshkin, V., Plutonium and ^{210}Pb northeast Atlantic sediments: subsurface anomalies caused by non-local mixing. Earth Planet Sci.Lett., 1986/1987, **81**, 15-28.
14. Goldberg, E.D., and Koide, M., Geochronological studies of deep-sea sediments by the ionium/thorium method. Geochim. Cosmochim. Acta, 1962, **26**, 417-450.
15. Balsam, W.G. and McCoy, F.W. Jnr. Atlantic sediments: Glacial/interglacial comparisons. Paleoceanogra., 1987, **2**, 531-542.
16. Ritchie, G.D. and Shimmield, G.B.,1991 This Volume
17. Wheatcroft, R.A., Jumars, P.A., Smith, C.R. and Nowell, A.R.M., A mechanistic view of the particulate biodiffusion coefficient: Step lenghts, rest periods and transport directions. Journ. of Marine Res. 1990, **48**, 177-207.
18. Boudreau, B.P., Mathematics of tracer mixing in sediemnts: II. Nonlocal mixing and biological conveyor-belt phenomena. Amer. Journ. Sci., 1986, **286**, 199-238.
19. Rice, D.L., Early diagenesis in bioadvective sediments: Relationships between the diagenesis of beryllium-7, sediment reworking rates and the abundance of conveyor-belt deposit-feeders. Journ. Marine Res., 1986, **44**, 149-184.
20. Robbins, J.A., A model for particle-selective transport of tracers in sediments with conveyor-belt deposit feeders. Journ. Geophysical Res., 1986, **91**, No. C7, 8542-8558.
21. Boudreau, B.P., Mathematics of tracer mixing in sediments: I Spatially dependent, diffusive mixing. Amer. Journ. Sci., 1986, **286**, 161-198.
22. Boudreau, B.P. and Imboden, D.M., Mathematics of tracer mixing in sediments: III. The theory of non-local mixing within sediments. Amer. Journ. Sci., 1987, **287**, 693-719.
23. Gooday, A.J., and Turley, C.M., Responses by benthic organisms to inputs of organic material to the ocean floor: a review. Phil. Trans. R. Soc. Lond. A, 1990,**331**, 199-138.
24. Turley, C.M. and Lochte, K., Microbial response to the input of fresh detritus to the deep-sea bed. Palaeogeography, Palaeoclimatology, Palaeoecology, 1990, **89**, 3-23.
25. Moore, W.S., and Dymond, J., Correlation of ^{210}Pb removal with organic carbon fluxes in the Pacific Ocean. Nature, 1988, **331**, 339-341
26. Graf, G., Benthic-pelagic coupling in a deep-sea benthic community. Nature 1989, **341**, No. 6241, 437-439.
27. Smith, K.L.Jnr. and Baldwin, R.J., Seasonal fluctuations in deep-sea sediment community oxygen consumption: central and eastern North Pacific Nature, 1984, **307**, 624-626.
28. Martin, W.R. and Bender, M.L., The variability of benthic fluxes and sedimentary remineralisation rates in response to seasonaly variable organic carbon rain rates in the deep sea: A modelling study, Amer. Journ. Sci., 1988, **228**, 561-574.

THE TOTAL ALPHA PARTICLE RADIOACTIVITY FOR SOME COMPONENTS OF MARINE ECOSYSTEMS

ERIC I. HAMILTON
Phoenix Research Laboratory
Penglebe, Dunterton, Tavistock, Devon PL19 OQJ, England

ROBERT WILLIAMS
Plymouth Marine Laboratory
West Hoe, Plymouth, Devon PL1 3DH, England

PETER J. KERSHAW
Ministry of Agriculture, Fisheries and Food
Directorate of Fisheries Research
Fisheries Laboratory, Lowestoft
Suffolk NR33 OHT, England

ABSTRACT

Examples are provided to illustrate how unique information can be obtained by an examination of the specific surface alpha particle radioactivly (for thick or thin sources) of marine systems (ARMS Project) using the dielectric detector CR-39. Alpha particle radiographs of sediment cores can be used to determine features associated with bioturbation; a depth resolution of about 100 microns is practical and offers considerable advantages over conventional 1 cm core sections when useful information is lost. Alpha particle distributions in sediments can identify changes in seasonal deposition of sediment debris and sources.

For the coastal environment examples are provided to illustrate the method for sediment cores. In the Irish Sea the spatial distribution of alpha emitters shows enhanced levels during periods of peak productivity; an extension of this study into the English Channel and North Sea provides data on the provenance of the alpha emitters. In the Atlantic Ocean alpha particle distributions associated with suspended particulate matter can identify nepheloid layers in the water column; also, in some areas the presence of Mediterranean water at depth. Radiographic study of biota identifies sites for the uptake of alpha emitters (mainly ^{210}Po and ^{226}Ra), associations with faecal debris, organs and tissues from surface layers to a depth of 3000 m.

INTRODUCTION

Cherry and Shannon (1) identified the need for comprehensive and co-ordinated knowledge concerning the distribution of alpha-radioactive nuclides in marine ecosystems. There is a need to include all the significant alpha-emitting nuclides and a satisfactory selection of samples from the marine environment; the distinction between natural background levels and artificially introduced radionuclides; and, the fine details of the distribution of alpha emitters in marine organisms. Examples of the distribution of alpha emitters in marine samples as determined by the dielectric detector CR-39 are presented in order to illustrate the flexibility of the method.

Apart from providing pictures of the distribution of alpha emitting radionuclides which are produced in the CR-39, from a consideration of the range of alpha particles (4-8 MeV) and the geometry of track pits Hamilton and Clifton (2), the energy of individual alpha emitters can be determined. Fews and Henshaw (3) provided a comprehensive evaluation of high resolution alpha spectroscopy using CR-39 and Hatzialekou *et al.* (4) described a fully automated image analysis system for the analysis of alpha-particle autoradiographs and energy determinations. The objective of this work is simply to illustrate some possibilities in the use of CR-39 techniques in order to describe and understand the distribution of alpha emitters in marine materials. A full scientific discussion of the selected examples will be published elsewhere. The method complements conventional analyses for short to medium radioactive half-life emitters by surface barrier spectroscopy and thermal ionisation mass spectrometry for long lived radionuclides. For many marine samples the alpha particle activity is usually dominated by ^{226}Ra and its daughters, especially ^{210}Po. Knowledge of the distribution of alpha emitters in a sample can provide valuable information which is not obtained by conventional destructive methods of analysis.

MATERIALS AND METHODS

Details of the method have been described by Hamilton and Clifton (2) and Hamilton (5). The distribution of alpha emitters in diverse materials is described for samples from estuarine, coastal and oceanic systems. For example: sediments (surficial and cores); suspended particulates, following the filtration of sea water (1-2 litres); plankton net samples; together with the histological examination of thick sections for a variety of marine fauna. Alpha particle emissions are calculated for thin or thick geometry (CR-39, 0.008-1 mm thick for either 16 or 36 hour curing). A detection limit for alphas is about 0.05 Bq kg^{-1}. Measurements are made using a binocular microscope with x10 wide angle occulars and x5/0.15, x16/0.32 or x40 (dry)/0.57 objectives. A typical CR-39 background alpha track value of <1x10^{-6} alphas cm^{-2} s^{-1} is obtained. For filter residues, organic filter papers are used (e.g. Millipore HA, 0.45 μm) which have a similar background activity to that of CR-39; glass fibre filters have values of $>6\times10^{-5}$ alphas cm^{-2} s^{-1} and organic gridded filter papers have values of $>8\times10^{-5}$ alphas cm^{-2} s^{-1} rendering both useless for low level samples. Exposure periods from a few minutes to 10 years at 4°C have been used with negligible fading of the latent image. Rare interference from radon emanation products can occur and is monitored by determining the track density in the CR-39 on the side not exposed to the sample. Typical exposure times are: for coastal materials 3 months; 1 year for sediment cores; and, between 9 months to 2 years for low level biological samples. Albeit exposure times are long, once a back log of samples has been provided information is steadily produced, hence a need for careful organisation of work schemes. While exposures are usually made in the laboratory they have also been successful in the field, e.g. emplacement in fine grained estuarine and lacustrine sediments. From an examination of the geometry of alpha tracks, the shape and size of the emitting body can be determined. When a single representative sample, with a mass of a few milligrams, can be provided (e.g. individual hot particles, copepods or faecal pellets), the elemental separation of most alpha emitters on 3"x1" plastic slides coated with thin layer chromatography media can be accomplished (2). The CR-39 detector is then exposed against the chromatograph and the alpha tracks from specific elements are recorded. Larger samples can be processed provided that the point of application to the chromatographic media is not saturated. The specific determination of ^{210}Po can be made using micro thermal volatilisation techniques.

RESULTS AND DISCUSSION

Hamilton and Clark (6) described a sedimentation record for the Esk estuary, Cumbria, which receives low level radioactive waste from a nuclear fuel reprocessing plant at Sellafield, some 10 km to the north. A variable sedimentation rate of between 3-6 cm y^{-1} over 20 years was derived with a transport time from the source of 10 months for ^{137}Cs and 5 months for ^{241}Am, $^{240+239+238}$Pu. For a core obtained from Senhouse Dock, 24 km to the north of the Sellafield plant, Kershaw et al., (7) obtained an overall sedimentation rate of 6.4 cm y^{-1} and a transport time from the source of 1-1.5 years. The dock was last dredged in the early 1950's since when it has not been used commercially. As an independent evaluation 76 sections of the core, cut at nominal 1 cm increments, were analysed by the CR-39 technique. From an evaluation of the distribution of alpha emitters a mean sedimentation rate of 5.0 cm y^{-1} was obtained and a transport time of 4 months. Figure 1 compares a section of the CR-39 alpha core data with the monthly discharge data for alpha emitters by the plant. The shapes of both curves are very similar; the lower values in the later 1980's reflect considerably reduced discharges by the plant. In some sections of the core CR-39 alpha peaks were not obvious in the discharge data which can be explained as a consequence of transient changes in the direction of transport of the radioactive wastes; changes in the chemistry of the receiving sediments for the near-field environment; and, fluctuations in the rate of release of radionuclides together with other natural or industrial sources of alpha emitters, such as ^{210}Po (7).

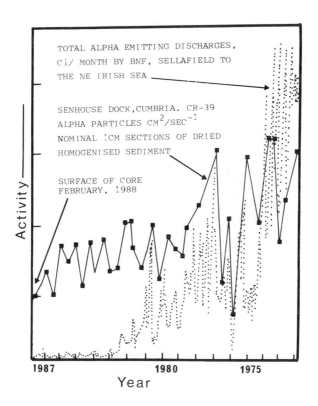

Figure 1. A comparison between alpha discharges (Ci/month) by British Nuclear Fuels plc into the North East vIrish Sea and the distribution of alpha activity (cm^{-2} s^{-1}) from 1 cm sections of core from Senhouse Dock, Cumbria.

Figure 2 illustrates a high resolution alpha particle profile of a 12 cm core taken at 2000 m in the Atlantic Ocean. The top 9 cm were oxic. CR-39 was placed against a median section of the core, hence, the record is continuous, with the practical resolution with depth being < 500 μm. Figure 2 represents 0.9 mm increments. Position A marks the region where ^{234}Th is no longer detected by gamma spectrometry with maximum alpha activity occurring at a depth of 3 cm. At position B there is a pronounced decrease in alpha activity which coincides with the absence of excess ^{210}Pb. The ^{226}Ra concentrations in the core vary between 63-88 Bq kg^{-1}. The insert on Figure 2 shows that at position B the sediment is drier and this coincides with a significant inflexion in the surface area of aliquots of dry sediment. Element data provided by neutron activation analysis show that K and Al tend to increase with depth, Ca, REE, Sb and V tend to be constant, but at 0.5-0.7 cm (point A in Fig. 2) there is a localised decrease (e.g. >50%) in the abundance of terrigenous-derived elements such as Co, Cr, Fe, Hf, Hg, Rb, Sc, Ta, and Mn; the concentrations of Zr, Th and U tend to vary slightly throughout the core.

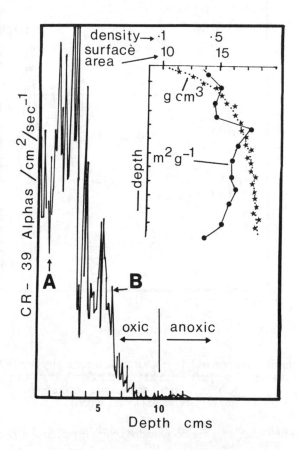

Figure 2. Atlantic Ocean core; length 10.5 cm, water depth 2000 m, position 51°0.5'N, 12°39.4'W, 1987. High resolution CR-39 data from an intact core, together with density and surface area profiles. A - Max. depth where ^{234}Th detected; B - Max. depth where excess ^{210}Pb detected.

Figure 3 illustrates the use of the CR-39 method for determining the alpha particle radioactivity (alphas cm^{-2} s^{-1}) emitted from surfaces of various sediments and particulate debris filtered from 2 litre samples of seawater from the North East Irish Sea, collected during a period of low productivity in December, 1989. The highest values are found very close to the British Nuclear Fuels plc effluent outfall off Sellafield, Cumbria; the alpha particle activity from bottom sediments tends to be significantly higher than that for suspended particulates. The data indicates that, at the time of sampling, significantly enhanced levels of alpha activity were restricted to the close proximity of the outfall. Table 1 illustrates similar data from the Esk estuary, and elsewhere, and serves to place the alpha particle activity for various environmental materials into perspective, i.e. materials can contain higher concentrations of alpha emitters from natural sources than from contamination by radionculides from nuclear fuel reprocessing.

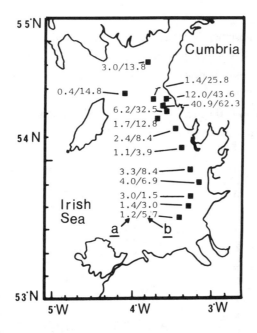

Figure 3. CR-39 alpha particle activity for suspended particulates from 2 litres of seawater (a), and bottom sediments of the Irish Sea (b).
a - Thin source cm^{-2} s^{-1} x 10^{-4}
b - Thick source cm^{-2} s^{-1} x 10^{-4}.

The CR-39 method has also been used to determine the temporal and spatial distribution of alpha activity in the open ocean as an indicator of the distribution of biota, suspended particulates and sediments, with an emphasis upon provenance and fluxes of

TABLE 1
Characteristic alpha particle activity from the surfaces of various materials determined by the CR-39 method.

Sample	Locality	alphas cm^2 s^{-1}
Silt	Esk, Cumbria	$1-12 \times 10^{-3}$
Sand	Esk, Cumbria	$2-20 \times 10^{-5}$
Gravels	Esk, Cumbria	$1-3 \times 10^{-6}$
Silts	Senhouse Dock	$1-5 \times 10^{-3}$
Silts	Maryport Harbour	$10^{-3} - 10^{-4}$
Silts	Tamar estuary, Devon	$4-8 \times 10^{-4}$
Granites	United Kingdom	$10^{-1} - 10^{-5}$
Particulates	Atlantic Ocean	$10^{-4} - 10^{-5}$
Acantharia	Atlantic Ocean	$6-8 \times 10^{-3}$
Plankton		
Fish	Atlantic Ocean	$10^{-4} - 10^{-7}$

materials. Figure 4 illustrates the alpha profile for suspended particulates from the Porcupine Sea Bight, North East Atlantic Ocean for 20-200 μm net samples after filtration of between 50-500 litres of seawater. Of particular interest is the association with organic carbon between 120-280 m and the presence of a nepheloid layer derived from the continental shelf at 800 m (Figure 5). The presence of Mediterranean water off the North East coast of Portugal is indicated at a depth of 700-900 m.

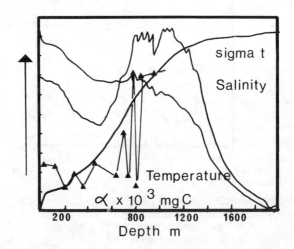

Figure 5. Cr-39 activity x 10^{-3} mg^{-1} C., salinity, temperature, sigma t data for the Porcupine Sea Bight, North East Atlantic Ocean; data for particulate debris in 20-200 μm net mesh from between 50-5000 litres of seawater. Data indicate the presence of Mediterranean water at a depth of about 800 m.

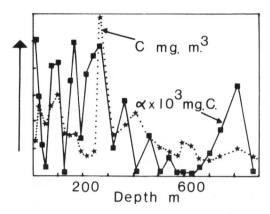

Figure 4. An indication of the presence of a nephoid layer at about 800 m from the Porcupine Sea Bight. The carbon at ≈ 800 m is in a degraded form.

The distribution of alpha emitters obtained by autoradiographic analysis of zooplankton and fish is given in Tables 2 and 3, together with the alpha particle activity which is associated with faecal debris from pellets of gut contents about to be released,

TABLE 3
Alpha particle activity emitted from thick sources, $cm^2 s^{-1} \times 10^{-4}$ for various marine animals from the Porcupine Sea Bight, 12 April-12 May 1986, pre bloom period, 51.36'N-13.09'W
(hepato. - hepatopancreas; E - Euphausiid)

Water depth (m)	Species	Faeces	Whole animal	Organ/tissue				
34-60	*Salpa fusiformis* (S)	5.0	0.9	-			-	
50	*Meganyctiphanes norvegica* (E)	8.3	0.09	-			-	
252-268	*Cyclothone* sp. (F)	8.2						
	Radiolaria (sample, a)	26.0	-	-			-	
336-365	Radiolaria (sample, b)	23.0	-	-			-	
	Small Euphausiid	-	4.1	-			-	
400-500	Decapod	59.0	-	muscle (fore)	0.8	(tail)	1.7	
	Euclio pyramidata	-	9.0	-			-	
	Radiolaria	-	12.0	-			-	
	Myctophid (F)	33.0	-	vertebrae	36,	eye	0.4	
	Aygyropelecus hemigymnus (F)	-	-	muscle	9.1,			
	Ostracod *Gigantocypris* sp.	-	-	tissue	0.04,	shell	0.9	
800	*Notoscopelas notoscopelus*	210.0	-	muscle	3.9,	vertebrae	29	
	Kroeyerii (F)			brain	0.05			
860-1050	*Acanthaphyra purpurea* (D)	40.0	-	muscle ext. shell	0.2, 0.3	hepato	33,	
	Myctophid juvenile (F)	-	-	muscle	1.4,	stomach	10	

and depurated faeces from some species kept alive on board ship. Faecal debris contains major amounts of ^{210}Po, ^{210}Pb and ^{226}Ra. There does not appear to be any progressive enrichment in the alpha activity of faeces with depth, probably a reflection of rapid transport from the surface layers to bottom sediments. The concentration of alpha emitters in marine organisms is usually significantly higher during periods of peak productivity compared to post bloom situations. Nyctoepipelagic fish can contain significant levels of alpha activity which is associated with melanocratic skin.

TABLE 2
Alpha particle activity emitted from thick sources, $cm^2 s^{-1} \times 10^{-4}$ for various marine animals from the North East Atlantic Ocean 47°15.2'N-19°29.7'W, June 1988
(hepato. - hepatopancreas; C - Copepod, D - Decapod, F - Fish, H - Hyperiid, S - Salp)

Water depth (m)	Species	Faeces	Whole animal	Organ/tissue				
20	Radiolaria	-	6.5-13	-		-		
	Parathemisto gaudichaudi (H)	-	-	exoskeleton	0.4			
0-100	Myctophid (F)	-	0.2	-		-		
0-200	Copepods	1.5	2.5	-		-		
	Parathemisto gaudichaudi (H)	1.8	-	exoskeleton	32.0			
200-500	Radiolaria	-	82.0	-		-		
	Copeopods	8.8	5.2	-				
	Argyropelecus hemigymnus (F)	12.0	-	muscle	0.2,	gills	3.3	
				skin	2.7			
300-350	Parathemisto gaudichaudi (H)	1.7	-	skin	0.5,	gills	1.8	
	Euphausiid	11.0	-	muscle	0.3,	gut	1.3	
	Sergestes arcticus (D)	11.0	-	muscle	0.3,	hepato.	5.7	
280-350	Euphausiid	-	0.3	muscle	0.1			
360-530	Chauliodus sloani (F)	0.5	-	muscle	0.05,	vert.	3.7	
520	Benthosema glaciale (F)	3.5	-	muscle (tail)	0.1,	vert.	3.7	
600-760	Euphausiid	0.3	-	muscle (body)	0.5,	(tail)	0.1	
	Euphausiid	1.0	-	muscle	0.05			
	Acanthephyra sp. (D)	-	-	muscle	0.07,			
				exoskeleton	0.3,	hepato	5.5	
800-950	Sergestes arcticus (D)	-	-	muscle	0.03,	hepato	34.0	
	Decapods	12-13.0	-	-		-		
1000	Ostracod Gigantocypris sp.	-	0.8	-		-		
940-1050	Acantharia	16.0	-	-		-		
	Acanthephyra pelagica (D)	2.7	-	muscle (tail	0.2			
	Gonostoma bathyphilum (F)	17.0	-	muscle	0.01,	mid-gut	2.0	
1000-1500	Mixed copepods (C)	-	0.4	-		-		-
	Metrida lucens	-	0.4	-		-		-
120	Bathycalanus sp. (C)	2.0	-	exoskeleton	0.6			
	Argyropelecus gymnus	0.7	-	muscle	0.1,	eye	0.4,	
				skin (black)	1.1,			
				black skin gut	69.0			
	Acantharia mixed	-	7.2	-				-
	Acanthephyra pelagica (D)	2.5	-	muscle	0.07			
	Myctophid	5.1	-	muscle	0.6			
	Euphausiid	2.0	-	muscle	0.4			
	Eucopia sp. (D)	0.5	-	muscle	0.4			
12-1400	Parapasiphaeo sp. (D)	1.0	-	-		-		

Figure 6 illustrates the distribution of alpha activity for suspended particulate matter for UK coastal waters together with data for the distribution of uranium determined by conventional surface barrier alpha spectroscopy. The distributions correlate well with the known distribution of contaminants as controlled by water movements.

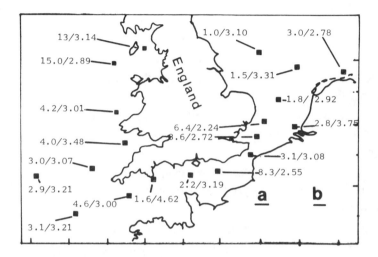

Figure 6. The distribution of alpha particle radioactivity (α cm^{-2} s^{-1} x 10E^{-4}) for particulates filtered from 2 litres of seawater (a), and the concentration of dissolved uranium (μg.l^{-1} in seawater for UK coastal waters. (b)

The brief descriptions provided here illustrate the usefulness of the CR-29 approach, when combined with other conventional methods, in furthering our knowledge of oceanic processes. Methodology is inexpensive and large numbers of samples can be processed, hence providing a reliable framework for judicious sampling prior to the use of more conventional methods. Current applications concern the use of the method in investigating the temporal and spatial composition of marine detritus in relation to carbon dioxide cycling in the oceans, and studying the dynamics and processes associated with the transfer of alpha radionuclides and stable elements in the water column. The method is proving useful in determining the turnover rate and removal rate of very fine and composite particles in the top 200 m of the water column, and the selective feeding behaviour of marine organisms.

ACKNOWLEDGEMENTS

This work has been partially funded, under contract to E.I.H., by the CEC-B16-038-UK and MAFF. CSG. 1527, who is also grateful for the cooperation and assistance provided by D.V.P. Conway and N.J.P. Owen, and also a number of ocean going yachtsman, in providing samples of surface oceanic waters, and, to E. Sabbioni CEC-JRC Ispra for radioactivation measurements.

REFERENCES

1. Cherry, R.D. and Shannon, L.V., The alpha radioactivity of marine organisms. Atomic Energy Rev., 1974, **12**, 3-45.
2. Hamilton, E.I. and Clifton, R.J., CR-39, a new alpha-particle sensitive polymeric detector applied to investigations of environmental radioactivity. Int. J. Appl. Rad and Isotopes, 1981, **32**, 313-324.
3. Fews, A.P. and Henshaw, D.L., High resolution alpha-spectroscopy using CR-39 plastic track detector. In Proc. 11th Int. Conf. Solid State Detectors, Bristol, 1982, pp. 641-645.
4. Hatzialekou, U., Henshaw, D.L. and Fews, A.P., Automated image analysis of alpha-particle autoradiographs of human bone. Nuc. Inst. & Meth. in Phys. Res., 1988, **A263**, 504-514.
5. Hamilton, E.I., The disposal of radioactive wastes into the marine environment: the presence of hot particles containing Pu and Am in the source term. Min. Mag., 1985, **49**, 177-194.
6. Hamilton, E.I. and Clarke, K.R., The recent sedimentation history of the Esk estuary, Cumbria, U.K.: the application of radiochronology. Sci. Total Environ. 1984, **35**, 325-386.
7. P.J. Kershaw, Woodhead, D.S., Malcolm, S.J., Allington, D.J. and Lovett, M.B. A sediment history of Sellafield discharges. J. Environ. Radioactivity, 1990, **12**, 201-241.

TRANSURANICS CONTRIBUTION OFF PALOMARES COAST: TRACING HISTORY AND ROUTES TO THE MARINE ENVIRONMENT

L.Romero, A.M.Lobo, E.Holm[1] and J.A.Sánchez[2]
PRYMA-CIEMAT, Av.Complutense 22, 28040-Madrid SPAIN
[1]Department of Radiation Physics, Lund University SWEDEN
[2]Dept.Física de las Radiaciones, UAB, Barcelona SPAIN

ABSTRACT

Following the Palomares accident, 1966, a land monitoring program has been running up to now. The study of the possible transfer of the residual transuranics contamination on land to the marine environment began in 1986, under a project sponsored by CEE. This paper presents the use of artificial and natural radionuclides as valuable tools for the evaluation of the possible pathways by which the transfer to Mediterranean sea could occur.

INTRODUCTION

The major source of transuranium elements to the Mediterranean Sea is fallout from nuclear weapons tests. Minor contributions originate from runoff from rivers contaminated by nuclear facilities. The eventual contribution from the Palomares accident (southern coast of Spain) in 1966, began to be studied in 1986 under a project sponsored by CEE focused on the depositional history recorded in sediments.

The Palomares accident occurred during a mid-air refueling operation of two USAF planes, one of them was carrying four thermonuclear bombs. Two of the bombs whose parachutes did not deploy experienced non-nuclear explosion, resulting in the spreading of their fissile material; a land area of 2.3 km^2 was contaminated with transuranic elements. After the remedial operations a transuranics contamination remainded in the area.

Previous conclusions on the land to sea transfer of this contamination stated that it was mainly confined to an area south of the mouth of the Almanzora river (see Fig.1) [1]. To evaluate inventories and distribution of the transferred contamination, additional sediment samples were analyzed. During the analysis three hot particles were found at different locations.

^{210}Pb dating and artificial radionuclide ratios (^{137}Cs, ^{238}Pu, $^{239+240}$Pu, ^{241}Pu and ^{241}Am) were employed to investigate the possible pathways of the transfer of transuranics from land to sea.

MATERIALS AND METHODS

Sediment samples were collected using a box-corer during a sampling expedition in 1985, surveying the Vera Gulf (Southeastern Spanish coast, Fig. 1). Cores were extruded in 1 cm thick slices, discarding the outer centimetre of each section to avoid the mixing of contiguous sections. The samples were dried at 110°C to constant weight, ball-milled and sieved through 1 mm mesh size for radionuclide analysis. This paper presents the results for three sediment cores collected at the continental shelf and deep sea floor.

Samples were analyzed by gamma spectrometry and radiochemically treated to obtain the plutonium and americium activities.

A 45 g aliquot of each section was measured by gamma spectrometry. The use of a coaxial hyperpure n-type germanium detector allows the simultaneous determination of a wide range of gamma emitters [2] (from ^{210}Pb, E=46.5 keV to ^{40}K, E=1461 keV). Self-absorption effects were considered by using sediments with similar texture as the calibration samples; besides, the transmission method [3] was applied to ensure the equivalence between problem and calibration samples.

Briefly, Pu and Am were extracted from sediment by acid leaching and separated by sequential ion-exchange steps [4]. The isolated radionuclides were electroplated onto stainless steel discs and counted by alpha spectrometry with Silicon Surface Barrier detectors. Radiochemical yield was determined by using ^{242}Pu and ^{243}Am as tracers.

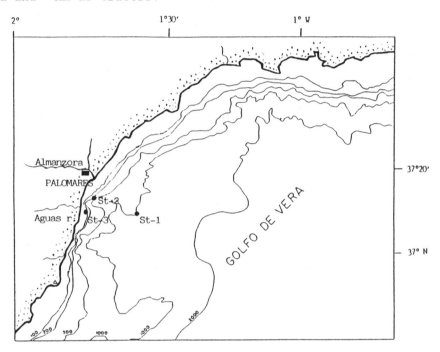

Figure 1. Map of the sampling area.

The accuracy of both alpha and gamma measurements was checked by participating in IAEA intercomparison exercises.

^{241}Pu was determined in only a few samples. The plutonium electroplated disk is directly measured in a low background liquid scintillation counter in order to record the low energy beta spectrum of ^{241}Pu [5].

The chronology of sediment cores was obtained by applying the CRS model [6] to unsupported ^{210}Pb profiles. The unsupported ^{210}Pb values in each core section were obtained by substracting the ^{226}Ra concentration from the total ^{210}Pb. Fig.2 shows the different profiles so obtained in cores St-1 (1000 m depth) and St-2 (94 m). This latter presents discontinue ^{210}Pb values, due to its proximity to the coast (resuspension phenomena) and also to the intermitent supply of fluvial sediments by the Almanzora river (which is a seasonal river).

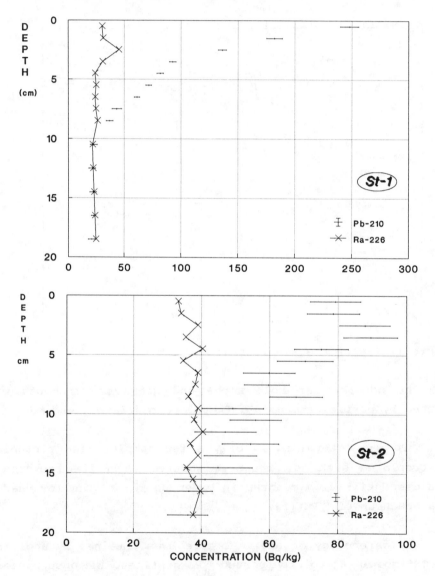

Figure 2. ^{226}Ra and total ^{210}Pb profiles in Cores St-1 and St-2

RESULTS

To evaluate the pathways of the transfer of transuranics from land to sea, as consequence of the referred accident, the chronologies of the sediment cores were identified and the isotopic ratios of Pu were used in order to characterize the sources of this contribution.

During these analysis a hot particle was found at 1000 m depth, far from the point of impact, 20 km aproximately. This fact has been observed before at the Thule accident, 1968, where hot particles were found up to 45 km from the point of impact [7]. The detailed study on this sediment core, St-1, has been published elsewhere [8]. Also the ^{241}Pu activity has been determined for this particle. The total activities of the isolated fraction (2.75 g) were the following:

$^{239+240}$Pu : 21.7 ± 1.1 Bq
^{238}Pu : 0.431 ± 0.050 Bq
^{241}Pu : 31.2 ± 9.8 Bq
^{241}Am : 9.70 ± 0.76 Bq

The activity ratios ^{238}Pu/$^{239+240}$Pu and ^{241}Pu/$^{239+240}$Pu correspond to typical values for weapons grade Pu [7], but the ^{241}Am/$^{239+240}$Pu ratio corresponds more to that found today in integrated nuclear test fallout after build up from ^{241}Pu.

The dating results of this core are shown in Table 1. The core St-1 presents a very low sedimentation rate and dates the second centimetre, where the hot particle was found, to 1967 with an uncertainty ranging from 1966 to 1969. It could be deduced that this particle was transported by aeolian distribution at the time of the accident. The isotopic ratios ^{241}Pu/$^{239+240}$Pu and ^{238}Pu/$^{239+240}$Pu, (see Table 4) confirm this assumption.

Analysis of duplicate samples [9] showed two new hot particles at the continental shelf, south of the Almanzora river mouth. The concentrations of ^{137}Cs and unsupported ^{210}Pb as well as the dating results of cores, St-2 and St-3, are listed in Tables 2 and 3.

TABLE 1

Dating results of core St-1. Coordinates 37°09.0 N 01°36.8 W; depth 1000 m.

SECTION (cm)	^{137}Cs Bq/kg ± 1σ	^{210}Pb$_{ex}$	DATE	Sed.rate g/cm^{2}y
0-1	9.18 ± 0.42	255. ± 11.	1975 (1974-76)	0.058
1-2	5.22 ± 0.27	172. ± 8.0	1967 (1966-69)	0.064
2-3	3.77 ± 0.25	104.1 ± 4.4	1957 (1954-59)	0.080
3-4	2.43 ± 0.22	71.8 ± 3.6	1948 (1945-51)	0.085
4-5	2.02 ± 0.22	65.6 ± 3.3	1937 (1933-41)	0.069
5-6	1.12 ± 0.23	53.5 ± 3.1	1923 (1918-30)	0.058
6-7	1.17 ± 0.19	43.1 ± 2.8	1901 (1891-16)	0.042
7-8	< 0.6	21.9 ± 5.0	1883 (1866-20)	0.043
8-9	< 0.6	13.5 ± 3.6	-	-
9-10	< 0.6	8.3 ± 3.5	-	-
10-11	< 0.6	3.1 ± 3.5	-	-
11-12	< 0.6	1.7 ± 3.5	-	-

The cores closer to the coast at the continental shelf in the southern part of Almanzora river mouth have higher sedimentation rates, being lower for St-2 than St-3. This latter one receives the terrigenous outflow from two rivers, the Almanzora and the Aguas (see Fig.1) which leads to an increase in sedimentary deposited material.

The 10 g of sediment sample from core St-2, were the particle was found, have the following activities:

$^{239+240}$Pu : 0.625 ± 0.037 Bq
^{238}Pu : 0.0175 ± 0.0021 Bq

The chronology obtained to the St-2 core dates the first centimetre at 1981, ranging between 1977 and 1986. The pathway for this hot particle could be via an extensive big flood occurred in 1973 at the Almanzora river, which could have washed out the contaminated river bed and remainder contamination at river surroundings, later to be deposited by descending currents. The higher value of sedimentation rate obtained to this core section confirms the assumption.

TABLE 2

Dating results of core St-2. Coordinates 37°10.7 N 01°46.3 W; depth 94 m.

SECTION (cm)	^{137}Cs Bq/kg ± 1σ	^{210}Pb$_{ex}$	DATE	Sed.rate g/cm^{2}y
0-1	4.85 ± 0.46	49.7 ± 8.1	1981 (1977-86)	0.314
1-2	5.59 ± 0.50	48.2 ± 8.4	1978 (1973-83)	0.290
2-3	6.52 ± 0.49	52.7 ± 8.1	1973 (1968-79)	0.234
3-4	7.32 ± 0.52	58.5 ± 8.6	1967 (1962-74)	0.179
4-5	6.69 ± 0.48	37.6 ± 8.6	1962 (1956-70)	0.235
5-6	5.44 ± 0.52	38.9 ± 9.0	1955 (1948-66)	0.189
6-7	4.00 ± 0.47	22.6 ± 8.3	1951 (1942-63)	0.272
7-8	3.43 ± 0.38	23.2 ± 7.2	1945 (1935-60)	0.226
8-9	3.88 ± 0.51	34.3 ± 8.5	1935 (1922-58)	0.120
9-10	2.61 ± 0.44	11.1 ± 13.	-	-
10-11	1.96 ± 0.43	19.7 ± 8.2	-	-
11-12	1.76 ± 0.45	2.8 ± 12.	-	-
12-13	1.17 ± 0.47	19.3 ± 9.1	-	-
13-14	1.17 ± 0.45	10.4 ± 8.4	-	-
14-15	1.07 ± 0.49	10.2 ± 16.	-	-
15-16	1.10 ± 0.32	4.0 ± 12.	-	-
16-17	1.15 ± 0.41	-3.0 ± 7.2	-	-

The plutonium activities found in the 10 g of analyzed sediment from core St-3 are:

$^{239+240}$Pu : 0.1960 ± 0.0065 Bq

^{238}Pu : 0.00353 ± 0.00040 Bq

In the St-3 core the hot particle was found at 4-5 centimetre depth, corresponding to 1973 (range between 1965-90). The high uncertainty interval is due to the low resolution of the ^{210}Pb profile and allows to assign its transfer pathway to both aerial and fluvial routes. The ^{137}Cs dating method was applied to try to elucidate its actual transport route. The maximum radiocaesium concentration in the profile appears in the 6-7 cm section; which should be assigned to 1963, the year of maximum fallout concentration. In this way, an average sedimentation rate of 0.40 g/cm^{2}y was estimated. As the hot particle appears in the 4-5 cm section, its age corresponds to 1969. It is still doubtful to distinguish between the two proposed routes, although the radiocaesium chronology points more likely towards an aerial transfer at the time of the accident and subsequent transport by descending currents.

TABLE 3

Dating results of core St-3. Coordinates 37°08.1 N 01°48.5 W; depth 64 m.

SECTION (cm)	^{137}Cs Bq/kg ± 1σ	^{210}Pb$_{ex}$	DATE		Sed.rate g/cm^{2}y
0-1	-	-	1983	(1984-89)	0.408
1-2	5.37 ± 0.47	37.2 ± 8.9	1980	(1979-90)	0.423
2-3	-	-	1978	(1974-90)	0.453
3-4	4.63 ± 0.53	26.5 ± 9.4	1975	(1970-90)	0.505
4-5	-	-	1973	(1965-90)	0.475
5-6	5.77 ± 0.50	25.0 ± 9.2	1970	(1960-91)	0.453
6-7	7.22 ± 0.52	35.0 ± 9.3	1965	(1952-92)	0.289
7-8	6.29 ± 0.58	29.4 ± 10.	1961	(1944-92)	0.299
8-9	-	-	1956	(1937-93)	0.274
9-10	4.51 ± 0.50	26.8 ± 9.2	-		-
10-11	-	-	-		-
11-12	2.52 ± 0.53	27.2 ± 9.4	-		-
12-13	-	-	-		-
13-14	< 0.6	13.9 ± 8.9	-		-
14-15	-	-	-		-
15-16	1.00 ± 0.40	9.8 ± 8.5	-		-
16-17	< 0.6	8.0 ± 9.4	-		-

TABLE 4

Isotopic ratios of the hot particles. ^{241}Pu activity referred to 1/1/89.

	St-1	St-2	St-3
^{238}Pu/$^{239+240}$Pu	0.0199 ± 0.0030	0.0280 ± 0.0037	0.0180 ± 0.0021
^{241}Pu/$^{239+240}$Pu	1.42 ± 0.46	-	-
^{241}Am/$^{239+240}$Pu	0.447 ± 0.042	-	-

It should be remarked that gamma spectrometry was performed on different sediment subsamples to those used to radiochemical analysis. In cores St-2 and St-3 the presence of a hot particle was detected by processing the sample radiochemically; in the case of St-1 it was detected during gamma spectrometry, so that a deeper study has been possible.

It seems obvious that the inhomogeneity present in the samples is not typical of the current fallout situation in Mediterranean sediments. It appears evident that the origin of the inhomogeneity found in the studied samples originates from the Palomares accident.

CONCLUSIONS

The use of both natural and artificial radionuclides to estimate the chronology of recent sediments is a valuable tool when evaluating transfer processes in the marine environment.

Isotopic ratios of radionuclides constitute a tool for identifying the different sources of radionuclides (plutonium in this case) found in the samples.

Their application to the marine environment in the Palomares area has shown the different pathways by which the aerosol and residual contamination in soil could reach the Mediterranean marine environment.

REFERENCES

1. Gascó C. Romero L. & Iranzo E.: J. Radioanal. Nucl. Chem. in press

2. Lobo A.M. Romero L. & Palomares J.: An. Fis. (Spain) in press.

3. Cutshall N.H. Larsen I.L. & Olsen C.R.: Nucl. Instrum. Meth., 1983, **206,** 309.

4. Holm E. Fukai R. & Ballestra S.: Talanta, 1979, **26,** 791.

5. Mitchell P.I.: personal communication (1990).

6. Appleby P.G. & Oldfield F.: <u>Catena</u>, 1978, **5**, 1.

7. Aarkrog A. Dahlgaard H. Nilsson K. & Holm E. <u>Health Phys.</u>, 1984, **46**, 29.

8. Romero L. Lobo A.M. & Holm E.: Proc. ANS International Topical Conference MARC-II, Kona 1991.

9. Jennings C.D.: personal communication (1988).

THE USE OF RADIONUCLIDES (unsupported ^{210}Pb, ^{7}Be AND ^{137}Cs) IN DESCRIBING THE MIXING CHARACTERISTICS OF ESTUARINE SEDIMENTS

ROBERT J. CLIFTON
Plymouth Marine Laboratory
Prospect Place, The Hoe, Plymouth PL1 3DH, UK.

ABSTRACT

By studying sediment depth profiles of radioisotopes having different sources and half-lives we differentiate between steady-state accretion and mixing. The effect of ^{7}Be and ^{137}Cs source variability on the sediment-depth profiles of these isotopes is discussed. Mixing rates throughout any estuarine sediment deposit cannot be assumed to be constant. At sites identified as 'mixed' we investigate different mixing components and boundary conditions and in the upper estuary, we quantify bulk sediment movement.

INTRODUCTION

Estuarine sediments are often a sink for contaminants released into estuarine and coastal waters and the subsequent release of adsorbed substances from these sediments is influenced by a variety of factors including:

(i) **Physical Disturbance**: Resulting in the release of pore waters, together with material desorbed from sediment surfaces due to increased exposure to overlying waters.

(ii) **Biological Disturbance**: Due to sediment reworking and irrigation resulting in the release of pore waters and adsorbed species as described in (i).

Large areas of intertidal sediments of the Tamar Estuary are colonised by a variety of benthic macrofauna, 40% of which comprise of 4 main species: *Nereis diversicolor, Arenicola marina, Scrobicularia plana* and *Nephtys hombergi* [1]. Previous work has demonstrated increased fluxes of metals and nutrients at bioturbated sites, over and above those recorded at sites identified as purely diffusional. [2]

(iii) **Diffusion**: Concentration gradients of contaminants generated in the pore waters resulting in a transfer of material across the sediment water interface.

Our present programme of work is very wide ranging and is investigating sediment/water exchange processes at sites throughout the Tamar estuary. In this paper we focus on just 3 sites to illustrate the use of the radioisotopes ^{210}Pb, ^{137}Cs and ^{7}Be to gain insight into the mixing and bulk movement characteristics of the bed sediments of the Tamar estuary as a prerequisite for assessing the role of mixing processes on the transfer of contaminants and nutrients across the sediment–water interface.

MATERIALS AND METHODS

Intertidal sites were selected to cover the tidal length of the river Tamar and to include areas which were known to be exposed primarily to physical disturbance as well those subject to bioturbation [2].

Sediment samples were collected by 'scraping' 1cm deep layers from within a $0.1 m^2$ stainless steel frame, to a depth of 6cm. Samples below 6cm were obtained by taking an 18cm diameter core and sectioning it at 2–4cm intervals to a total depth of 25–60 cm, depending on the unsupported ^{210}Pb (xs^{210}Pb) profile.

Wet sediment material (700–900g) was loaded into Marinelli beakers and sealed. The samples were counted using a high purity germanium coaxial photon detector system (HPGe). The ^{7}Be, ^{137}Cs and ^{210}Pb were determined as soon as possible after collection, using the 477.6, 661.6 and 46.5keV photopeaks respectively. The sample was counted again at least 30 days after sealing in the Marinelli beaker and the ^{226}Ra content estimated from the ^{214}Bi(609.3keV) and ^{214}Pb(352.0keV) photopeaks.

Atmospheric fluxes of ^{210}Pb and ^{7}Be were determined at 4–8 weekly intervals by collecting precipitation in a 2000cm^2 polyethylene tray containing 1NHNO$_3$ to a depth of 1cm. The total precipitation (rain plus particulates) was transferred to a beaker, Pb, Al, and Fe carriers were added and the whole adjusted to pH 11 with NH$_4$OH. The resulting precipitate plus particulate material was filtered on a GFC filter and counted on the HPGe detector to determine the ^{210}Pb and ^{7}Be. Yields, determined gravimetrically, on filtered rain–water samples containing Be as well as Pb yield tracers were consistently greater than 90% for both elements. ^{7}Be and ^{210}Pb atmospheric fluxes were determined regularly at the Plymouth laboratory site and less frequently at two other sites: one situated approximately mid–estuary and the other towards the tidal limit.

RESULTS AND DISCUSSION

Mixing, mixing depth and accretion.
Sediment mixing is frequently treated as a solid particle diffusion process [6–12] which affords a relatively simple solution of the steady–state relationship between depth (z) and the activity of a radioactive indicator. For example [12]:

$$A_{(z)} = A_0 \exp(\alpha z) \qquad (1)$$

where: A is the activity concentration of the indicator isotope (Bq/cm^3).

$\alpha = [S - SQR(S^2 + 4\lambda K)]/2K$ (2)

K is the particle mixing coefficient (cm^2/s).
λ is the radioactive decay constant (s^{-1})
S is the sediment accumulation rate (cm/s)

If $S^2 << 4\lambda K$ then: (3)

$K = -\lambda/\alpha^2$ (4)

However, the validity of this relationship is based on a variety of assumptions

(a) Steady-state accretion is negligible compared with mixing.

Apart from areas subject to large bulk-sediment movement, all of the sediments-depth profiles, analysed so far, exhibit a decrease of both ^{7}Be and xs ^{210}Pb activity with depth. These profiles are generated by steady-state accretion, sediment mixing or a combination of the two.

For values of K in the region of 10^{-7} to 10^{-6} cm^2 and a sedimentation rate of 1 cm/y

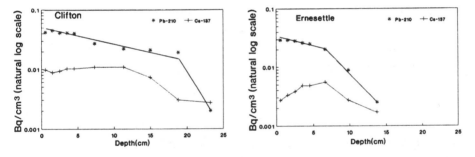

Figure 1. Depth profiles of ^{137}Cs and xs ^{210}Pb for Clifton and Ernesettle.

then condition (3) holds for ^{7}Be but not xs^{210}Pb. However, for a sedimentation rate of 0.1 cm/y and a mixing coefficient of 5 x10^{-7} cm^2 s^{-1}, 4λK is more than 2 orders of magnitude greater than S^2. Assuming condition 3 to be valid in this case would result in an over estimate of K by less than 10%. There is strong evidence to suggest mixing is the dominant process, defining the sediment-depth profile of both ^{7}Be and xs ^{210}Pb. For example:

(i) In the case of ^{7}Be the derived sedimentation rate from these profiles would indicate extremely high sedimentation rates (typically > 7cmy^{-1}).

(ii) If these profiles were primarily a result of steady-state accretion then sediment horizons dated according to the classical ^{210}Pb method [3–5] would imply that ^{137}Cs was present at a time which predates the first known introduction of this isotope into the environment by, in some cases, 25 years (14). This discrepancy may be the result of diffusion but the diffusion coefficient of ^{137}Cs, in marine sediments, is 1 to 2 orders of magnitude lower

[13] than the mixing coefficients determined in this work. Also there is strong evidence (Fig. 1) to suggest that ^{210}Pb and ^{137}Cs are governed by a similar process in that they both approach zero concentrations at a common depth in the sediment–implying a boundary condition or a finite mixing depth.

(iv) The observed density of benthic species colonising many of the intertidal sediment deposits of the Tamar estuary [2] imply a high degree of biological mixing.

(b) The porosity of the sediment is constant.

The porosity of some sediment deposits may change significantly over the top few cm but normally reaches a constant value at a depth of 5–10cm. We have addressed the problem of changes in porosity over the first few cm by normalising all concentrations (Bq/cm^3) to 50% water content. In fact this normalisation procedure has little effect on derived mixing coefficient based on either ^{7}Be or ^{210}Pb data but it does improve the depth correlation and reduces the residual bias of surface data points (top 5cm) to one side of the regression line.

(c) Mixing rate is a constant throughout the sediment.

Mixing rates cannot be assumed to be constant in any estuarine deposit of sediment. These sediments are subjected to a variety of mixing processes, driven by both physical (e.g. wave motion) and biological turbations and any conclusions based on the assumption of uniform mixing may be seriously in error.

(d) The source of the indicator is a constant.

The atmospheric flux of both ^{7}Be and ^{210}Pb (Bq/cm^2) has been determined over the period 11/88 to 11/90 (Fig. 2). While yearly averages of the fluxes of both these isotopes appear to be reasonably constant (Table 1) there are variations over the short–term.

TABLE 1

Annual Fluxes of ^{7}Be and ^{210}Pb measured in 1988–1990

YEAR	^{7}Be Flux Bq cm^{-2}y^{-1} x 10^3	^{210}Pb Flux Bq cm^{-2}y^{-1} x 10^3
11/88–11/89	74.0 ± 9.5	6.3±1.1
11/89–11/90	80.7 ± 9.2	7.2 ± 1.2

Although it is reasonable to anticipate that small, short–term variations in the flux of ^{210}Pb (half–life=22years) will have minimal effect on the sediment depth profile of this isotope, this

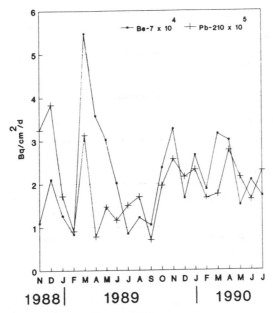

Figure 2   Atmospheric deposition of ^{7}Be and ^{210}Pb.

cannot be assumed in the case of ^{7}Be which has a much shorter half–life (53.3 days) and over 90% of its sediment inventory is deposited over the previous 4 months.

Simulation

To model sediment mixing it would be unrealistic to assume that all of the parameters governing the validity of equation 4 are constant throughout the whole of the sediment depth. We are developing a simple advective model which accommodates variable mixing and input rates.

Mixing coefficients are derived from equation 4, where α is determined as the slope of the line representing the decrease in activity (ln Bq cm^{-3}) of the indicator isotope (^{7}Be or xs^{210}Pb) over discrete sediment horizons, having a constant slope and representative of areas of uniform mixing. These areas are delineated by points of inflection in the line.

Apart from variable mixing, the reflection of a source of contaminantion in the sediment depth profile will depend on the magnitude and variability of the source, its affinity for particulate material and the dynamics of the overlying water and the bed sediments:

(a) The fact that different bed sediments respond very differently to a particular source has been referred to as the focussing ability of that sediment [14] where:

$$\text{Focussing ability (F)} = I_S/I_T \tag{5}$$

The sediment inventory (I_S) of an indicator is calculated by summing the specific activity A_i (Bq g^{-1}) of all of the depth intervals i with detectable ^{7}Be or xs^{210}Pb.

$$I_S = \sum_{i=1}^{N} A_i \rho_s (1-\varphi_i) \qquad (6)$$

The particle density ρ_s is taken to be 2.3g cm^{-3} and the porosity φ_i is calculated from the water loss on freeze drying.

$$I_T = V/\lambda \qquad (7)$$

where V is the atmospheric flux (Table 1) and λ is the decay constant of the indicator (y^{-1}).

(b) The proportion (P) of the contaminant flux associated with particulate material is estimated from:

$$P = R/(R+1) \qquad (8)$$

where $R = K_d \times L \times 10^{-3}$; K_d is the partition coefficient and L is the particulate loading of the water (g/l).

(c) Loss of a contaminant at the sediment water interface is expressed as a surface loss coefficient (K_s) where $K_s = 0$ for 100% retention.

Examples of two simulation runs are presented here:

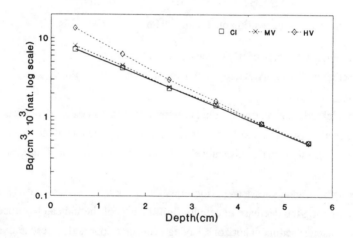

Figure 3. Simulated ^{7}Be depth profiles. (Constant input = CI, Medium variability input = MI, High variability input = HV).

(i) We have simulated the input of a variable source of ^{7}Be to the sediments surface over two periods (Fig. 3): the first was run for 500 days up to the peak of the November 1989 deposition (medium variability) and the second for 500 days up to the peak of the April 1984 deposition (high variability). Assuming F = 1, $K_s = 0$ and K = 5x10^{-7} cm^2 s^{-1}, the depth profiles resulting from variable inputs are compared to those generated from a continuous input of ^{7}Be of 2.2 x 10^{-4} Bq cm^{-2} d^{-1} (Fig. 2). The slopes of the lines (natural log scale) resulting from the variable input simulations are not linear (Fig. 3) and the magnitude of the differences between mixing coefficients derived from these profiles relative to those derived from the constant input simulation will depend on the depth over which they are compared. Over 3 cm these differences are 46% and 33% in the case of the high and medium variability peaks respectively and drop to 22% and 10% respectively if the profiles are compared over a depth of 6 cm.

(ii) The bed sediments at Ernesettle and Clifton have quite different properties:

(a) Ernesettle. The sediment in this area has a relatively high sand content and is colonised mainly by *Arenicola marina* and *Nephtys hombergi*. The derived mixing coefficients for these sediments are low (e.g. 7.5×10^{-8} cm^2.s^{-1})

(b) Clifton. This station is further up the estuary and is characterised by fine silts colonised by both *Hediste* and *Scobicularia*. The mixing coefficients derived for this area are very variable (Range 1.5×10^{-7} to 1.5×10^{-6} cm^2.s^{-1}).

TABLE 2

^{137}Cs – Simulation Parameters

STATION	MIXING DEPTH (cm)	MIXING COEFFICIENT (cm^2 s^{-1})	FOCUS FACTOR (F)
Clifton	20	1.76×10^{-7}	2.05
Ernesettle	8	0.75×10^{-7}	1.30

This simulation uses the criteria in Table 2 and assumes that the input pattern of ^{137}Cs, to the Tamar region, reflects the global fallout pattern (N. Hemisphere) over the past 35 years [15]. At depths greater than the mixing depth it is assumed that the ^{137}Cs dispersion in the sediment is a function of molecular diffusion only. The diffusion coefficient in this case is assumed to be 10^{-8} cm^2 s^{-1} [13]. Assuming a particulate loading of 0.5 g l^{-1} and a K$_d$ for ^{137}C$_s$ of 2×10^3 [16], P was calculated to be 0.5. The best fit of the simulation profiles to the data points (Fig. 4) corresponds to a surface loss factor (K$_s$) of 3.1×10^{-8} and 2.8×10^{-8} cm^2 s^{-1} at Clifton and Ernesettle respectively.

Bulk Movement of sediments

Unlike the deposits at Ernesettle and Clifton some surface sediments in the upper estuary are subjected to periodic movement–controlled primarily by water run–off [17] which can account for changes in sediment height of over 25 cm. Although it would not be appropriate

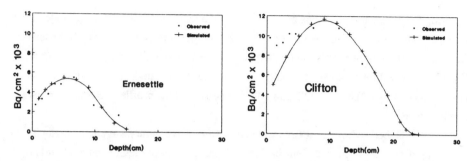

Figure 4 Comparison of observed and simulated depth profiles of ^{137}Cs at Ernesettle and Clifton.

to determine mixing coefficients under these circumstances, the ^{7}Be profiles are a useful guide to the amount of sediment deposited or eroded over short time periods. For example, sediment cores taken from the Clamoak site in Aug '89 show a deposit of highly mixed sediment approximately 10cm. above that recorded 6 months earlier in Feb.'89 (Fig. 5).

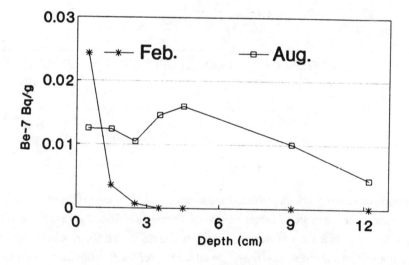

Figure 5 Depth profiles of ^{7}Be observed at Clamoak during periods of high (Feb.) and low (Aug.) 'run–off'.

CONCLUSIONS

At all sites investigated so far, the sediment–depth profiles of ^{7}Be, ^{137}Cs, ^{210}Pb are generated predominantly by mixing processes rather than steady–state accretion and it must be concluded, therefore, that this applies to all sediment associated contaminants.

Variations in the atmospheric flux of ^{210}Pb have minimal effect on the determination of sediment mixing coefficients. However, the greatest variation in the ^{7}Be flux, so far observed, would introduce an error of over 40%.

Mixing of estuarine sediments is complex and cannot be defined by a single mixing coefficient spanning, in some cases, over 50 cm. Variable porosity, mixing and flux all have to be considered together with boundary conditions which delimit physical, biological and diffusional processes.

ACKNOWLEDGEMENTS

The author would like to thank M. Carr, J.T. Davey, P.J. Radford, D. Turner and P.G. Watson for their comments concerning the chemical, biological and simulation aspects of this work.

This work is partly supported by the National Rivers Authority (NRA R&D project A5/104) and also forms part of the Interfacial Exchanges project of the Plymouth Marine Laboratory – a component body of the U.K. Natural Environment Research Council.

REFERENCES

1. Spooner, G.M. and Moore, H.B., The ecology of the Tamar estuary VI. An account of the macrofauna of the intertidal muds. J. Mar. Biol. Assoc. 1939, (U.K.), 24, 283–330.
2. Watson, P.G. and Davey, J.T.,1988. Estuarine sediment bioturbation and metal distribution. Final Report to British Department of the Environment, Contract No. 7/7/163 (unpublished): 61 pp.
3. Koide, M., Souter, A. and Goldberg, E.D., Marine geochronology with Pb–210. Earth Planet Sci. Lett., 1972. **14**, pp. 442–446.
4. Krishnaswani, S., Lah, D., Martin, J.M. and Maybeck, M., Geochronology of lake sediments. Earth Planet Sci. Lett., 1971, 11, pp. 407–414.
5. Robbins, T.A. and Edgington, D.H., Determination of recent sedimentation rates in Lake Michigan using Pb–210 and Cs–137. Geochim. Cosmochim. Acta., 1975, **39**, pp. 285–304.
6. Aller, R.S. and Cochran, J.K., ^{234}Th/^{238}U disequilibrium in near shore sediment: particle reworking and diagenetic time scales. Earth Planet Sci. Lett., 1976, **29**, 37.
7. Boudreau, B.P., Mathematics of Tracer Mixing in Sediments: I. Spatially–dependent, diffusive mixing. Am. J. Sci, 1986, **286**, pp. 161–198.
8. Christensen, E.R., A model for radionuclies in sediments influenced by mixing and compaction. J. Geophys. Res., 1986, **87**, pp. 566–572.
9. Kershaw, P.J., Garbutt, P.A., Young, A.K. and Allington, D.J., Scavenging and bioturbation in the Irish Sea from measurements of ^{234}Th/^{238}U and ^{210}Pb/^{226}Ra

disequilibria. In Radionuclides: A Tool for Oceanography, ed. J.C. Guary, P. Guegueniat and R.J. Pentreath, Elsevier Applied Science Publishers, London and New York, 1988, pp. 131–142.

10. Officer, C.B., Mixing, Sedimentation rates and age dating of sediment cores. Mar. Geol., 1981, **42**, pp. 261–276.

11. Schink, D.R. and Guinasso, N.L. Redistribution of dissolved adsorbed materials in abyssal marine sediments undergoing biological stirrling. Am. J. Sci., 1978, **278**, pp. 687–702.

12. Krishnaswami, S., Benninger, L.K., Aler, R.C. and Vonn Damon, K.L. Atmospherically derived radionuclides as tracers of sediment mixing and accumulation in near–shore marine and lake sediments: Evidence from ^{7}Be, ^{210}Pb and ^{239,240}Pu. Earth Planet. Sci. Lett. **47** pp. 307–318.

13. Duursma, E.K. and Bosch, C.J. Theoretical, experimental and field studies concerning diffusion of radio isotopes in sediments and suspended particles of the sea. Part B: Methods and experiments. Netherlands J. Sea Res. 1970, **4**, pp. 395–469.

14. Dibb, E.J. and Rice, D.L. Temporal and spatial distribution of Beryllium–7 in the sediments of Chesapeake Bay. Estuar. coastal and shelf sci. 1989, **28**, pp. 395–406.

15. Cambray, R.S., Playford, K. and Lewis, G.N.J. Radioactive fallout in air and rain: Results to the end of 1984. Atomic Energy Research Establishment report. AERE–R 11915. Oct. 1985.

16. Sediment K_ds and concentration factors for radionuclides in the marine environment. Technical reports series No. 247. International Atomic Energy Agency, 1985.

17. Bale, A.J., Morris, A.W. and Howland, R.J.N., Seasonal Sediment Movement in the Tamar estuary. Oceanol. Acta., 1985, **8**, pp. 1–6.

PLUTONIUM, AMERICIUM AND RADIOCAESIUM IN SEA WATER, SEDIMENTS AND COASTAL SOILS IN CARLINGFORD LOUGH

P.I. Mitchell, J. Vives Batlle and T.P. Ryan
Department of Experimental Physics, University College, Dublin 4.

C. McEnri, S. Long, M. O'Colmain and J.D. Cunningham
Nuclear Energy Board, Dublin 14.

J.J. Caulfield, R.A. Larmour and F.K. Ledgerwood
Environmental Protection Division, Department of the Environment (N.I.),
Belfast BT1 1FY.

ABSTRACT

Artificial radioactivity within the sediment-water system in Carlingford Lough has been studied and the extent of transport of plutonium from sea to land has been assessed by analysing soil samples taken at selected inland sites close to the shore of the Lough. Samples of sea water, suspended particulate and sediment have been analysed for ^{134}Cs, ^{137}Cs, ^{238}Pu, ^{239,240}Pu and ^{241}Am and the speciation of both plutonium and americium determined in Lough waters. The association of transuranium nuclides with colloids has also been examined and it has been shown that a significant proportion of the plutonium and almost all of the americium in the filtrate fraction (<0.22 μm) is in a colloidal form. In general, concentrations of radiocaesium and the transuranics were found to be very low and showed little or no enhancement when compared to similar materials sampled elsewhere near the north-east coast of Ireland. Finally, the sea-to-land transfer of plutonium and, by inference, other actinides was found to be insignificant in the vicinity of the Lough and the adjacent coastline.

INTRODUCTION

Carlingford Lough, located on the east coast of Ireland (Figure 1), is a shallow estuary approximately 13 km in length and 5 km at its widest, with a relatively narrow fetch of about 1.7 km at the entrance. In recent years important developments in the field of mariculture have taken place in the Lough. There is now a significant mussel (*Mytilus edulis*) farming industry located near Rostrevor while the Lough has become the largest producer of Pacific oysters (*Cressostres gigas*) in Ireland [1].

Considerable water quality data are now available following studies conducted by the Carlingford Lough Marine Laboratory [1, 2]. Recorded pH values in the Lough vary from 7.1 to 8.8 but, in general, exceed 8.0 with a mean value of 8.2 [1]. The waters are, therefore, slightly alkaline. Salinity variations in the Lough are relatively small, though there is a greater variation along the northern shore and lowest values occur near Narrow Water due to the influence of the Newry River. Little difference is evident between surface and bottom waters in the Lough and suggests that the volume of fresh water being discharged into the Lough by the inflowing streams has only a localised effect [1]. Dissolved oxygen levels in the water are generally above 10 mg l^{-1}. The Lough appears to be slightly depleted in nitrate and phosphate nutrients, whereas there is some evidence of silicate enrichment [1, 2].

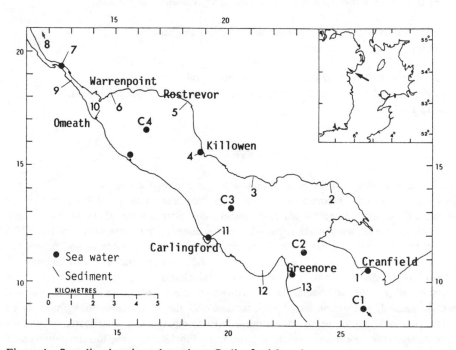

Figure 1. Sampling locations throughout Carlingford Lough.

An important objective of this study was to assess whether artificial radioactivity concentrations in environmental materials from the Lough were enhanced above general levels found elsewhere along the north-east coast of Ireland. Other objectives included an examination of the speciation of plutonium and americium in Lough waters, including the association of these elements with colloids, and an analysis of coastal soils to determine the extent of 'sea-to-land' transfer of plutonium in the vicinity of the Lough. An assessment of the chemical speciation of plutonium was considered to be of particular importance, as it is now well established that plutonium can exist in the water column in a number of different oxidation states simultaneously and that the nature of these states has a very significant influence on the environmental behaviour of this element.

METHODS AND ANALYSES

Samples of surface (~1 m depth) sea water (50 l) and sediment (0-10 cm) were taken from aboard the R.V. *Lough Beltra* at four stations along the axis of Carlingford Lough on the 25-27 September 1990 (Figure 1). Additional samples of sea water and sediment (0-5 cm) were also collected from the inter-tidal zone at various sites around the Lough. Furthermore, samples of undisturbed soil were taken at separate locations 10 metres and 50 metres from the high-water mark for the purpose of assessing the extent of the transfer of plutonium from sea to land.

Standard sample handling and radioanalytical techniques were used to determine the concentrations of radiocaesium and the transuranics in these samples, the details of which have been described elsewhere [4, 5, 6, 7]. Separation of the reduced and oxidised states of plutonium, i.e., Pu(III,IV) from Pu(V,VI), was carried out using the neodymium fluoride co-precipitation method of Lovett and Nelson [8]. The fraction of the plutonium and americium in Lough waters which was colloidally bound, was determined by passing each sample through a set of four aluminium oxide beds arranged in series after the method of Nevissi and Schell [9, 10]. In this technique the 'colloidal' fraction is defined as that which can pass through a 0.22 μm filter, but is efficiently sorbed on the first aluminium oxide bed and may be quantified by the appreciably higher collection efficiency on the first bed compared with the second and third beds [10].

RESULTS AND DISCUSSION

Sea water

Concentrations of ^{134}Cs, ^{137}Cs, ^{238}Pu, ^{239,240}Pu and ^{241}Am in sea water sampled at three centrally-located stations inside Carlingford Lough and one station outside the entrance to the Lough, are given in Table 1. Little variation is evident in the ^{137}Cs levels in filtered (<0.22 μm) water at any of the stations, the mean value being 93 ± 4 Bq m^{-3}. Caesium-137 levels in filtered water sampled at six shore-line stations around the Lough were also found to show little variation and, with one exception, are indistinguishable from the levels recorded at the centre-line stations. In all cases, ^{137}Cs levels on the suspended particulate fraction (1.0-1.5 mg l^{-1}) were below the detection limits, the same being true for ^{134}Cs. Data on ^{137}Cs concentrations in the open waters of the western Irish Sea (private communication) indicate that the ^{137}Cs levels within the Lough are approximately 10-15% higher than those prevailing outside.

Concentrations of plutonium and americium in Lough waters were between two and three orders of magnitude lower than those of radiocaesium. Little variation is evident in the plutonium and americium concentrations in filtered water at the three central stations inside the Lough. However, outside the Lough (station C1) and further afield, concentrations of plutonium in filtered sea water were considerably higher, whereas those of americium were very much lower [11]. Between 18% to 22% of the ^{239,240}Pu and 17% to 25% of the ^{241}Am was associated with the particulate phase in Lough waters, whereas to seaward (station C1) the corresponding values were 8% and 49%, respectively (Table 2). The latter are in good agreement with data from studies carried out in the open waters of the western Irish Sea in 1988 and 1989 [11].

TABLE 1

Radionuclide concentrations in surface seawater sampled at central stations in Carlingford Lough, September 1990

[Station] Co-ordinates	Fraction	(Bq m^{-3})		(mBq m^{-3})			^{238}Pu/^{239,240}Pu	^{241}Am/^{239,240}Pu	^{137}Cs/^{239,240}Pu
		^{134}Cs	^{137}Cs	^{238}Pu	^{239,240}Pu	^{241}Am			
[C2] 54° 01.76' 06° 07.04'	Filtrate	<2	91 ± 4	41 ± 2	212 ± 4	64 ± 7	0.195 ± 0.007	0.30 ± 0.03	430 ± 20
	Particulate			13 ± 2	59 ± 2	21 ± 3	0.22 ± 0.03	0.36 ± 0.05	-
[C3] 54° 03.31' 06° 10.00'	Filtrate	<2	91 ± 3	40 ± 2	205 ± 4	75 ± 9	0.196 ± 0.010	0.37 ± 0.04	440 ± 20
	Particulate			11 ± 9	45 ± 1	15 ± 2	0.24 ± 0.02	0.33 ± 0.04	-
[C4] 54° 04.40' 06° 13.50'	Filtrate	<2	98 ± 4	42 ± 4	182 ± 12	74 ± 8	0.23 ± 0.02	0.41 ± 0.05	540 ± 40
	Particulate			9 ± 2	40 ± 2	20 ± 3	0.23 ± 0.05	0.50 ± 0.08	-
Mean (C2-C4)	Filtrate	<2	93 ± 4	41 ± 1	200 ± 16	71 ± 6	0.21 ± 0.02	0.36 ± 0.06	430 ± 90
Mean (C2-C4)	Particulate			11 ± 2	48 ± 10	19 ± 3	0.23 ± 0.01	0.40 ± 0.09	-
[C1] 53° 59.76' 06° 00.56'	Filtrate	<2	88 ± 4	58 ± 2	283 ± 5	18 ± 6	0.204 ± 0.006	0.06 ± 0.02	310 ± 15
	Particulate			6 ± 1	26 ± 2	17 ± 1	0.22 ± 0.05	0.65 ± 0.06	-

Uncertainties quoted are ± 1 S.D.

The ^{238}Pu/^{239,240}Pu ratios in filtered sea water and suspended particulate, at 0.21 ± 0.02 and 0.23 ± 0.01, respectively, were identical and entirely consistent with the corresponding ratio in radioactive waste discharges from Sellafield. Outside the Lough, the ^{241}Am/^{239,240}Pu ratio in filtered water and particulate differed sharply, whereas in Lough waters they were similar, at 0.36 ± 0.06 and 0.40 ± 0.09, respectively.

Significant variations were observed in the percentage of oxidised plutonium in filtered water at the three stations within the Lough (Table 2). However, the mean value for ^{239,240}Pu (V,VI), at 30 ± 15%, is considerably smaller than the mean observed in the open waters of the western Irish Sea namely, 83 ± 7% [11], and readily explains why plutonium concentrations in filtered water within the Lough are considerably lower than those outside. Although the chemical speciation of americium in filtered water at these stations was not examined, it is reasonable to assume that it is almost entirely in a reduced form, Am(III), with little or none of the oxidised species, Am(V), present [12].

TABLE 2

Percentage of plutonium and americium on suspended particulate and percentage of Pu(V,VI) in filtered (<0.22 μm) surface water from Carlingford Lough, September 1990 (corresponding percentages for surface waters of the Irish Sea in 1989 are also given).

Station	% on suspended particulate			% Pu(V,VI) in filtrate	
	^{238}Pu	^{239,240}Pu	^{241}Am	^{238}Pu	^{239,240}Pu
C2	24 ± 4	22 ± 1	25 ± 3	32 ± 3	23 ± 2
C3	22 ± 18	18 ± 1	17 ± 3	46 ± 5	50 ± 4
C4	18 ± 4	18 ± 1	21 ± 4	22 ± 4	17 ± 1
Mean (C2-C4)	21 ± 3	19 ± 2	21 ± 4	33 ± 14	30 ± 15
C1	9 ± 2	8 ± 1	49 ± 9	-	-
Western Irish Sea (9/89)[a]	7 ± 6 [2-15]	7 ± 6 [2-15]	45 ± 13 [30-63]	-	-
Western Irish Sea (12/89)[a]	16 ± 5 [12-23]	21 ± 5 [15-26]	56 ± 21 [30-70]	85 ± 7 [80-93]	83 ± 7 [80-92]
Eastern Irish Sea (12/89)[a]	60 ± 20 [45-88]	60 ± 20 [47-88]	90 ± 2 -	84 ± 5 [77-90]	85 ± 4 [80-89]

Uncertainties quoted are ± 1 S.D.   [a] Source: Mitchell et al. [11].

The partition of plutonium and americium between particulate, 'colloidal' and soluble fractions of sea water sampled at stations C2 and C3 is summarised in Table 3. At both stations the percentage of plutonium in filtered water which was associated with the 'colloidal' fraction was about 50%, whereas almost all of the americium appeared to be so bound. Given that at least 50% of the plutonium and almost all of the americium [12] in filtered water from the Lough are in a reduced chemical form, it is not unreasonable to infer that the bulk of both (reduced) fractions is colloidally bound.

TABLE 3

Distribution of plutonium and americium between particulate (>0.22 μm), 'colloidal' and soluble fractions of sea water sampled at stations C2 and C3 in Carlingford Lough.

Station	Fraction	Concentration (mBq m^{-3})		% of total	
		^{239,240}Pu	^{241}Am	^{239,240}Pu	^{241}Am
C2	Particulate	33 ± 2	15 ± 1	22 ± 5	56 ± 6
	'Colloidal'	56 ± 12	12 ± 2	36 ± 9	44 ± 8
	Soluble	65 ± 17	<3	42 ± 12	
C3	Particulate	24 ± 2	15 ± 2	15 ± 2	16 ± 2
	'Colloidal'	70 ± 10	81 ± 5	45 ± 10	84 ± 7
	Soluble	70 ± 20	<8	45 ± 10	

Uncertainties quoted are ± 1 S.D.

Sediment/water distribution coefficients (K_ds) for plutonium and americium in Carlingford Lough have been deduced from the data and are given in Table 4. Total K_d values for both elements at Station C1, just outside the Lough, differ by an order of magnitude and are fully consistent with the K_d values reported for the western Irish Sea elsewhere in these proceedings [11]. However, within the Lough, total K_d values for both elements are indistinguishable with means of $(2.1 \pm 0.4) \times 10^5$ and $(2.3 \pm 0.6) \times 10^5$, respectively. The mean K_d for reduced plutonium, Pu(III,IV), is also very similar at $(3 \pm 1) \times 10^5$.

The similarity between the total K_d for plutonium and americium is readily explained by the fact that the bulk of the plutonium in the filtrate is in a reduced state and, consequently, the oxidised fraction contributes only marginally to the total K_d. Furthermore, competition between suspended solid and other species likely to be present in estuarine waters, such as colloidal organic carbon, will tend to diminish measured K_ds [13] and may explain the lower values (about a factor of two) observed in Lough waters.

Sediments
Radionuclide concentrations in estuarine sediments were determined at 13 inter-tidal sites around the Lough and 3 sites near the centre of the Lough (Table 5). Caesium-137 concentrations in the range 8-200 Bq kg^{-1} were observed, while ^{239,240}Pu and ^{241}Am levels were at least an order of magnitude lower at 1.4-18 Bq kg^{-1} and 0.4-8.8 Bq kg^{-1}, respectively. These concentrations are broadly similar to those observed elsewhere along

the north-east coast of Ireland [14, 15]. The results of a particle-size distribution and classification of the sediment samples showed wide variations within the sand fractions, with several samples having significant amounts of gravel. Only a few contained significant amounts of silt with small or virtually no clay material being associated with these. There was a clear correlation between radioactivity concentration and particle size, finer particles having a larger surface area per unit mass and, thus, adsorbing more activity.

TABLE 4

K_d values (Bq kg^{-1} dry weight suspended load /Bq l^{-1} of sea water filtered through a 0.22 μm filter) for plutonium and americium in Carlingford Lough (corresponding values for surface waters of the Irish Sea in 1989 are also given).

Station	K_d (Total) ^{239,240}Pu	^{241}Am	K_d (Reduced Pu)
C2	(2.5 ± 0.9)x10^5	(3.0 ± 1.2)x10^5	(3.6 ± 1.3)x10^5
C3	(2.2 ± 0.9)x10^5	(1.8 ± 0.7)x10^5	(4.1 ± 1.7)x10^5
C4	(1.7 ± 0.7)x10^5	(2.1 ± 0.9)x10^5	(2.1 ± 0.8)x10^5
Mean (C2-C4)	(2.1 ± 0.4)x10^5	(2.3 ± 0.6)x10^5	(3 ± 1)x10^5
C1	(5.7 ± 1.5)x10^4	(5.9 ± 2.5)x10^5	-
Western Irish Sea[a]	(5 ± 3)x10^4 (n = 14)	(5 ± 3)x10^5 (n = 11)	(6 ± 4)x10^5 (n = 7)
Eastern Irish Sea[a]	(4 ± 2)x10^5 (n = 6)	(3 ± 1)x10^6 (n = 2)	(3 ± 2)10^6 (n = 6)

Uncertainties quoted are ± 1 S.D.   [a]Source: Mitchell et al. [11].

The mean ^{238}Pu/^{239,240}Pu ratio for sediments from Carlingford Lough shows little variation at 0.176 ± 0.008 and is consistent with ratios reported by others for sediments along the north-east coast of Ireland [14, 15]. The ^{241}Am/^{239,240}Pu ratio also shows little scatter, the mean being 0.45 ± 0.04. The ^{137}Cs/^{134}Cs ratio does show considerable variation due largely, we believe, to the imprecision of our ^{134}Cs determinations resulting from the very low concentrations of this nuclide present on sediment. Nevertheless, the mean value of 80 ± 30 is consistent with values reported elsewhere throughout the Irish Sea [14].

^{238}Pu/^{239,240}Pu ratio in coastal soils

In an attempt to establish whether a significant transfer of plutonium from Lough waters to land had taken place, the ^{238}Pu/^{239,240}Pu isotopic ratio was examined in coastal soils sampled at various locations in the vicinity of Carlingford Lough (Table 6). In a

TABLE 5

Radionuclide concentrations in inter-tidal sediments (0 - 5 cm) in Carlingford Lough, September 1990

| Site No. | Grid Ref. | Bq kg^{-1} (dry weight) | | ^{137}Cs/^{134}Cs | mBq kg^{-1} (dry weight) | | | ^{238}Pu/^{239,240}Pu | ^{241}Am/^{239,240}Pu |
		^{134}Cs	^{137}Cs		^{238}Pu	^{239,240}Pu	^{241}Am		
[1]	262 107	<0.2	14.1 ± 0.4	—	198 ± 9	1389 ± 29	360 ± 24	0.143 ± 0.007	0.26 ± 0.02
[2]	246 143	1.1 ± 0.2	49 ± 2	45 ± 8	—	—	1353 ± 496[a]	—	—
[3]	211 145	0.8 ± 0.2	77 ± 2	96 ± 18	1611 ± 29	8572 ± 116	3890 ± 99	0.188 ± 0.03	0.45 ± 0.01
[4]	189 155	2.0 ± 0.3	150 ± 3	75 ± 10	—	—	5333 ± 657[a]	—	—
[5]	182 178	2.5 ± 0.8	173 ± 6	69 ± 23	2129 ± 37	11,987 ± 160	5774 ± 165	0.178 ± 0.002	0.48 ± 0.02
[6]	148 181	0.9 ± 0.2	68 ± 2	76 ± 17	—	—	6516 ± 1040[a]	—	—
[7]	125 194	2.7 ± 0.4	162 ± 4	60 ± 9	2073 ± 78	12,323 ± 307	4820 ± 151	0.168 ± 0.006	0.39 ± 0.02
[8]	102 233	4.4 ± 0.3	179 ± 4	41 ± 3	—	—	8835 ± 1140[a]	—	—
[9]	130 188	2.7 ± 0.2	174 ± 4	64 ± 5	—	—	7853 ± 837[a]	—	—
[10]	142 170	1.1 ± 0.2	66 ± 2	60 ± 11	916 ± 18	5360 ± 72	2524 ± 82	0.171 ± 0.003	0.47 ± 0.02
[11]	192 118	2.5 ± 0.4	200 ± 4	80 ± 12	3193 ± 106	18,050 ± 400	8620 ± 260	0.177 ± 0.005	0.48 ± 0.02
[12]	216 104	<0.2	18.5 ± 0.5	—	—	—	1671 ± 329[a]	—	—
[13]	225 095	0.1 ± 0.1	13.1 ± 0.4	131 ± 66	—	—	659 ± 170[a]	—	—
53°59' 06°00'[b]		0.2 ± 0.1	16.6 ± 0.4	83 ± 21	—	—	—	—	—
54°03' 06°10'[b]		<0.2	7.7 ± 0.3	—	—	—	616 ± 184	—	—
54°04' 06°13'[b]		0.6 ± 0.2	74 ± 2	123 ± 31	—	—	2440 ± 400	—	—
Mean				80 ± 30 (n = 13)				0.176 ± 0.008 (n = 5)	0.45 ± 0.04 (n = 5)

[a]Measured by high resolution gamma spectrometry   [b]Decca co-ordinates

previous survey carried out in 1987-89 the mean ratio for inland sites throughout Ireland was found to be 0.034 ± 0.004 [6]. This value is exceeded at both sites on the northern shore of the Lough, though in each case only at a point 10 metres from the high-water mark. On the southern shore of the Lough no evidence of any distortion in the ratio was apparent. Fifty metres inshore, the ratio at all four sites sampled was identical to the global fallout ratio of 0.034. Two other sites, one situated 17 km N (Newcastle) and the second 20 km S (Clogher Head) of the Lough were also examined. Only at Clogher Head - a particularly exposed site of low relief - was there clear evidence of perturbation in the ratio. Sites with elevated ratios at the 10-metre mark compared to the 50- or 100-metre mark are indicative of localities where some degree of sea-to-land transfer of plutonium has taken place, given that the $^{238}Pu/^{239,240}Pu$ ratio in Lough waters and the western Irish Sea is typically in the range 0.17-0.23.

That only sites on the northern shore of the Lough appear to be affected, is not unexpected as the prevailing wind is south-westerly and almost directly on-shore. Other evidence of a weak sea-to-land transfer of actinides along the north-east coast of Ireland has previously been reported by Garland et al. [15], who pointed out that only at some sites on the east coast could the resulting accumulation in soil close to the beach be distinguished from global fallout.

TABLE 6

Pu-238/Pu-239,240 isotopic ratio in samples of undisturbed soil (0-5 cm) from the vicinity of Carlingford Lough and further afield.

Site (Grid Ref.)	Distance from high water mark (m)	$^{238}Pu/^{239,240}Pu$
Carlingford Lough:		
Mill Rock	10	0.071 ± 0.004
(246 143)	50	0.034 ± 0.003
Carrigaroan	10	0.069 ± 0.003
(211 145)	50	0.037 ± 0.004
Omeath	10	0.036 ± 0.004
(137 177)	50	0.037 ± 0.003
Carlingford	20	0.033 ± 0.002
(206 109)	50	0.036 ± 0.004
Other sites:		
Newcastle	10	0.037 ± 0.002
	100	0.043 ± 0.003
Clogher Head	10	0.078 ± 0.004
	100	0.046 ± 0.004
Irish Inland Survey	-	0.034 ± 0.004[a]

[a]mean (± 1 S.D.)

CONCLUSIONS

Generalised radionuclide concentrations in sea water and sediments from Carlingford Lough are broadly similar to those observed elsewhere along the east and north-east coasts of Ireland [4, 14, 15]. There is, therefore, little evidence to suggest that the Lough is a more effective 'sink' for artificial radioactivity in the western Irish Sea than other regions along the north-east coast.

The percentage of reduced plutonium in the filtered water fraction was found to be much higher in the Lough than in the open waters of the Irish Sea. This is not unexpected and may be related to the fact that dissolved organic carbon levels are also considerably higher in Lough waters. The fact that most, if not all, of the reduced plutonium and americium in filtered Lough water appears to be colloidally bound, is a further indication that colloidal organic carbon is competing with the particulate phase for the available plutonium and americium. This hypothesis is supported by the observation that the ^{241}Am/^{239,240}Pu ratio in filtered water from the Lough was considerably higher than that observed in the open waters of the Irish Sea. Given these considerations, measured K_d values in Lough waters are fully consistent with those reported for the western Irish Sea [11].

Finally, the marginal distortion in the ^{238}Pu/^{239,240}Pu ratio in soil sampled close (10 m) to the high-water mark on the northern shore of the Lough suggests that a detectable transfer of plutonium (and by inference other actinides) from sea to land has taken place. However, the effect is a marginal one and has clearly no radiological significance whatsoever.

ACKNOWLEDGEMENTS

The facilities provided by the staff of the Carlingford Lough Marine laboratory and, in particular, the help given by D.J. Douglas and S. McQuaid, is gratefully acknowledged. The authors also wish to thank J.S. Batch and J.E. Pinkerton (Industrial Science Centre, Northern Ireland) who carried out the particle-size analysis of the sediment samples and the captain and crew of the R.V. *Lough Beltra* for their generous support.

REFERENCES

1. Douglas, D.J. and Conlon, J., Physico-chemical analysis of the water in Carlingford Lough. In Carlingford Lough Marine Laboratory, Bulletin No. 6, Nov. 1990, Published by Carlingford Lough Aquaculture Association., Omeath, Ireland, ISSN 0791-4512, pp. 20-8.
2. Douglas, D.J., Conlon, J. and McQuaid, S., Water analysis in Carlingford Lough. In Carlingford Lough Marine Laboratory, Bulletin No. 3, May 1990, Published by Carlingford Lough Aquaculture Association, Omeath, Ireland, ISSN 0791-4512, pp. 14-23.
3. Caulfield, J.J. and Ledgerwood, F.K., Terrestrial gamma-ray dose rates out of doors in Northern Ireland. Environmental Monitoring Report No. 2, Department of the Environment (Northern Ireland), Belfast, November 1989, 23 pp.

4. O'Grady, J. and Currivan, L., Radioactivity monitoring of the Irish marine environment, 1987. Nuclear Energy Board, Dublin, Ireland, 1990, 28 pp.
5. Crowley, M., Mitchell, P.I., O'Grady, J., Vives, J., Sanchez-Cabeza, J.A., Vidal-Quadras, A. and Ryan, T.P., Radiocaesium and plutonium concentrations in *Mytilus edulis* (L.) and potential dose implications for Irish Critical Groups. *Ocean and Shoreline Management*, 1990, **13**, 149-61.
6. Mitchell, P.I., Sanchez-Cabeza, J. Ryan, T.P., McGarry, A.T. and Vidal-Quadras, A., Preliminary estimates of cumulative caesium and plutonium deposition in the Irish terrestrial environment. *J. Radioanal. Nucl. Chem. (Articles)*, 1990, **138**, No. 2, 241-56.
7. Mitchell, P.I., Vives Batlle, J., O'Grady, J., Sanchez-Cabeza, J.A. and Vidal-Quadras, À., Critical group doses arising from the consumption of fish and shellfish from the western Irish Sea. In *New Developments in Fundamental and Applied Radiobiology: Proc. 23rd Annual Meeting of the European Society for Radiation Biology*, Dublin, 23-26 September 1990 (Eds. C.B. Seymour and Carmel Mothersill), Publ. Taylor and Francis, 1991, 380-91.
8. Lovett, M.B. and Nelson, D.M., Determination of some oxidation states of plutonium in sea water and associated particulate matter. In *Techniques for Identifying Transuranic Speciation in Aquatic Environments*, Panel Proc. Ser., Ispra, IAEA, Vienna (STI/PUB/613), 1981, pp. 27-35.
9. Nevissi, A. and Schell, W.R., Distribution of plutonium and americium in Bikini Atoll Lagoon. *Health Phys.*, 1975, **28**, 539-47.
10. Nevissi, A. and Schell, W.R., Efficiency of a large volume water sampler for some radionuclides in salt and freshwater. In *Proc. Fourth National Symposium on Radioecology*, Corvallis, Oregon, 12-14 May 1975, The Ecological Society of America, Spec. Publ. No. 1, (Ed. C.E. Cushing Jr.), Dowden, Hutchinson and Ross Inc., Stroudsburg, Pennsylvania, 1976, pp. 277-82.
11. Mitchell, P.I., Vives Batlle, J., Ryan, T.P., Schell, W.R., Sanchez-Cabeza, J.A. and Vidal-Quadras, A., Studies on the speciation of plutonium and americium in the western Irish Sea. *These Proceedings*.
12. Pentreath, R.J., Harvey, B.R.. and Lovett, M.B., Chemical speciation of transuranium nuclides discharged into the marine environment. In *Speciation of Fission and Activation Products in the Environment* (Eds. R.A. Bulman and J.R. Cooper), Elsevier, 1986, pp. 312-25.
13. Nelson, D.M., Larsen, R.P. and Penrose, W.R., Chemical speciation of plutonium in natural waters. In *Proc. Symposium on Environmental Research on Actinide Elements*, Hilton Head, South Carolina, November 7-11, 1983. US Department of Energy, August 1987, CONF-841142, pp. 27-48.
14. Hunt, G.J., Radioactivity in surface and coastal waters of the British Isles, 1989. Aquatic Environment Monitoring Report, No. 23, Ministry of Agriculture, Fisheries and Food Directorate of Fisheries Research, Lowestoft, 1990, 66 pp.
15. Garland, J.A., McKay, W.A., Burton, P.J. and Cambray, R.S., Studies of environmental radioactivity on the coasts of Northern Ireland. *Nucl. Energy*, 1990, **29**, No. 3, 205-23.

RADIOECOLOGY OF COBALT-60 UNDER TROPICAL ENVIRONMENTAL CONDITIONS

B. Patel and S. Patel
Health Physics Division
Bhabha Atomic Research Centre
Bombay - 400 085
India

ABSTRACT

Cobalt-60 and cesium-137 released into the near shore environment of a BWR (400 Mwe) nuclear power station at Tarapur, have been found to be readily sorbed by the sedimentary particles. Surveillance over the last two decades showed that beyond 2 km distance north and south from the discharge point concentrations were within background range despite the continued release of controlled low levels of radioactive wastes. Within 1 km of the effluent point of both the nuclides dropped exponentially with time, yielding effective half-lives of 1.5 and 2.0 y against their respective physical half-lives of 5.2 and 30 y. Under laboratory-simulated conditions the sorption-desorption of cesium could be explained in terms of ambient salinity and other environmental parameters. It was, however, not possible to demonstrate similar behaviour for cobalt-60 in laboratory studies. The possible mechanism governing sorption-desorption of both radionculides under field and laboratory conditions are discussed.

INTRODUCTION

Sediment is a major sink for many kinds of waste arising from human activities. It has, therefore, been given a unique status by marine ecologists in attempts to understand many complex marine processes. The sorption of various kinds of anthropogenic pollutants, including radionuclides, by sedimentary material is governed by (i) the nature and chemical form of the contaminant (ii) the crystal structure and chemical composition of the sedimentary material and (iii) the physico-chemical composition of the recipient ecosystem. Furthermore, the rates of sorption-desorption of radionculides is controlled by the type of reaction mechanism these nuclides might undergo at the sediment-water interface. The distribution patterns in oceanic sediments of various radionuclides originating from global atmospheric fall-out and in the effluents from nuclear facilities, are well documented (RIME, 1971; Duursma, 1972; Heft *et al.*, 1973; Robertson *et al.*, 1973; Fukai *et al.*, 1974; Nakamura and Nagaya, 1975, 1977; Jinks and Wrenn, 1976; Patel *et al.*, 1975, 1978, 1985; Ueda *et al.*, 1979; Stanners and Aston, 1982; Zucker *et al.*, 1984; Nicholson and Mackenzie, 1988). However, it is not known with certainty whether the radionuclides once sorbed onto sediment could be desorbed under changing environmental conditions, and if so to what extent. In the case of river sediments the desorption of radionuclides has been observed when these are transported to the marine environment. No attempt seems to have been made to understand the possible desorption of various radionuclides except for cesium-134, 137, ruthenium-106, cerium-144 and americium-241 (Duursma and Eisma, 1973; Patel et al., 1978; Stanners and Aston, 1982) from near shore sediments under environmental conditions. During the last three decades controlled low levels of radioactivity have been discharged into the coastal waters off Bombay. The distribution patterns and levels of radionuclides from fuel reprocessing and other facilities have been studied in the water, sediment and biota. The present paper, in continuation of our earlier work (Patel *et al.*, 1978) on sorption-desorption of cesium-137, ruthenium-106 and cerium-144 from the Bombay harbour sediment, discusses the distribution and sorption-desorption of cobalt-60 and cesium-137 present in the controlled low level aqueous radioactive wastes from a BWR (400 Mwe) power at Tarapur about 100 kms north of Trombay (Fig. 1). The major radionuclides present in the effluents are cesium-134, 137; iodine-131 and cobalt-60 (Patel et al. 1973, 1975, 1980, 1985).

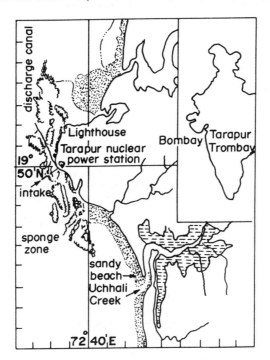

Fig. I. The intertidal zone off Nuclear power station at Tarapur.

MATERIALS AND METHODS

The topography and ecology of Tarapur waters have been discussed earlier (Patel and Ganguly, 1973; Patel et al., 1973, 1980, 1984, 1985; Bhat *et al.*, 1981).

The sediment samples (~1 kg) were scraped from coralline mud flats (~2 cm layer) at fortnightly intervals except during the monsoon period (July-September), when the land run offs were highest making coastal waters extremely turbid. Sediment samples were freeze dried, homogenised and about 20 g packed in a plastic vial and counted for gamma emitting radionuclides using a well-type NaI (T1) scintillation crystal (9 cm x 9 cm) coupled to a 1024 multichannel pulse height analyser. The counting time was adjusted between 20 and 40 min to obtain counts in excess of 95% confidence level. The mean annual budget of cesium and cobalt radionuclides was derived for the period 1977-1990.

To understand the possible mechanism of desorption/remobilisation of radionuclides from the sediment under environmental conditions, about 10 g sediment (8 samples) was mechanically shaken with sea water (250 ml) of various salinities in the range 9.5 - 38.0 ppt. The shaking method was preferred since it simulated very nearly the natural conditions prevalent in the coastal regions (Patel et al., 1978; Stanners and Aston, 1982). The water was changed daily and radioactivity desorbed was measured in decanted water, container washes and sediment after 1,2,4,6,8,10, days. The experiment was repeated at least five times and the mean activity of cesium and cobalt calculated. To measure the extent to which radionuclides were loosely bound, about 5 g sediment was treated with 100 ml aliquots of 1 N- ammonium acetate, pH 7.0; a mixture of 1 N-ammonium acetate and acetic acid (1:1), pH 5.4; and 5 % Na-EDTA pH 7.0 for 24 hours. Both the leaching agent and sediment were counted for gamma-activity.

RESULTS AND DISCUSSION

The distribution pattern of cesium and cobalt radioactivity was measured from the outfall to 5 km due north and south. Radioactivity, however, could be detected only in the sediment deposited within 2 km of the discharge point. The annual mean levels of cesium and cobalt activity in sediment deposited within 1 km radius of the outfall varied from 0.4 to 7.0 and 0.3 to 5.0 Bq g^{-1} dry respectively (Fig. 2).

Furthermore, it will be seen that on reaching peak sorption with the releases, concentrations of both the nuclides dropped exponentially with time under environmental conditions despite continued releases yielding effective sedimentary half-times of about 2.0 and 1.5 y (E) for cesium and cobalt against their respective physical half-lives (P) of 30 and 5.2 y. The ecological half life (e) could be calculated from the relationship $e = (E) \times (P)/(P-E)$. This works out to ca 3.2 y for cobalt-60 and ca 1.6 y for cesium. This closely followed the effective half life of both the nuclides in the releases (Fig. 2). Furthermore, the maximum radioactivity was sorbed on the fine fractions of the sediment (Table 1).

TABLE 1
Distribution of cesium-137 and cobalt-60 as a function of size fraction of sediment from Tarapur intertidal zone

Size fraction	radionuclide Bq g^{-1}	
μm	Cesium-137	Cobalt-60
Over 64	0.72	0.71
32-64	0.51	6.00
16-32	7.30	3.90
8-16	5.60	6.30
Under-8	17.40	7.70

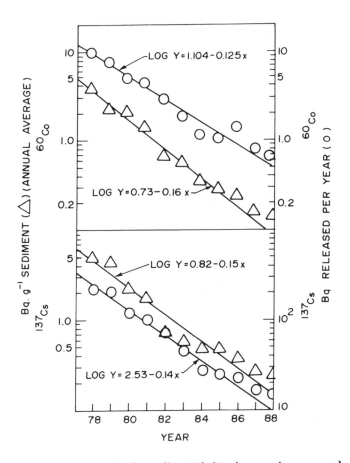

Fig.2, Levels of radioactivity (annual average) in the sediment and effluent.

The turnover patterns of both cobalt-60 and cesium-137 could be expressed by the following polynomial, which takes into account inputs of both the radionuclides and their behaviour within the two environmental compartments: water and sediment:

$$C_n = F\,(D_n + RfD_{n-1} + (Rf)^2 D_{n-2} + \cdots\cdots + (Rf)^{n-1} D_1)$$

where 'C_n' is the concentration of radionuclide at the end of nth year, 'F' is the fraction of radionuclide adsorbed by the sediment, 'D_n' is the quantity of radionuclide discharged during the nth year, 'R' is the fraction of radionuclide left over after one year of its decay and 'f' is the fraction retained in the sediment (Bq g^{-1} dry sediment) within the discharge zone.

Knowing the total inputs (D_n) and the observed inventory of the radionuclide over the period of monitoring- (C_n- cumulative), it was found that if on a yearly basis about 30% and 2%(f) of cobalt-60 and cesium-137 respectively, is retained during the following year, in addition to what may be aquired from current discharges, then the estimated environmental budget closely matched the actually observed levels. This, however, does not take into account the possible changes in levels due to the movement of sediment and influx of fresh sediment (~20%) during monsoon wash outs (annual rain fall ~250 cm). This suggests that not only the information on the chemical form of cobalt-60 in various abiotic and biotic matrices is desired (Patel et al., 1984), but also that it is most important to understand the possible mechanism of its sorption and desorption under environmental conditions.

The percentage of cesium-134, 137 desorbed after 10 days of shaking at various salinities is recorded in Fig 3. The maximum desorption, 40%, occurred at 38 ppt and the minimum, 25%, at the lowest salinity 9.5 ppt. The desorption of cobalt-60, however, was difficult to demonstrate with certainty, and did not appear to be as dependent upon ambient salinity as observed for cesium. Maximum desorption of about 15% was observed at the lowest salinity (9.5 ppt).

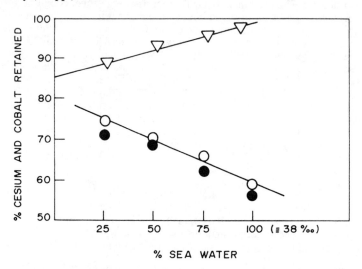

Fig. 3, Effect of salinity on the desorption of cesium -134 (●), cesium - 137 (○) and cobalt - 60 (▽) from the sediment after 10 d of shaking under simulated laboratory conditions.

About 13-25% of cesium was found to desorb on treatment with ammonium acetate/acetic acid and EDTA. On the other hand hardly any cobalt activity was released on treatment with the first of these mild leaching agents. However, it was complexed (15%) on treatment with EDTA. This shows that significant cesium activity was more or less loosely bound to the sedimentary material in an exchangeable form. The sorption of Cs-137 activity has also been suppressed at higher salinity due to increased concentration of the exchangeable ions (Na^+, K^+ and Mg^{++}). The mechanism of desorption of Cs-137 under both field and laboratory conditions can be most simply explained in terms of the law of mass action describing the competition for binding sites between Cs-137 and either its stable counterpart or other major ions (Na^+, K^+ and Mg^{++}) present in ambient seawater.

The desorption of Co-60, however, was more difficult to demonstrate with certainty, and did not appear to be as dependent upon ambient salinity as for cesium. This suggests that Co ions absorbed onto the sedimentary particles are not exchangeable with Co ions in solution, nor do they compete with Na^+, K^+ or Mg^{++} for binding sites. Thus the behaviour of Co-60 cannot simply be explained in terms of ion-exchange reaction. Under environmental conditions the Co ions may be complexed, perhaps as a hydroxycarbonate, rendering the system more or less irreversible. It is, therefore, difficult to predict the desorption behaviour of cobalt complexes from the sedimentary particles. (Duursma and Eisma, 1973; Robertson et al., 1973)

The present study suggests that under the environmental conditions sorption of cobalt-60 on the sediment was more or less labile (exchangeable?). This is rather unusual and contrary to its behaviour in the Urozoko Bay where it was shown that the sedimentary load of cobalt-60 remained practically the same over a period of 5 years even though the discharges had decreased by almost two orders of magnitude (Nakamura and Nagaya, 1975, 1977). This indicates that the heavy initial load of cobalt-60 was retained in the Urozoko Bay as a result of deposition and irreversible accumulation; its remobilisation (desorption) being almost negligible (very slow). This further suggests that there was hardly any influx of fresh sediment and/or churning and/or transport of sediment due to tidal currents. In contrast to this, in the surface deposited sediment from Tarapur discharge zone, cobalt-60 radioactivity decreased by a factor of seven with a drop in the release by same magnitude over the last two decades. Thus unlike simulated laboratory conditions, under environmental conditions cobalt-60 followed sorption-desorption patterns similar to cesium-137 in sediment along Trombay coast (Patel et al., 1978). Only a small fraction of cobalt-60 has been found bound with naturally active chelators- humic and fulvic acids - present in the sediment. However, it is tightly bound with synthetic chelating agents like EDTA, used for decontamination during nuclear operations. Furthermore, the mobility of cobalt-60 in sandy sediment has been found to be dependent upon the pH, redox conditions and the ion strength of the ambient water (Mahara and Kudo, 1981). The fact that the environment decontaminates itself as has been observed during present study offers a possibility to revise the present levels of maximum permissible discharges, to measure rates of siltation and influx of fresh sedimentary material through land run off and to trace the movement of bed material.

REFERENCES

Bhat, I.S., Patel, S., Patel, B. and Kamath, P.R. (1981). Cycling of radionuclide and impact of operation releases in the nearmecosystem off the west coast of India. In: Impact of radionuclide releases into the marine environment. IAEA, Vienna, STI/PUB/565, 431-438.

Duursma, E.K. (1972). Geochemical aspects and applications of radionuclides in the sea. In: Oceanography and Marine Biology Annual Review. Barnes, H., (ed.) Vol. 10, 137-223.

Duursma, E.K. and Eisma, D. (1973). Theoretical, experimental and field studies concerning reactions of isotopes with sediments and suspended particles of the sea. Part C: Applications to field studies. Netherland Journal of Sea Research. 6, 265-324.

Fukai, R., Statham, G. A., Ballestra, S. and Asari, K. (1974). Intercalibration of methods for radionuclide measurements on a marine sediment sample. In: Environmental surveillance around nuclear installations, IAEA, Vienna, STI/PUB/353, 313-335.

Heft, R.E., Phillips, W.A., Ralston, H.R. and Steele, W.A. (1973). Radionuclide transport studies in the Humboldt bay marine environment. In: Radioactive Contamination of the Marine Environment. IAEA, Vienna, STI/PUB/313, 595-614.

Jinks, S.M. and Wrenn, M.E. (1976). Radiocesium transport in the Hudson river estuary. Chapter 11, 207-227. In Miller, M.W. and Stannard, J.N. (Eds): Environmental Toxicity of Aquatic Radionuclides, Models and Mechanisms. Ann Arbor Sci. Pub. Inc. Michigan.

Mahara, Y. and Kudo, A. (1981). Fixation and mobilisation of cobalt-60 on sediments in coastal environments. Health Physics 41+. 645-655.

Nakamura, K. and Nagaya, Y. (1975). Dispersion and accumulation of radionuclide in sediment of Urazoka Bay-I. J. Oceanogra. Soc. Japan, 31, 145-153.

Nakamura, K. and Nagaya, Y. (1977). Dispersion and accumulation of radionuclide in sediment of Urazoko Bay-II. J.Oceangra. Soc Japan. 33. 1-5.

Nicholson, S. and Mackenzie, J. (1988). The remobilisation of radionuclides from marine sediments: implications for collective dose assessment. UK AEAS and R Directorate, SRD R453 Publi. 1-36.

Patel, B. and Ganguly, A.K. (1973). Occurrence of selenium-75 and tin-113 in oysters. Health Physics. 24. 599-562.

Patel, B., Mulay, C.D. and Ganguly, A.K. (1975). Radioecology of Bombay harbour - a tidal estuary. Estuarine and Coastal Marine Science. 3. 13-42.

Patel, B., Patel, S. and Balani, M.C. (1985). Can sponge fractionate isotopes? Proc. R. Soc. Lond. B.224. 23-41.

Patel, B., Patel, S. and Pawer, S. (1978). Desorption of radioactivity from near shore sediment. Estuarine Coastal Marine Science. 7, 49-58.

Patel, B., Patel, S. and Taylor, D.M. (1984). The chemical form of cobalt-60 in the marine sponge, Spirastrella cuspideifera. Mar. Biol. 80, 45-48.

Patel, B., Pawer, S.S., Balani, M.C. and Patel, S. (1980). Microalgae as sentinel of trace and heavy metals in the management of environment. In: Management of Environment. Patel, B. (ed). Wiley Eastern, New Delhi, 371-388.

Patel, B., Valanju, P.G., Mulay, C.D., Balani, M.C. and Patel, S. (1973). Radioecology of certain molluscs in Indian coastal waters. In: Radioactive contamination of the marine environment, IAEA, Vienna, STI/PUB/313, 307-330.

RIME (1971). Radioactivity in the marine environment. (Chairman, Seymour, H). US National Academy of Sciences publication, 272 pp.

Robertson, D.E., Silker, W.B., Langford, J.C. Petersen, M.R. and Perkins, R.W. (1973). Transport and depletion of radionuclides in the Columbia river. In: Radioactive Contamination of the Marine Environment. IAEA, Vienna, STI/PUB/313, 141-158.

Stanners, D.A. and Aston, S.R. (1982). Desorption of Ru-106, Cs-134, 137, Ce-144 and Am-241 from intertidal, sediment contaminated by nuclear fuel reprocessing effluents. Estuarine Coastal Marine. Science. 14, 687-691.

Ueda, T., Suzuki, Y. and Nakamura, R. (1979). Radioecology of cobalt-60 in Urazoko Bay. In: Proc. of third NEA seminar on Marine Radioecology, NEA- OCED Publi. Tokyo, 265-273.

Zucker, C.L., Olson, C.R., Larson, I.L. and Cutshall, N.H. (1984). Inventories and sorption-desorption trends of radiocesium and radiocobalt in James river estuary sediments. Environ. Geol Water Sci. 6, 171-181.

Radionuclides in Biological Systems

Thorium isotopes as tracers of particle dynamics and carbon export from the euphotic zone

J. K. Cochran[1], K. O. Buesseler[2], M. P. Bacon[2] and H. D. Livingston[2]

[1] Marine Sciences Research Center, SUNY, Stony Brook, NY 11794-5000
[2] Woods Hole Oceanographic Institution, Woods Hole, MA 02543

Several thorium isotopes (Th-234, Th-228, Th-230) are produced from decay of dissolved parents in sea water and associated strongly with particles. We have used measurements of these isotopes in solution and on suspended (>0.5 μm) and sinking particles, coupled with the known rates of their production and decay, to determine radionuclide (and by extension carbon) fluxes in the surface ocean. The short half-life of Th-234 (24 days) makes it a particularly appropriate tracer for variations in particle fluxes over time scales of weeks to months, and shipboard measurements of The-234 were made by gamma spectrometry during the JGOFS North Atlantic Bloom Experiment (April-May, 1989). The results show increasing association of Th-234 with particles and consequent removal from surface water as the spring phytoplankton bloom progressed. Depth profiles of Th-234 determined four times during the course of the bloom permit calculations of the net export of Th-234 at several depths in the water column. These Th-234 fluxes may be converted to carbon fluxes if the C/Th-234 ratio of material sinking out of the euphotic zone is known. Using the C/Th-234 ratio measured in suspended particles and sediment trap material gives a range of euphotic zone carbon export of 15-42 mmol C/m^2/d and 6-20 mmol C/m^2/d, respectively, over the course of the bloom. Thorium-234 and carbon export vary during the bloom, while primary production remains relatively constant. Thus, the f-ratio (new production/primary production) ranges from 0.16 to 0.43 or 0.07 to 0.20, depending on the choice of C/Th-234 ratio. Data from the other thorium isotopes will be discussed in the context of the Th-234 results.

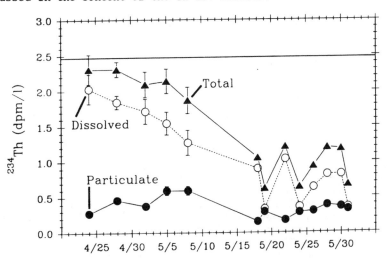

Temporal variation of Th-234 on particles and in solution in surface seawater during the spring bloom (1989) in the NE Atlantic (47°N, 20°W).

INTERANNUAL VARIATION IN TRANSURANIC FLUX AT THE VERTEX TIME-SERIES STATION IN THE NORTHEAST PACIFIC AND ITS RELATIONSHIP TO BIOLOGICAL ACTIVITY

S.W. FOWLER[1], L.F. SMALL[2], J. LA ROSA[1], J.-J. LOPEZ[1] & J.-L. TEYSSIE[1]
1. International Laboratory of Marine Radioactivity, IAEA,
19, Avenue des Castellans, MC98000 MONACO
2. College of Oceanography, Oregon State University,
Corvallis, OR 97331, USA

ABSTRACT

From October 1986 through May 1988, seven cruises took place at approximately three-month intervals at the VERTEX time-series station (33°N, 139°W) in the North Pacific. Data on plutonium ($^{239+240}$Pu) and, in some cases, ^{241}Am in unfiltered sea water, suspended particulate matter, sediment trap material, and zooplankton and their fecal pellets were collected over a 2000 m water column in order to examine temporal variations in particulate-associated transuranic flux and its relationship to biological activity in the upper water layers. Some seasonality in $^{239+240}$Pu flux was noted with the highest fluxes corresponding to the lowest mass fluxes below 400 m in summer through early winter (possibly due to scavenging by fine particles), and the lowest $^{239+240}$Pu fluxes corresponding to the highest mass fluxes above 150 m at the same time of year (due apparently to minimal scavenging in the euphotic zone). Little apparent difference was noted in the relationship between $^{239+240}$Pu flux and mass flux during late winter and spring at all depths, suggesting minimal scavenging at depth and generally small mass fluxes out of the euphotic zone. Salps which feed on tiny particles have relatively high concentrations of $^{239+240}$Pu in their fecal pellets while copepods which feed more selectively on larger particles have lower transuranic concentrations in their pellets. Nevertheless, copepods being more abundant and ubiquitous within the zooplankton than salps, contribute heavily, via feces production, to total transuranic flux out of the euphotic zone. Large particles in sediment traps (often fecal pellets and amorphous flocs) showed increasing $^{239+240}$Pu and ^{241}Am concentrations with depth up to at least 1200 m suggesting transuranic scavenging occurs as the particles sink through the water column. Biogenic particulates from zooplankton are closely linked to the downward flux of $^{239+240}$Pu at this site and there is some evidence for a direct coupling between Pu flux and carbon flux (carbon-specific "new production") through the base of the euphotic zone; however, no clear relationship was noted between the Pu flux and either the rate of primary production in the overlying waters or the fraction of primary production that was exported from the euphotic zone. Calculated $^{239+240}$Pu residence times in the upper 150 m water column for each deployment result in an average of approximately four years over the 18-month period.

INTRODUCTION

The transuranium nuclides plutonium and americium enter the marine environment primarily via atmospheric fallout from previous nuclear tests [1]; however, in addition, significant amounts have been discharged directly into the coastal zone from nuclear fuel processing activities [2] or accidentally from episodic events such as the bomber crash in Thule, Greenland [3] and the Chernobyl reactor fire [4,5]. Once introduced to the ocean, plutonium and americium, which display differing chemistries with respect to soluble and particulate phases, have proven to be useful tracers for gaining insight into particulate biogeochemical cycles [6-9].

Earlier work has identified the ocean sediments as the ultimate repository for $^{239+240}$Pu and ^{241}Am, and particle transport has been suggested as the prime vector for moving these surfaced-introduced transuranium nuclides to depth [10-12]. More recently, detailed studies of radionuclide flux have demonstrated that biological activity in the upper water column exerts a strong control on the vertical flux of $^{239+240}$Pu and ^{241}Am [5,13,14]. For example, biogenic particles such as phytoplankton cells, marine snow aggregates and zooplankton fecal pellets from a variety of oceanic regimes have been found to contain relatively high concentrations of Pu and Am [8,15]. These large particulate aggregates are known to sink at speeds of 10s to 100s of meters per day and are largely responsible for the downward transport of primary particulate organic carbon (POC) produced in the euphotic zone [See 15 for review]. During their descent, they can also scavenge reactive elements such as Pu and Am and thereby transport these radionuclides to depth. Experiments at the VERTEX I, II, III and IV sites in the northeastern Pacific have further shown that transuranic flux is highest and residence time shortest in the upper layers of eutrophic regimes compared to oligotrophic regions [13,14].

Although the data base is not extensive, there appears to be a close coupling between transuranic flux and resultant residence time and the export of carbon from the euphotic zone as a measure of "new production". Such observations have lead to the development of models which can be used to predict the flux of radionuclides and other elements in oceanic regimes of varying productivity [16,17].

Despite the evidence pointing to a strong coupling between primary productivity in surface waters and element or radionuclide flux to the deep sea, very few studies have attempted to quantify this relationship or discern how the relationship varies over time. Bacon et. al. [18] measured fluxes of $^{239+240}$Pu and other radionuclides at 3200 m using a three-year collection of sediment trap samples (<37 μm sieved fraction) from a deep ocean station near Bermuda. They found that variations in Pu flux were in phase with variations in total mass flux, which in turn were closely coupled to the annual cycle of primary production in the surface waters at that site. The close co-variance of Pu flux at great depth with the primary productivity cycle (based on a historical data set) strongly suggested rapid vertical transport via fast-sinking biogenic aggregates. Their study focused on transuranic flux in the deep sea and, thus, shed little light on the mechanisms involved in the initial processes of particle formation and transuranic scavenging which occur in the biologically active, upper layers of the water column. Furthermore, their conclusions were somewhat limited by lack of real-time primary production measurements with which to directly compare their radionuclide flux data.

In the fall of 1986, an 18-month times-series study was initiated in the northeast Pacific subtropical gyre as part of the NSF-sponsored Vertical Transport and Exchange Study (VERTEX). One of the primary objectives of this study was to measure seasonal trends in carbon export (as an estimate of new production) and its relationship to the primary productivity cycle [19]. During the experiment, we measured the transuranic nuclides $^{239+240}$Pu and ^{241}Am in sediment trap samples as well as in plankton and a variety of biogenic particles sampled in the overlying waters in order to better understand transuranic scavenging and removal processes in the upper 2000 meters. Our main aim in the study was to extend measurements of plutonium and americium flux on a temporal scale, and to determine the relationship between transuranic fluxes and seasonal changes in primary productivity and resultant POC export from the euphotic zone.

METHODS AND MATERIALS

Sediment Trap Deployments

The flux experiments took place during an approximately 18-month period between October 1986 and May 1988 at the VERTEX time-series station in the central northeast Pacific Ocean (33°N 139°W). On five occasions, replicate, cylindrical free-floating multi-traps [20] with a 0.0039 m^2 surface area were deployed at a series of depths between 40 and 2100 m for approximately three months each. The set of sediment traps (from two to six at each depth) used for transuranic analyses contained a NaCl density gradient solution plus 2% formalin, and were unscreened. Upon recovery, the contents of each replicate trap at a given depth were pooled and the samples stored in a small amount of this supernatant at 4°C until used.

Trap Sample Preparation

Bulk sediment trap samples were sequentially sieved through a series of nylon screens (1000, 600 and 150µm mesh size) to remove zooplankton "swimmers" which actively entered the traps. Zooplankton bodies were carefully removed from each screen and any non-living particulate matter adhering to the screen was replaced in the sample. After sieving through 150µm, the remaining sample was carefully examined with a dissecting microscope (x36 magnification) under dark-field in an attempt to remove with forceps and needle the less obvious forms of "swimmers" and their mucous products [21]. The final sample was filtered onto 1.2 µm Millipore filters, rinsed briefly with distilled water and oven-dried at 60°C to constant weight.

Sea Water and Suspended Particulates

During each cruise at the station, approximately 180 l of unfiltered sea water were collected from 1, 80 and 200 m depths. The sea water was precipitated onboard following the procedure described by Fowler et. al. [8]. To collect suspended particles, several hundred liters of surface sea water were pumped through a series of 0.45 µm Nuclepore polycarbonate filters (293 mm dia.) packed in a deck filtration unit. Filters were immediately frozen and later dried at 100°C for analyses.

Biogenic Material

To compare with large particles trapped at depth, samples of fresh biogenic material were also collected at the seasonal station. Live zooplankton were sampled at night by making 30-minute horizontal tows at 15 m and 80 m depths with a 1-meter opening-closing net (200 µm mesh). In order to isolate three different size-fractions of the mixed zooplankton, each collection was first sieved through large mesh (2000 μm) netting to remove any large organisms and detritus, and then sequentially through 1000, 500 and 200 μm mesh size. The fractions retained on those nets were carefully transferred to large volume fecal pellet collectors [8] and left for several hours to produce natural fecal pellets. Following the incubation, the pure fecal pellet sample and the zooplankton producing them were removed by filtration and frozen. In cases where sample size was small, either pellets from the different size-fractions or the organisms were combined to form a pooled sample. In the laboratory, all samples were dried at 60°C to constant weight.

Radioanalysis

The analytical procedures for separating, purifying and measuring $^{239+240}$Pu and ^{241}Am in similar samples have been reported in detail elsewhere [8,22]. Briefly, the separated and purified Pu and Am fractions were electro-deposited and counted by alpha-spectrometry using silicon surface barrier detectors. Unfortunately, a problem with ^{241}Am contamination of several of the samples rendered the ^{241}Am results unreliable and, therefore, they are not reported. Because of the low levels associated with many of the small samples, counting times of several days to weeks were often necessary. Counting errors associated with all measurements were propagated and are reported at the 1σ level.

RESULTS AND DISCUSSION

Transuranic Concentrations in Sea Water

Concentrations of transuranics in sea water from three depths at the time-series station are given in Table 1. During the 18-month study, concentrations of both Pu and Am were remarkably constant varying by only a factor of 5 in surface waters and approximately 2-3 at depth. The marked

TABLE 1

Plutonium-239+240 and 241Am concentrations (mBq m-3) in unfiltered sea water from the VERTEX time-series station in the northeast Pacific (33°N, 139°W).

Sea Cruise	Depth (m)					
	Surface		80 m		200 m	
	Pu	Am	Pu	Am	Pu	Am
1 (Oct./Nov.'86)	3.60±0.40	4.0±3.0	1.49±0.34	-	-	-
2 (Jan.'87)	1.17±0.21	1.0±0.2	1.53±0.23	0.72±0.15	9.20±0.70	2.24±0.34
3 (May '87)	1.20±0.20	0.9±0.2	1.10±0.20	1.2±0.3	4.60±0.60	1.2±0.3
4 (Jul.'87)	0.70±0.10	0.8±0.2	1.40±0.20	0.7±0.2	4.70±0.50	1.6±0.3
5 (Oct.'87)	0.70±0.20	0.9±0.2	0.80±0.20	0.7±0.2	3.30±0.40	1.0±0.2
6 (Feb.'88)	0.70±0.10	-	0.80±0.20	-	7.80±0.90	-
7 (May '88)	1.30±0.20	-	0.80±0.20	-	3.60±0.40	-

increase in Pu concentration noted at 200 m is an effect of the subsurface Pu maximum which is a characteristic feature ranging between 250 to 750 m at these latitudes in the north Pacific [8,11,23,24]. Americium also displays a subsurface concentration maximum but it is not as prominent as that of Pu [8, 24, 25]. The Am/Pu activity ratio was the highest (x = 1.0) in surface waters and lowest (x = 0.29) at 200 m. Seasonal variations in the ratio were never more than approximately a factor of 2 during the study.

Particulate-Associated Transuranics
The fractions of transuranics associated with 0.45 μm-filtered particles from surface waters were generally very low and varied little throughout the study, suggesting that most of the Pu and Am is present in soluble form (Table 2). Such low percentages of particulate-associated transuranics at this station appear to be typical of north Pacific central gyre waters [24]. Americium/Pu activity ratios ranging from 0.6 to 3.6 were considerably higher than the ratio normally found in fallout (≈0.3) and indicate that Am is preferentially taken up by suspended particles relative to plutonium. This is in keeping with the general hypothesis that Am is more particle-reactive than plutonium [7-9].

TABLE 2
Plutonium-239+240 and 241Am concentrations (mBq m-3) in filtered particles (0.45 μm filters) and percentages of transuranics (in parentheses) associated with the particulate phase in surface sea water at the VERTEX time-series station in the northeast Pacific (33°N, 139°W).

Sea Cruise	Pu	Am	Am/Pu Activity Ratio
1 (Oct./Nov.'86)	0.025±0.01 (0.7%)	0.09±0.02 (2.3%)	3.6
2 (Jan.'87)	0.04±0.01 (3.4%)	0.06±0.04 (6%)	1.5
3 (May '87)	0.05±0.03 (4.2%)	-	-
4 (Jul.'87)	0.028±0.015 (4.0%)	0.019±0.008 (2.4%)	0.67
5 (Oct.'87)	0.07±0.02 (10%)	0.05±0.01 (5.6%)	0.71
6 (Feb.'88)	0.037±0.007 (5.3%)	0.023±0.005	0.62
7 (May '88)	0.019±0.006 (1.5%)	-	-

Concentrations and vertical flux of $^{239+240}$Pu and ^{241}Am associated with particles during the 18-month study are given in Table 3. The data indicate some seasonality in $^{239+240}$Pu flux with the highest radionuclide fluxes corresponding to the lowest particulate mass fluxes below 400 m during the summer through the winter period (Fig. 1). This relationship is particularly evident if the data from the time of year when the water column is stratified are plotted (Fig. 2a). In this case, the lowest Pu fluxes correspond to the highest mass fluxes out of the euphotic zone (upper 150 m) at the same time of year. On the other hand, during late winter and spring when the upper water column is well-mixed, there is no apparent relationship between $^{239+240}$Pu flux and mass flux (Fig. 2b). It is also evident that despite decreasing or nearly constant mass flux with depth, there is generally a concomitant slight increase in transuranic flux with depth at least down to 1200 m (Table 3, Fig. 1). This is due to an

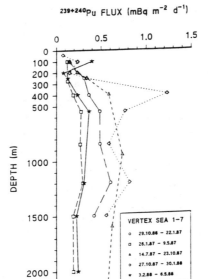

Figure 1. Vertical flux of particulate $^{239+240}$Pu at the VERTEX time-series station (33°N 139°W) during the period October 1986 - May 1988.

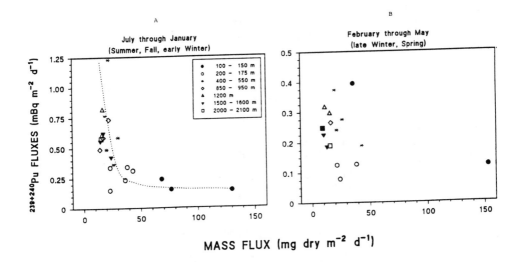

Figure 2. Relationship between $^{239+240}$Pu fluxes and total mass flux at the VERTEX time-series station (33°N 139°W) for the period a) summer through early winter and, b) late winter through spring.

overall increase in transuranic concentration in particles which most likely have taken up plutonium as they sank through the layers that are

TABLE 3

Transuranic concentrations, fluxes and activity ratios at the VERTEX time-series station, October 1986 - May 1988.

Depth (m)	Mass Flux mg dry m^{-2}d^{-1}	$^{239+240}$Pu mBq g^{-1} dry±1sd	Pu Flux mBq m^{-2}d^{-1}	^{241}Am mBq g^{-1}dry±1sd	Am Flux mBq m^{-2}d^{-1}	Am/Pu Ratio
		SEA 1-2 (29 Oct. 86 - 22 Jan. 87)				
100	130.0	1.1 ± 0.3	0.143	0.5 ± 0.2	0.065	0.455
250	43.1	7.2 ± 1.2	0.310	1.1 ± 0.6	0.047	0.153
400	26.2	13.9 ± 1.9	0.364	4.6 ± 1.5	0.121	0.331
550	20.0	24.3 ± 2.4	0.486	11.1 ± 1.6	0.222	0.457
850	14.2	34.3 ± 4.6	0.487	13.6 ± 2.6	0.193	0.397
1200	16.1	37.5 ± 4.4	0.604	18.9 ± 2.7	0.304	0.504
1500	24.0	17.6 ± 2.1	0.422	6.9 ± 1.0	0.166	0.392
		SEA 2-3 (26 Jan. 87 - 9 May 87)				
100	153.0	0.8 ± 0.2	0.122	0.5 ± 0.1	0.077	0.625
275	38.5	3.3 ± 0.7	0.127	1.8 ± 0.4	0.069	0.545
400	42.9	4.4 ± 0.4	0.189	2.4 ± 0.4	0.103	0.545
550	26.3	10.5 ± 1.0	0.276	7.0 ± 1.0	0.184	0.667
850	16.3	16.3 ± 1.1	0.266	8.1 ± 0.9	0.132	0.497
1200	15.4	19.2 ± 1.6	0.296	10.7 ± 1.6	0.165	0.557
1500	12.9	14.5 ± 2.4	0.187	8.0 ± 1.0	0.103	0.552
2000	15.6	12.2 ± 1.9	0.190	9.5 ± 1.8	0.148	0.779
		SEA 3-4 (No data)				
		SEA 4-5 (14 July 87 - 23 Oct. 87)				
100	76.3	2.0 ± 0.2	0.153			
200	36.4	6.4 ± 1.0	0.233			
250	38.3	8.9 ± 1.4	0.341			
400	30.3	19.3 ± 2.8	0.585	--	--	--
950	22.0	33.3 ± 2.3	0.733			
1600	17.2	36.1 ± 2.7	0.621			
2100	15.4	37.4 ± 2.7	0.576			
		SEA 5-6 (27 Oct. 87 - 30 Jan. 88)				
40	106.0	0.7 ± 0.2	0.074			
100	67.9	3.5 ± 0.6	0.238			
200	22.9	6.5 ± 1.5	0.149			
250	23.1	14.6 ± 2.3	0.337			
400	21.8	56.4 ± 6.1	1.230	--	--	--
550	18.5	41.3 ± 4.4	0.764			
850	17.1	34.2 ± 4.4	0.585			
1200	16.6	49.0 ± 5.5	0.813			
1500	14.7	37.9 ± 4.4	0.557			
2000	6.9	- lost -	-			
		SEA 6-7 (3 Feb. 88 - 6 May 88)				
40	37.1	- lost -	-			
100	35.9	11.0 ± 1.8	0.395			
200	24.1	3.3 ± 1.4	0.079			
250	21.5	5.8 ± 1.2	0.125			
400	21.3	11.3 ± 1.7	0.241	--	--	--
550	19.5	19.2 ± 2.5	0.374			
1200	11.2	28.2 ± 4.0	0.316			
1500	10.1	22.5 ± 2.3	0.226			
2000	9.2	27.1 ± 3.7	0.248			

highly enriched in plutonium, i.e., the subsurface maximum. There is also some evidence which suggests that large particles are releasing radionuclide in waters deeper than 1200 m. This pattern, which occurred throughout the study, is identical to that noted at other VERTEX stations in the north Pacific [8,14,24]. Only two ^{241}Am flux profiles were obtained from the study (Table 3). While the highest ^{241}Am concentrations and fluxes occurred at depth, there was no clear trend over the depth range examined. Furthermore, there was no consistent increase in Am/Pu ratio with depth as has been observed at other VERTEX stations, although ratio values generally greater than 0.3-0.4 indicate a similar enrichment of ^{241}Am relative to $^{239+240}$Pu in these particles.

It is noteworthy that plutonium Kd values for large particles in the subtropical gyre, computed from the 80 and 100 m data in Tables 1 and 3, respectively, are all on the order of 10^6. These values are at least one order of magnitude higher than those which have been measured in similar particles collected in the California Current [8] and two southern California basins [26]. Such large differences in Kd may be a result of different distributions of the reduced and oxidized states of plutonium in the different water masses.

The particles in the traps are primarily biogenic and, in an attempt to determine what factors might be controlling transuranic flux, various size-fractions of zooplankton and their particulate products were also analyzed for Pu and Am (Table 4). Plutonium concentrations in fecal pellets from mixed zooplankton (primarily copepods) sampled from the euphotic zone averaged 1.2 mBq g^{-1} dry over all VERTEX SEA cruises while the 100-150 m sediment trap contents (minus one very high value) averaged 1.85 mBq g^{-1}. Furthermore, $^{239+240}$Pu concentrations in fecal pellets were higher (1.5 mBq g^{-1}) during the summer-early winter period (when Pu flux through the euphotic zone was highest) than during the later winter-spring months (0.79 mBq g^{-1}) when Pu fluxes were relatively low (see Fig.2). No other biogenic fraction came close to the sediment trap value. For example, zooplankton of all sizes combined, averaged only 0.046 mBq g^{-1}. Moreover, no obvious differences in concentration among the different-sized groups of zooplankton were evident; most likely, any small differences in transuranic concentration between zooplankton of different sizes would be masked by the variability in the data (Table 4).

The average value for two marine snow samples (Table 4 and unpublished results from VERTEX IV station) is 0.53 mBq g^{-1}, only about one-third of the concentration in the sediment trap samples. Similarly, mixed microplankton (64-200 µm) and pure phytoplankton (Coscinodiscus sp.) from other VERTEX stations also contained relatively low levels of $^{239+240}$Pu, 0.42 and 0.35 mBq g^{-1}, respectively [24]. On the other hand, salp fecal pellets at the time-series station were highly enriched in Pu (8.9 mBq g^{-1}) further suggesting that pellets, indeed, are an important contributor to the transuranic flux. Thus, even though Pu fluxes through the top 150 m are not particularly high (Table 3), the particulate Pu fraction leaving the euphotic zone appears to be mainly in the form of fecal pellets.

Although the data are sparse, salp feces contains as much as ten times more Pu and Am than copepod pellets (Table 4). Salps are known to be indiscriminate filterers of minute particles (presumably those highly enriched in transuranics) whereas copepods feed on larger particles including animal matter. It seems clear that at times and in areas where salp swarming occurs, particle-associated transuranics will be effectively packaged into fast-sinking fecal pellets and removed from the euphotic zone. Nevertheless, the fact that copepods are ubiquitous and normally

TABLE 4

$^{239+240}$Pu and ^{241}Am (values in parentheses) concentrations (mBq g^{-1} dry wt.) in zooplankton, fecal pellets and marine snow from the VERTEX time-series station.

Cruise	Collection Depth (m)	Zooplankton (mBq g^{-1} dry wt.)				Animal ID	Fecal Pellets (mBq g^{-1} dry wt.)			
		200-500 μm	500-1000 μm	1000-2000 μm	All sizes combined		200-500 μm	500-1000 μm	1000-2000 μm	All sizes combined
Sea-1 Oct./Nov. 86	15-25m	0.04±0.02 (0.03±0.02)	0.04±0.03 (0.024±0.014)	0.09±0.03 (0.032±0.010)	-	Copes	-	-	-	0.5±0.2
	80-100m	-	-	-	0.05±0.03	Copes	-	-	-	1.21±0.96 (1.7±0.8)
Sea-2 Jan. 87	20-50	-	-	-	0.049±0.008	Copes	-	-	-	0.9±0.5
	50-60	-	-	-	0.06±0.01	Copes	-	-	-	-
Sea-3 May 87	20-50	-	-	-	0.062±0.0094 (0.012±0.006)	Copes	-	-	-	0.85±0.29 (0.27±0.15)
Sea-4 July 87	15-25m	0.030±0.008	0.029±0.007 (0.015±0.007)	0.034±0.007	-	Copes	-	2.77±0.76	1.03±0.29	-
	80-100m	0.032±0.010	0.025±0.010 (0.023±0.009)	0.030±0.009 (0.012±0.007)	-	Copes	1.11±0.78	1.69±0.90 (1.35±0.64)	1.40±0.47 (0.8±0.3)	8.9±1.1 (5.0±0.5)
						Salps	-	-	-	-
Sea-5 Oct. 87	20-50	-	-	0.016±0.006	-	Copes	-	1.40±0.80	-	-
	80-100	-	0.035±0.008	0.034±0.008	-	Copes	-	-	-	-
Sea-6 Feb. 88	20-40m	*0.11±0.01	-	0.03±0.01	-	Copes	*0.80±0.40	-	1.60±0.8	-
	80-100m	*0.11±0.01 (0.016±0.006)	-	0.039±0.009 (0.013±0.006)	-	Copes	0.70±0.40	-	0.40±0.40	-
						Marine snow	-	-	-	0.37±0.06 (0.16±0.06)
Sea-7 May 88	20-40m	*0.040±0.010	-	0.032±0.009	-	Copes	*0.40±0.20	-	-	-
	80-100m	*0.036±0.008	-	0.048±0.007	-	Copes	-	-	-	-

* 200-500 μm and 500-1000 μm fractions were combined.

more abundant than salps indicates that copepods contribute heavily, via fecal pellet production, to Pu flux out of the euphotic zone.

If much of the Pu in the surface layers is indeed leaving the euphotic zone in the form of fecal pellets and other biogenic particles, then a close relationship between Pu flux and new production in the form of carbon export might be expected. Estimates of primary production and new production for the periods in which Pu flux was measured are shown in Table 5. No relation is apparent between Pu flux and the percentage of primary production which is exported from the euphotic zone. Furthermore, no solid relation exists between Pu flux and average primary production. On the other hand, the two highest Pu fluxes do correspond to the periods when new production in terms of carbon flux was also highest. This observation is in agreement with the general oceanic radionuclide flux model of Fisher et. al. [17]; however, far more data are needed to establish whether there is a real relationship between these two parameters. Taken together, our results suggest that Pu flux is not directly tied to seasonal changes in primary production but rather, is more closely related to the processes by which zooplankton transform the particle field and resultant fluxes, and may even be a function of zooplankton type and biomass.

Estimates of $^{239+240}$Pu residence times in the euphotic zone during the 18-month period are also shown in Table 5. They range between 2.6 and 6.8 years with an average of 4.1 ± 1.8 years. These residence times are slightly longer than the single measurement of 2.4 years reported for the upper 80 m at the VERTEX I station in the California Current [8], and most likely, reflect the relatively lower carbon export from the euphotic zone which occurs in oligotrophic waters of the North Pacific central gyre [19,27].

TABLE 5
Primary production, new production (in terms of carbon flux), $^{239+240}$Pu flux out of the euphotic zone and Pu residence times in the upper 150 m at the VERTEX time-series station.

Sea Cruise (Deployment Period)	Primary Production* PP (mgC m^{-2}d^{-1})	Trap Flux as New Production^c NP (mgC m^{-2}d^{-1})	% NP of Total PP	$^{239+240}$Pu Flux** (mBq m^{-2}d^{-1})	Pu Residence Time^{JJ} (yr)
1-2 (Oct.86-Jan.87)	245	42	17%	0.200	5.2
2-3 (Feb.87-May 87)	215	32	15%	0.124	6.8
3-4 (May 87-July 87)	380	-	-	-	-
4-5 (July 87-Oct.87)	468	30	6%	0.193	3.0
5-6 (Oct.87-Feb.88)	325	29	9%	0.194	3.0
6-7 (Feb.88-May 88)	343	53	16%	0.238	2.6

* Estimated from Fig. 2b in Ref. [19].
^{c} Data from Ref. [19].
** Estimated for 150 m depth from Table 3.
^{JJ} Pu inventories (mBq m^{-2}) for the upper 150 m were computed from data in Table 1 using an interpolated value for 150 m depth. For Pu at 200 m during Sea Cruise 1, an average value (5.5 mBq m^{-3}) computed from all other cruises was used.

ACKNOWLEDGEMENTS

The International Laboratory of Marine Radioactivity (ILMR) operates under an agreement between the International Atomic Energy Agency and the Government of the Principality of Monaco. This work was part of the VERTEX programme and was supported by NSF grant OCE 86-00460 to LFS. We thank S. Moore and our colleagues in VERTEX who helped with sampling at sea and V.E. Noshkin for critical comments on the manuscript.

REFERENCES

1. Perkins, R.W. and Thomas, C.W., Worldwide fallout. In Transuranic Elements in the Environment, ed. W.C. Hanson, U.S. Department of Energy TIC-22800, Washington, D.C., 1980, pp. 53-85.

2. Hetherington, J.A., The behaviour of plutonium nuclides in the Irish Sea. In Toxicity of Aquatic Radionuclides: Models, and Mechanisms, ed. M.W. Miller and J.N. Stannard, Ann Arbor Science, Ann Arbor, Michigan, 1976, pp. 81-106.

3. Aarkrog, A., Dahlgaard, H., Nilsson, K. and Holm, E., Further studies of plutonium and americium at Thule, Greenland. Health Phys., 1984, 46(1), 29-44.

4. Holm, E., Aarkrog, A., Ballestra, S. and Lopez, J.-J., Fallout deposition of actinides in Monaco and Denmark following the Chernobyl accident. In Proceedings 4th International Symposium on Radioecology, Centre d'Etudes Nucléaires de Cadarache, France, 1988, pp. A-22.

5. Fowler, S.W., Ballestra, S., La Rosa, J., Holm, E. and Lopez, J.-J., Flux of transuranium nuclides in the northwestern Mediterranean following the Chernobyl accident. Rapp. Comm. int. Mer Médit., 1990, 32(1), 317.

6. Fukai, R., Holm, E. and Ballestra, S., A note on vertical distribution of plutonium and americium in the Mediterranean Sea. Oceanol. Acta, 1979, 2, 129-132.

7. Koide, M., Williams, P.W. and Goldberg, E.D., Am-241/Pu-239+240 ratios in the marine environment. Mar. Environ. Res., 1981, 5, 241-246.

8. Fowler, S.W., Ballestra, S., La Rosa, J. and Fukai, R., Vertical transport of particulate-associated plutonium and americium in the upper water column of the northeast Pacific. Deep-Sea Res., 1983, 30, 1221-1233.

9. Livingston, H.D., Mann, D.R., Casso, S.A., Schneider, D.L., Suprenant, L.D. and Bowen, V.T., Particle and solution phase depth distributions of transuranics and ^{55}Fe in the north Pacific. J. Environ. Radioactivity, 1987, 5, 1-24.

10. Noshkin, V.E. and Bowen, V.T., Concentrations and distributions of long-lived fallout radionuclides in open ocean sediments. In Radioactive Contamination of the Marine Environment, International Atomic Energy Agency, Vienna, 1973, pp. 671-686.

11. Bowen, V.T., Noshkin, V.E., Livingston, H.D. and Volchok, H.L., Fallout radionuclides in the Pacific Ocean: Vertical and horizontal distributions, largely from GEOSECS stations. Earth Planet. Sci. Lett., 1980, 49, 411-434

12. Livingston, H.D. and Anderson, R.F., Large particle transport of plutonium and other fallout radionuclides to the deep ocean. Nature, 1983, 303, 228-231.

13. Fowler, S.W., VERTEX: Factors affecting the vertical flux of $^{239+240}$Pu and ^{241}Am in the upper layers of the northeast Pacific Ocean. In Colloque Internationale d'Océanologie. Ecosystèmes de Marges Continentales, C.I.E.S.M., Monaco, 1987, p. 82.

14. Fowler, S.W., Ballestra, S., La Rosa, J. and Gastaud, J., Biological control of transuranic flux through the upper water column of the northeast Pacific Ocean. Rapp. Comm. int. Mer Médit., 1985, 29, 189-193.

15. Fowler, S.W. and Knauer, G.A., Role of large particles in the transport of elements and organic compounds through the oceanic water column. Prog. Oceanogr., 1986, 16, 147-194.

16. Fisher, N.S. and Fowler, S.W., The role of biogenic debris in the vertical transport of transuranic wastes in the sea. In Oceanic Processes in Marine Pollution, Vol. 2, Physico-chemical Processes and Wastes in the Ocean, eds. T.P. O'Connor, W.V. Burt and I.W. Duedall, R.E. Krieger Publ. Co., Malabar, Florida, 1987, pp. 197-207.

17. Fisher, N.S., Cochran, J.K., Krishnaswami, S. and Livingston, H.D., Predicting the oceanic flux of radionuclides on sinking biogenic debris. Nature, 1988, 335, 622-625.

18. Bacon, M.P., Huh, C.-A., Fleer, A.P. and Deuser, W.G., Seasonality in the flux of natural radionuclides and plutonium in the deep Sargasso Sea. Deep-Sea Res., 1985, 32, 273-286.

19. Knauer, G.A., Redalje, D.G., Harrison, W.G. and Karl, D.M., New production at the VERTEX time-series site. Deep-Sea Res., 1990, 37, 1121-1134.

20. Knauer, G.A., Martin, J.H. and Bruland, K.W., Fluxes of particulate carbon, nitrogen and phosphorus in the upper water column of the northeast Pacific. Deep-Sea Res., 1979, 26, 97-108.

21. Michaels, A.F., Silver, M.W., Gowing, M.M. and Knauer, G.A., Cryptic zooplankton "swimmers" in upper ocean sediment traps. Deep-Sea Res., 1990, 37, 1285-1296.

22. Ballestra, S. and Fukai, R., An improved radiochemical procedure for low-level measurements of americium in environmental matrices. Talanta, 1983, 30, 45-48.

23. Noshkin, V.E., Wong, K.M., Jokela, T.A., Eagle, R.J. and Brunk, J.L., Radionuclides in the marine environment near the Farallon Islands. University of California Lawrence Livermore Laboratory Report UCRL-52381, 1978, pp. 1-17.

24. Fowler, S.W., Small, L.F., La Rosa, J. and Ballestra, S., Flux of transuranics at VERTEX stations in the northeast Pacific Ocean. MS in preparation.

25. Bowen, V.T., Element specific redistribution processes in the marine water column. In _Processes determining the input, behavior and fate of radionuclides and trace elements in continental shelf environments,_ U.S. Department of Energy CONF 790382, Washington, D.C., 1980, pp. 29-38.

26. Wong, K.M., Jokela, T.A., Eagle, R.J., Brunk, J.L. and Noshkin, V.E., Radionuclide concentrations, fluxes and residence times at Santa Monica and San Pedro basins. Prog. Oceanogr., (in press).

27. Knauer, G.A., Martin, J.H. and Karl, D.M., The flux of particulate organic matter out of the euphotic zone. In _Global Ocean Flux Study,_ Nat. Acad. Press, Washington, D.C., 1984, pp. 136-150.

ENHANCED DEPOSITION OF CHERNOBYL RADIOCESIUM BY PLANKTON IN THE NORWEGIAN SEA: EVIDENCE FROM SEDIMENT TRAP DEPLOYMENTS

MARION BAUMANN and GEROLD WEFER
Fachbereich Geowissenschaften, Universität Bremen
D-2800 Bremen 33, F.R.G.

ABSTRACT

Chernobyl contaminated sediment trap material from the Norwegian Sea revealed a strong influence of the food web interactions in controlling the transport of radiocesium to deeper waters. The production of copepod fecal pellets from suspended matter increased the ^{137}Cs-concentration about ten times in the sediment trap material. The vertical flux of ^{137}Cs via sinking fecal pellets to deeper water depths was at least ten times higher than during periods of biological recycling in the upper parts of the water column. High ^{137}Cs-specific activities in the sinking particles from sediment trap deployments were propably due to the particulate form of Chernobyl-^{137}Cs and to high radiocesium concentrations in the sea-water shortly after deposition.

INTRODUCTION

The reactor accident at Chernobyl released $3,8 \times 10^{16}$ Bq ^{137}Cs from late April to May 1986 [1]. Contaminated air reached the Norwegian coast on the 28th and 29th of April and on the 5th of May 1986 and within several days Chernobyl isotopes were deposited to the sea-surface by washout. In the Norwegian Coastal Current 130 Bq m^{-3} ^{137}Cs were detected shortly after the accident. For over a year after deposition ^{137}Cs concentrations were 1,5 to 3 times higher than to those measured before the Chernobyl accident [2].

Marine organisms concentrate radionuclides from seawater by adsorption, adsorption into tissues, and by the ingestion of contaminated organic matter and detritus [3]. Zooplankton fecal pellets have been shown to be an especially efficient means of transporting Chernobyl radionuclides to greater water depths [4]. In this study we will focus on the ^{137}Cs-accumulation and sedimentation processes in a pelagic realm that also has been contaminated by Chernobyl-derived radiocesium. The seasonal radiocesium flux detected in the sediment trap materials (10% of the fallout) was temporally and quantitatively related to the radiocesium inventories in the surface sediments [5]. Radionuclide investigations combined with detailed biological investigations of three short-term and two long-term sediment trap deployments provided the unique opportunity to elucidate the role of plankton in the sedimentation process of radiocesium.

MATERIALS AND METHODS

Locations and positions of the sediment traps and deployments are shown in Table 1. The detailed sediment trap parameters have been published elsewhere [5]. The samples were analysed by non-destructive gamma ray measurements with a 64 cm³ Ge(Li)-detector. Cesium activities of VP2 and NB2 sediment trap material were analysed in the original trap cups. Efficiency calibrations for the trap cups of VP2 were performed with a standard solution. A 1/8 split of each sample of the VP1 traps was filtered onto a 0.45 μm filter prior to analysis. All reported activities represent decay corrected values to May 1st, 1986 for purposes of comparision.

The observed $^{134}Cs/^{137}Cs$ ratios of about 0.4 to 0.6 are typical for Chernobyl fallout [6]. Cesium ratios from the discharges of the reprocessing plants at Sellafield and La Hague are lower because the primary ratio at the time of discharge is smaller and ^{134}Cs (t/2 = 2.04 yrs) decays much faster compared to ^{137}Cs (t/2 = 30.2 yrs). Due to the time of the study and the fact that only small amounts of sediment were recovered, no other gamma-emmitting radioisotopes could be detected in the traps. Since all detected ^{137}Cs originated from the Chernobyl accident, only ^{137}Cs activity is reported here.

TABLE 1

Location of sediment trap deployments and depths of the sediment traps at the Vøring Plateau (VP), and Norwegian Basin (NB).

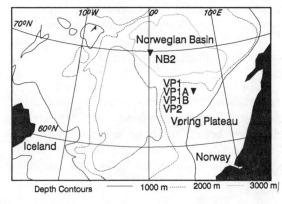

Trap	Water depth	Trap depth
VP1	1244 m	250 m
VP1.2		1000 m
VP 1A		140 m
VP 1B		140 m
		350 m
		1000 m
VP2	1429 m	700 m
NB2 1	3295 m	750 m

RESULTS

The results are shown in Figures 1-3. In Figures 2 and 3, time-scales are the same for better comparision. On the Vøring Plateau total fluxes of 103-581 mg m^{-2}d^{-1} (VP1A) and 375-1095 mg m^{-2}d^{-1} (VP1B 1) were measured to 140 m water depth six weeks after the accident at Chernobyl. The ^{137}Cs-fluxes were lower than 1.8 Bq m^{-2}d^{-1}, and the specific ^{137}Cs-activities were less than 3 Bq g^{-1} (Fig. 1).

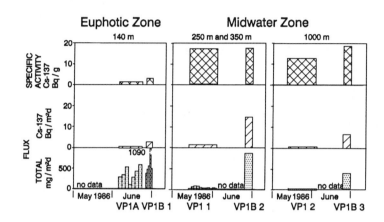

FIGURE 1. Analytical and gamma-spectroscopic results of the sediment trap material of the deployments VP1, VP1A, and VP1B.

In the midwater zone the total flux was 28.1-89.3 mg m^{-2}d^{-1} at 250m (VP1 1) and 27.6 mg m^{-2}d^{-1} at 1000 m (VP1 2) during the last two weeks of May. The ^{137}Cs-fluxes were lower than 0.9 Bq m^{-2}d^{-1} to both traps. But specific ^{137}Cs-activities in the sediment trap material were 17.0 Bq g^{-1} at 250 m and 12.8 Bq g^{-1} at 1000 m. In the last week of June (VP1B 2 and 3) total-fluxes of 868 mg m^{-2}d^{-1} at 300 m and 369 mg m^{-2}d^{-1} at 1000 m were measured. ^{137}Cs-fluxes were 14.6 and 6.8 Bq m^{-2}d^{-1}, respectively. Specific ^{137}Cs-activities were 16.8 Bq g^{-1} at 300 m and 18.4 Bq g^{-1} at 1000 m.

The long term sediment trap on the Vøring Plateau VP2 (Fig. 2) showed declining total-fluxes at 700 m water depth from 287 mg m^{-2}d^{-1} in mid July to 44.5 mg m^{-2}d^{-1} in January 1987. The total flux correlated well with lithogenic particle-flux ($R^2 = 0.90$). The ^{137}Cs-flux correlated best with particulate organic matter flux ($R^2 = 0.71$) and lithogenic flux ($R^2 = 0.63$). The ^{137}Cs-flux declined from 2.8 Bq m^{2}d^1 in July to less than 0.1 Bq m^{2}d^1 in winter. ^{137}Cs-specific activities were highest in the first weeks of July (16.7 Bq g^{-1}) and decreased to 1.9 Bq g^{-1} in January 1986.

The long-term sediment trap deployment in the Norwegian Basin (Fig. 3) showed ^{137}Cs-values ten times lower than those in the Vøring Plateau sites. The total-flux at 750 m increased from 22.6 at the outset to 116.9 mg m^{-2}d^{-1} in late August and dropped to 4.7 mg m^{-2}d^{-1} in May 1987. A slight increase in flux was detected during June 1987. ^{137}Cs-specific activites reached 12.1 Bq g^{-1} in July and decreased to values of approximately 1.5 Bq g^{-1} from October 1986 to June 1987.

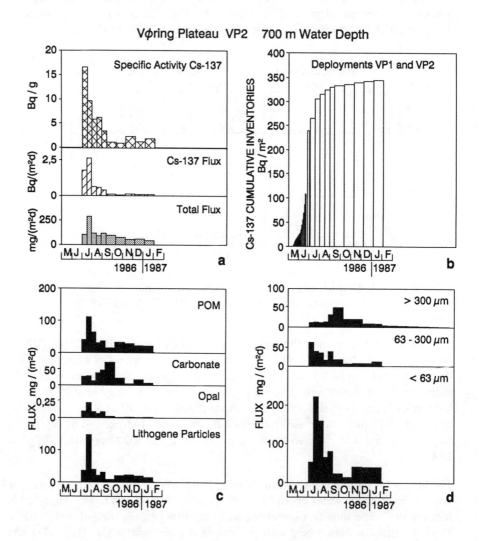

FIGURE 2. Gamma-spectroscopic results (a), cumulative ^{137}Cs-inventories of VP1 and VP2 traps (b), and analytical results (c,d) of the sediment trap material from the VP2 deployment.

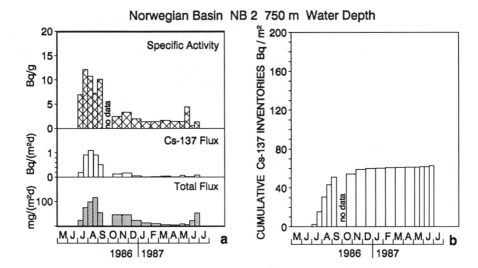

FIGURE 3. Analytical and gamma-spectroscopic results (a) and cumulative ^{137}Cs-inventories (b) of the sediment trap material from the NB2 deployment.

DISCUSSION

The relationship between ^{137}Cs-flux, ^{137}Cs-specific activity, and the biological situation encountered at the Vøring Plateau is shown in Figure 4. Low ^{137}Cs-fluxes and low ^{137}Cs-specific activities were observed in the euphotic zone (VP1A and VP1B) in May/June 1986 (Fig. 4a). In this time period clay minerals, organic matter and fecal pellets were recycled by grazing in the euphotic zone [7]. An equilibrium of aggregation, dissolution, breakup, coagulation resulted in low vertical fluxes of particulate matter and nutrients [8]. Also in the midwater zone, total-fluxes and ^{137}Cs fluxes were low, because during this time period Chernobyl-contaminated suspended matter was recycled in the uppermost part of the water column. Only small amounts of material was carried to greater water depths. This sinking matter composed of copepod fecal pellets [8] showed high ^{137}Cs-specific activities (VP1 1, VP1 2).

In late June (VB1B 2 and 3) copepods moved to deeper parts of the water column. Nutrients in the euphotic zone were depleted and the suspended particles were packaged in large fecal pellets [8]. Copepods metabolize about 10-90% of the organic matter during ingestion [9]. Thus inorganic matter such as clay minerals and undigestible biogenic debris such as silicate, carbonate, and chintin [10] become relatively enriched. Clay minerals at the Vøring Plateau site were mainly composed of illite [11] which can selectively adsorb radiocesium [12]. In the traps deeper than 140 m water depths (Fig. 4a) the fecal pellets were 6-10 times more enriched in ^{137}Cs compared to sinking matter at time of recycling (VP1A, VP1B 1). Due to the the sedimentation pulse, also ^{137}Cs-fluxes were 10 times higher than during time of recirculation. In July (VP2) sedimentation was dominated by minipellets of less than 50 m in length [8]. ^{137}Cs-specific activities of up to 16.8 Bq g^{-1} were detected in material dominated by small fecal pellets. But ^{137}Cs-fluxes were 10 times lower than in June propably because the main plankton bloom was already over.

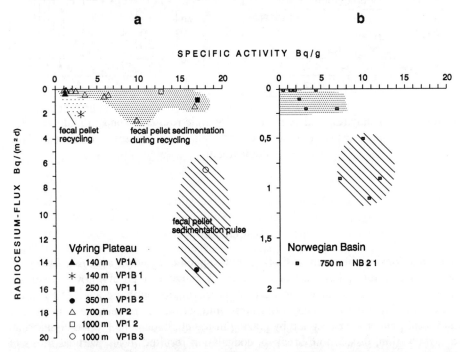

FIGURE 4. The relationship between ^{137}Cs-flux and ^{137}Cs-specific activity and the biological situation encountered at the Vøring Plateau Site (a) and the Norwegian Basin site (b). Note the larger scale for NB2 ^{137}Cs-flux.

In the deepest sediment trap VP1B 3 at 1000 m, ^{137}Cs-flux was about 50% lower than that in the VB1B 2 trap at 350 m. The loss may be attributed to either a primarily lower mass flux or to the dissolution of fecal pellets during sinking. But ^{137}Cs-specifiv activity increased by 10% from 350 m to 1000 m water depth. This may be due to repackaging of fecal matter in deeper parts of the watercolumn, due to the uptake and digestion of fecal pellets by ostracods [8].

In the Norwegian Basin, ^{137}Cs-fluxes were ten times lower than at the VP2 site, probably because saisonality was weaker, as well as the amount ^{137}Cs deposited to the seasurface was declining with distance from the coast [2]. The Norwegian Basin sediment traps (Fig. 4b) have not bee n investigated so far for biological composition. But the combination of high ^{137}Cs-specific activities and ^{137}Cs-fluxes for cups NB2 1.2-5 may also be caused by fecal pellet sedimentation as was observed in the Voring Plateau site VP1B.

Due to the sedimentation pulse, 70% of the total observed ^{137}Cs-trap-inventory at the Vøring Plateau sites was sedimentated within four (June/July) out of the 40 week deployment. Spring bloom developed 6-8 weeks later at the NB2 station compared to the Vøring Plateau site. But due to high specific activities, 60 Bq m^{-2} (95% of the total amount of the NB2 1 trap) have been transported to a depth of 750 m within three months.

^{137}Cs-concentrations observed for all sediment trap material in our study as well as for other sediment trap material contaminated by the Chernobyl accident (D1 Baltic Sea [13], NS III North Sea [14], MS northwestern Mediterranean Sea [4,15], BS VI Black Sea [16], VP2 Vøring Plateau) were extremely high (several hundred to 18200 Bq kg^{-1}). For comparision, in 1984 marine species and fine grained sediments of the Irish Sea close to Sellafield showed in maximum 4500 Bq kg^{-1} ^{137}Cs [17].

Figure 5 shows the calculated concentration factors C_f for Chernobyl contaminated sediment trap material from different regions based on published data. Seawater activities were taken from the respective trap literature or from the same marine region at the same time (ref. see Figure 5). Pre-Chernobyl C_f [18] are marked as solid lines. All post-Chernobyl sediment trap deployments not only show high ^{137}Cs-specific activities but also very high C_f (200-110.000). Since ^{137}Cs water and sediment trap material ^{137}Cs concentrations probably have not always been been determined for the same water mass, it may be that the water ^{137}Cs activities are underestimated. This is propable, because shortly after deposition, extremely high ^{137}Cs activities were restricted to the upper meter of the water column causing a strong contamination of the suspended matter. After several weeks ^{137}Cs concentration in the surface layer decreased by turbulent mixing [19], eventually resulting in a 'disequilibrium' between ^{137}Cs-concentration in suspended matter and in water, because desorption of radiocesium is a slower process than adsorption and not completely reversible [12].

^{137}Cs-determinations in sinking particles and seawater were undertaken in the northwestern Mediterranean Sea [4,15] in the same water masses. These data show the relatively low C_f in zooplankton and high enrichment factors for fecal pellets and unspecified sediment trap material. Furthermore other sediment trap investigations (Figure 5) show high C_f whenever fecal pellets appeared to sedimentation (VP2, possibly also NS III) or aggregation of phytoplankton (D1,

BS VI) was encountered [13] or hypothesized [16]. These observations suggest a scavenging of suspended matter from the water column by sinking blooms, that were very effective in accumulating and transporting Chernobyl-contaminated particulate matter.

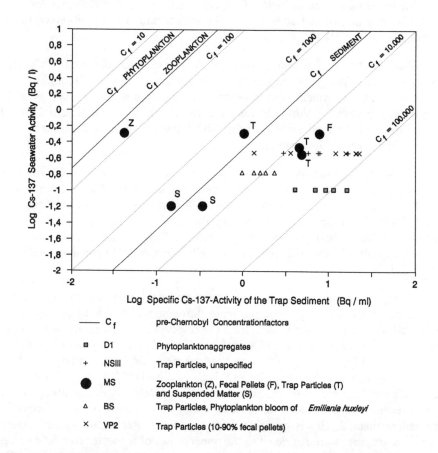

FIGURE 5 Concentration factors of Chernobyl contaminated sediment trap material from different marine regions in comparision with concentration factors compiled by IAEA [18], calculated from: D1 Baltic Sea [12,20], NS III Northern North Sea [14], MS Mediterranean Sea [4,15], BS VI Black Sea [16], VP2 Vøring Plateau. Data were taken for May-July 1986 only.

High ^{137}Cs-specific activities may have been caused also by the chemical form of ^{137}Cs washed out from the atmosphere. At the time of deposition to the sea surface, ^{137}Cs was mainly fixed in aerosol particles that dissolved in seawater only after several weeks [15]. These particles may have been attached to or incorporated in the suspended matter, resulting in higher specific activities than would have occurred by direct uptake of dissolved radiocesium.

CONCLUSIONS

In the Norwegian Sea, effective transport of radiocesium from the upper parts of the water column to the seafloor was driven by the plankton community. The production of fecal pellets concentrated ^{137}Cs tenfold in the sediment trap material and caused a strong vertical ^{137}Cs flux. The high specific actvities were caused by the relative enrichment of clay minerals in the fecal pellets that dominated the sinking particulate matter in this region. Relatively high specific activities in the sinking particles from the Norwegian Sea as well as in other post-Chernobyl sediment trap deployments were propably due to the particulate form of Chernobyl-^{137}Cs and to high radiocesium concentrations in the sea-water shortly after deposition.

ACKNOWLEDGEMENTS

Thanks are due to the crews of FS Meteor and FS Poseidon and to S. Middendorf, who carefully read the manuscript. This work was supported by the Universität Bremen (Tschernobyl-Projekt).

REFERENCES

1. IAEA (1986). International Atomic Energy Agency, Summary Report on the Post-Accident Review Meeting on the Chernobyl Accident, Report of the Nuclear Safety Advisory Group.- Safety Series 75-INSAG-1. Wien, IAEA, 106 pp..
2. Aakrog, A., Buch, E., Chen, Q.J., Christensen, G.C., Dahlgaard, H., Hansen, H., Holm, E. and Nielsen, S.P., Environmental radioactivity in the north Atlantic region including the Faroe Islands and Greenland Sea, 1987. Riso Reports, 1989, Riso-R-**564**.
3. Rice, T.R. and Wolfe, D.A., Radioactivity - chemical and biological aspects. In Impingement of Man on the Oceans, ed. D.W. Hood, Wiley, New York, 1971, pp. 325-380.
4. Fowler, S.W., Buat-Menard, P., Yokoyama, Y., Ballestra, S., Holm, E. and Nguyen, H.V., Rapid removal of Chernobyl fallout from Mediterranean surface waters by biological activity. Nature, 1987, **239**, 56-58.

5. Baumann, M., Segl, M., von Bodungen, B. and Wefer, G.. Chernobyl derived radiocesium in the Norwegian Sea: Flux between water column and sediment. In The Radioecology of Natural and Artificial Radionuclides, ed. W. Feldt, Proc. XVth Reg. Congr. IRPA, Visby, Schweden, September 1989, Köln, Verlag TÜV Rheinland, 1989, pp. 318-323.

6. Devell, L., Tovedal, H., Bergström, U., Applegreen, A., Chyssler, J. and Anderson, L., Initial observations of fallout from the reactor accident at Chernobyl. Nature, 1986, 321, 192-193.

7. Peinert, R., Bathmann, U., Bodungen, B.v., Noji, T., The impact of grazing on spring phytoplankton growth and sedimentation in the Norwegian Sea. Mittl. Geol. Pl. Inst. Univ. Hamburg, 1987, 62, 149-164.

8. Bathmann, U., Peinert, R., Noji, T.T. and Bodungen B. v., Pelagic origin and fate of sedimenting particles in the Norwegian Sea, Progr. Oceanogr., 1990, 24, 117-125.

9. Gaudy, R., Feeding four species of pelagic copepods under experimental conditions. Marine Biology, 1974, 25, 125-141.

10. Fowler, S.W. & Knauer, G.A., Role of large particles in the transport of elements and organic compounds through the oceanic water column. Progr. Oceanogr., 1990, 16, 147-194.

11. Berner, H., Mechanismen der Sedimentbildung in der Framstraße, im Arktischen Ozean und in der Norwegisch Grönlandischen See. Dissertation am Fachbereich Geowissenschaften, Universität Bremen, 1991, in prep..

12. Brouwer, E., Baeyens, B., Maes, A. & Cremers, A., Cesium and rubidium ion equilibria in illite clay. J. Phys. Chem., 1983, 87, 1213-1219.

13. Passow, U., Vertikalverteilung und Sedimentation von Phytoplankton in der mittleren Ostsee während des Frühjahres 1986. Berichte aus dem Institut für Meereskunde an der Christian-Albrecht-Universität, Kiel, 1990, 192.

14. Kempe, S. and Nies, H., Chernobyl nuclide record from a North Sea sediment trap. Nature, 1987, 329, 828-831.

15. Whitehead, N.E., Ballestra, S., Holm, E. and Huynh-Ngoc, L., Chernobyl radionuclides in shellfish. J. Environm. Radioact., 1988, 7, 107-121.

16. Buesseler, K.O., Livingston, H.D., Honjo, S., Hay, B.J., Manganini, S.J., Degens, E., Ittekkot, V., Izdar, E. and Konuk, T., Chernobyl radionuclides in a Black Sea sediment trap. Nature, 1987, 329, 825-828.

17. Hunt, G.-J, Radioactivity in surface and coastal waters of the British Isles, 1984. In Aquatic Environment Monitoring Report, ed. Ministry of Agriculture, Fisheries and Food, Directorate of Fisheries Research, 13, Lowestoft, Crown, 1985.

18. IAEA International Atomic Energy Agency, Sediment K_{ds} and Concentrationsfactors for Radionuclides in the Marine Environment. Technical Report Series, 1985, 247, Wien, IAEA, 71 pp..

19. Müller-Navara, S.H. and Mittelstaedt, E., Modelluntersuchungen zur Ausbreitung künstlicher Radionuklide in der Nordsee. Deutsches Hydrographisches Institut, Berichte, 1988, DHI Hamburg, 63 pp..

20. Deutsches Hydrographisches Institut, Die Auswirkungen des Kernkraftwerkunfalles von Tschernobyl auf Nord- und Ostsee. Meereskundliche Beobachtungen und Ergebnisse, Hamburg, DHI, 1987, 62.

POLONIUM-210 AND LEAD-210 IN MARINE ORGANISMS : ALLOMETRIC RELATIONSHIPS AND THEIR SIGNIFICANCE

R.D. CHERRY AND M. HEYRAUD
Department of Physics, University of Cape Town
Rondebosch, 7700 South Africa

ABSTRACT

Allometric relationships which indicate that Po-210 concentrations in marine organisms decrease with increasing organism mass have been reported previously in a few taxa. We report here the results of a study of nearly 400 data covering nine taxa of marine organisms. The data for each taxon are fitted to the allometric equation log Q = log a + b log M, where Q is the Po-210 concentration (mBq/g dry mass) and M is the dry mass per individual (g). For eight of the nine taxa the slope is negative, and for six the correlation is statistically significant. Data for Pb-210, the Po-210 grandparent, are available for some sixty per cent of the samples, covering eight taxa. For seven of these taxa b is again negative, and for four the correlation is statistically significant. The weighted mean of the nine Po-210 slopes is -0.24 ± 0.05, and of the eight Pb-210 slopes is -0.22 ± 0.05. These values are close to the slope of -0.25 frequently found in mass-specific allometric relationships in biology; an association between radionuclide concentration and food ingestion rate is indicated. The intertaxon variations in the intercept log a are large, nearly two orders of magnitude for Po-210, a fact which almost certainly reflects intertaxon differences in diet and/or assimilation. Within taxa, sub-groupings of the Po-210 data are found; these are discussed and an attempt is made to classify them statistically for the data as a whole.

INTRODUCTION

The importance of animal size in biology is well established. An "amazing number of morphological and physiological variables" [1] are scaled according to allometric relations of the form log y = log a + b log M, where y is the variable and M is the organism mass. The most extensively studied variable is metabolic rate; if this rate is expressed on a mass-specific basis, the slope is usually found to be about -0.25 or -0.33. This is true for a variety of organisms, both terrestrial and aquatic [1-4]. Theoretical arguments can be used to underpin either of these two values: McMahon's model, based on elastic similarity criteria, predicts -0.25, while the so-called "surface law" predicts -0.33 [1,3,5]. Opinion is divided as to which of the two is correct, with recent work tending to favour -0.25 [5-9]. Allometric relationships with similar slopes are found for other rate processes such as ingestion and growth rates [3,9].

Allometric relationships for trace element concentrations in marine organisms are often reported, but the slopes vary depending on the element and on the organism involved. They can be positive, negative or zero [10,11], and no simple or consistent picture is evident. Data for the concentration of the naturally-occurring alpha-radioactive element Po-210 in marine organisms have shown allometric relationships on several occasions [12-15], always with negative slopes but sometimes much more negative than -0.33. We have investigated these relationships further, for both Po-210 and its grandparent Pb-210, and we report here the results of the regression fits for several taxa of marine organisms.

RESULTS

Essentially all the Po-210 and Pb-210 data we have used have already been published [12-19] and in most cases the dry mass per individual was available in our laboratory records if not in the publication. We restricted our calculations to data representing concentrations in whole animals. We report in Tables 1 and 2 the regression parameters for all taxa for which at least six data points were available.

TABLE 1
Po-210 regression line summary

Taxon	n	Mrange	b	log a	r	Sig.
Carid shrimp	104	1000	-0.21 ± 0.07	2.27 ± 0.05	-0.28	.01
Penaeid shrimp	104	14000	-0.23 ± 0.05	3.03 ± 0.06	-0.45	.001
Fish	53	2200	-0.18 ± 0.08	2.46 ± 0.07	-0.29	.05
Euphausiids	47	190	-0.32 ± 0.10	1.36 ± 0.14	-0.42	.01
Amphipods	20	670	-0.08 ± 0.16	2.44 ± 0.22	-0.12	NS
Copepods	19	900	-0.27 ± 0.09	1.59 ± 0.28	-0.58	.01
Mysiids	18	9	$+0.21 \pm 0.14$	2.70 ± 0.60	+0.34	NS
Polychaetes	9	110	-0.63 ± 0.10	1.42 ± 0.10	-0.92	.001
Squid	9	4	-0.04 ± 0.29	2.04 ± 0.41	-0.06	NS

n = number of data; Mrange is the ratio maximum M to minimum M for the taxon concerned; errors on b and log a are standard errors; r is the correlation coefficient; NS = not significant at P < 0.05.

The regression line fitted was

$$\log Q = \log a + b \log M \qquad (1)$$

where Q is the concentration of Po-210 (or Pb-210) in mBq/g dry mass and M is the individual organism dry mass in g.

TABLE 2
Pb-210 regression line summary

Taxon	n	Mrange	b	log a	r	Sig.
Carid shrimp	74	260	-0.35 ± 0.09	0.66 ± 0.06	-0.44	.001
Penaeid shrimp	61	14000	-0.25 ± 0.06	1.01 ± 0.07	-0.50	.001
Fish	29	1400	-0.36 ± 0.11	0.64 ± 0.09	-0.55	.01
Euphausiids	30	140	-0.48 ± 0.16	0.02 ± 0.20	-0.49	.01
Amphipods	20	670	-0.06 ± 0.07	1.48 ± 0.10	-0.20	NS
Copepods	8	780	$+0.15 \pm 0.20$	2.06 ± 0.70	$+0.31$	NS
Mysiids	9	93	-0.02 ± 0.20	0.81 ± 0.27	-0.04	NS
Polychaetes	6	26	-0.57 ± 0.41	0.30 ± 0.25	-0.57	NS

Symbols have the same meaning as in Table 1.

REGRESSION LINE SLOPES

From Tables 1 and 2 we see that, for Po-210, eight of the nine taxa have a negative slope and that the correlation is statistically significant for six taxa; for Pb-210, seven of the eight taxa have a negative slope and the correlation is significant for four taxa. For both radionuclides, most of the correlations which are not significant are found in taxa which have a small number of data and/or a small mass range. The weighted means of the slopes are -0.24 ± 0.05 for Po-210 and -0.22 ± 0.05 for Pb-210. Both of these are consistent with the slope of -0.25 already referred to; insofar as individual taxa are concerned, only three of the seventeen regressions (the mysiid and polychaete lines for Po-210 and the amphipod line for Pb-210) have slopes which are significantly different from -0.25 (t-test, $P < 0.05$).

There is direct experimental evidence that food is the dominant source of Po-210 in marine organisms [15,16,20]. This being so, the Po-210 content of a marine organism will be governed by the rate of food ingestion, the Po-210 content of the food and the fraction of the ingested Po-210 which is assimilated. The mass-specific food ingestion rate in marine organisms is frequently found to follow an allometric relationship with slope -0.25 [3,9], and the allometric relationships for Po-210 concentrations are most likely a consequence of this scaling of the food ingestion rate. Information as to how the Po-210 content in the food scales with the mass of the consumer will be needed before the Po-210 allometry can be explained on a mathematically sound basis.

There is, to our knowledge, no direct evidence that food is the dominant source of Pb-210 in marine organisms. Nonetheless, the fact that the mean regression slopes are similar for both elements - despite the inherently less reliable nature of the Pb-210 data [19] - suggests that similar mechanisms operate for both elements and that food must, at least, be an important source of Pb-210.

One feature should be emphasized: the slopes of about -0.25 in Tables 1 and 2 are generally associated with regression lines calculated from data which cover a wide range of specimens comprising several species collected from different localities on different occasions, as well as a wide range of M of two orders of magnitude or more. When the data are less eclectic, and when in particular their coverage is limited to specimens collected in the same locality and on the same occasion, the slopes of the regression lines are generally steeper. We have reported two such cases previously: a slope of -0.98 ± 0.16 for a rather particular group of penaeid shrimp collected on one side of an oceanographic front in the Northeast Atlantic [19] and a slope of -0.59 ± 0.15 for a large collection of the euphausiid *Meganyctiphanes norvegica* collected during a period of less than six hours in the Mediterranean [12]. It is interesting to note that a laboratory determination of the food ingestion rate in *M.norvegica* [21] as a function of animal mass gave an allometric equation with a slope of -0.58 ± 0.08. This is remarkably close to the Po-210 allometric slope for the *M.norvegica* collection just referred to, but substantially different from the -0.25 slope usually associated with ingestion rates [3,9]. The discrepancy is perhaps a reflection of the difference between *inter*species and *intra*species allometric equations discussed by Feldman and McMahon [7].

The polychaetes in Tables 1 and 2 provide another example of more negative slopes, viz. -0.63 ± 0.10 and -0.57 ± 0.41 for Po-210 and Pb-210 respectively. They represent five species, and cover two orders of magnitude in mass, but they were all collected on the same day at the same site near the Cape of Good Hope.

REGRESSION LINE INTERCEPTS

The intercepts log a in Tables 1 and 2 are the values of log Q for organisms of 1g dry mass. The variations in intercept reflect differences, many of them significant, in the concentrations of the two nuclides in different taxa. Direct comparison between the different log a is, however, unrealistic, because the mass ranges in the different taxa do not bear the same relation to the (arbitrary) reference mass of 1g and because the regression slopes are unequal. To overcome this difficulty we made the assumption that the "true" slope for all taxa is indeed -0.25, and then used equation (1) to calculate the value of a "normalized" intercept, which we call log a*, for each data point. The merits of this procedure have been discussed recently in an investigation of general allometric equations in plankton organisms [9]. In view of the closeness of the weighted mean slopes to -0.25 for both Po-210 and Pb-210, we believe it is valid to apply the procedure here for the limited purpose of intercept comparison. The mean values of log a* for each taxon are given in Table 3. Intertaxa comparisons are now meaningful because the slopes are all the same and the arbitrary choice of reference mass is no longer a problem.

TABLE 3
Intertaxa comparisons of normalized intercepts, log a*

Taxon	Mean log a*,Po-210	Mean log a*,Pb-210
Carid shrimp	2.25±0.04	0.70±0.05
Penaeid shrimp	3.02±0.04	1.01±0.05
Fish	2.47±0.07	0.61±0.08
Euphausiids	1.45±0.05	0.31±0.07
Amphipods	2.25±0.13	1.27±0.07
Copepods	1.64±0.07	0.65±0.18
Mysiids	1.53±0.09	0.54±0.12
Polychaetes	1.70±0.10	0.42±0.17
Squid	2.50±0.05	-

Two features of Table 3 are immediately evident. (i) The intercepts of Po-210 are greater than those for Pb-210 by between about one and two for all taxa. This reflects the well-known fact that Po-210/Pb-210 ratios in marine organisms are typically between 10 and 100. (ii) Large intertaxon variations are observed. The range in the mean a* is a factor of 37 for Po-210 and 9 for Pb-210; for the data we have used, ranges in *individual* Po-210 and Pb-210 concentrations are, in comparison, factors of 2599 and 624 respectively.

Covariance analysis confirms the reality of the intertaxon variations and is shown in Table 4. These were performed on the individual data points and are not subject to the validity or otherwise of the common slope assumption used to produce the Table 3 intercepts.

TABLE 4
Covariance analysis of the log Q data

	Source of variation	SS	df	MS	F-ratio
Po-210	Covariate (logM)	3.43	1	3.43	20.58
	Group effect (taxon)	101.97	8	12.75	76.40
	Residual	62.22	373	0.17	
	Total	167.63	382		
Pb-210	Covariate (logM)	15.10	1	15.10	91.60
	Group effect (taxon)	17.32	7	2.47	5.01
	Residual	37.58	228	0.17	
	Total	70.00	236		

SS=sum of squares; df=degrees of freedom; MS=mean square.

All four F-ratios are highly significant (P<0.0001). The dominance of the intertaxon contribution to the variation is particularly striking for Po-210. For both nuclides, this contribution almost certainly reflects intertaxon differences in diet and/or assimilation of the nuclides.

SUB-GROUPS WITHIN TAXA

Our use of taxa constitutes an objective biological criterion. Within a taxon, sub-grouping of the Po-210 data has been reported on a few occasions [13,19]. Thus penaeid shrimp of the genera *Sergestes* and *Sergia* were lower in Po-210 than other penaeids of similar size (notably those of the genus *Gennadas*), while carid shrimp from the genera *Hymenodora*, *Meningodora*, *Pasiphaea* and *Parapasiphaea* were lower in Po-210 than the other carids. A sub-grouping in terms of a particular oceanic condition has also been observed: organisms which were collected on one side of an oceanographic front and which were associated with bloom conditions were found to be anomalously high in Po-210 in seven taxa. (This particular sub-grouping will be referred to as the "Southern Series" samples).

From our present data base we report a further sub-grouping, *viz.* in fish. This is shown in Figure 1, where we plot Po-210 data for common clupeoid fish [15] together with data for (smaller) meso- and bathypelagic fish [18, and unpublished data].

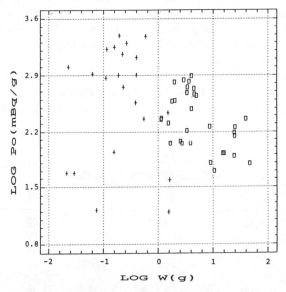

Figure 1. Po-210 data for clupeoid fish (squares) and meso- and bathypelagic fish (crosses). Clupeoid fish are mostly sardine and anchovy [15]. Meso- and bathypelagic fish are from the families Chauliodontidae, Eurypharyngidae, Gonastomatidae, Melamphaidae, Myctophidae, Opisthoproctidae, Serrivomeridae and Sternoptychidae.

Visual inspection suggest that most of the twenty-two meso- or bathypelagic fish lie on a common regression line with the clupeoids, but that six are below this line. Interactive outlier regression analysis rejects four of these six (95% confidence limits). All four are from the family Gonastomatidae (genera *Cyclothone* and *Gonastoma*), a family which is not represented in the other eighteen samples. The Po-210 data suggest that this family differs from the other fish in either diet or assimilation, and that further measurements on different genera of meso- and bathypelagic fish would be interesting.

We have used analysis of variance and multiple comparison statistical tests in an attempt to combine taxa and their sub-groups according to their Po-210 and Pb-210 contents. These procedures were carried out after excluding all data points from the anomalously high "Southern Series" samples already referred to. All data for amphipods, mysiids, polychaetes and squid were also excluded on the grounds that the data for these taxa are as yet too sparse. The spread in Po-210 concentrations is such that the test identifies four statistically separate levels, *viz.* high (comprising penaeid shrimp except those from the genera *Sergestes* and *Sergia*), medium high (comprising *Sergestes* and *Sergia* penaeids and fish except for those from the family Gonastomatidae), medium low (comprising carids except those from the genera *Hymenodora*, *Meningodora*, *Pasiphaea* and *Parapasiphaea*) and low (copepods, euphausiids and those genera of fish and carids excluded from the medium classifications). The smaller spread in Pb-210 concentrations results in the same tests producing only two clear-cut levels, *viz.* high, classified as for Po-210, and low, comprising the other three Po-210 classifications.

Regression fits over each level, for both Po-210 and Pb-210, give the results in Table 5.

TABLE 5
Po-210 and Pb-210 regression lines by level

Level	n	b	log a	r	Sig.
High Po-210	63	-0.30 ± 0.03	3.04 ± 0.04	-0.78	.001
Medium high Po-210	76	-0.30 ± 0.04	2.61 ± 0.04	-0.68	.001
Medium low Po-210	77	-0.25 ± 0.05	2.31 ± 0.04	-0.49	.001
Low Po-210	79	-0.24 ± 0.04	1.49 ± 0.07	-0.56	.001
High Pb-210	44	-0.31 ± 0.05	1.11 ± 0.06	-0.72	.001
Low Pb-210	134	-0.21 ± 0.05	0.57 ± 0.04	-0.36	.001

Symbols have the same meaning as in Table 1. See text for categorisation of Po-210 and Pb-210 levels.

The slopes remain close to -0.25, although several are also close to the alternative theoretical allometric value of -0.33. The greater range in log a for Po-210 as opposed to Pb-210 remains evident. The different Po-210 levels have been interpreted in terms of different diets; application of the Isaacs unstructured food-web model to Po-210 data [14] suggests that low, medium and high Po-210 levels tend to associate with herbivorous, omnivorous and detritivorous feeding respectively. More detailed considerations of individual taxa [13,15,20] support this tendency. Further application of the Isaacs model to the Po-210 data still awaits a solution to a fundamental difficulty: how to incorporate the mass M per individual organism into the model. A final *caveat*: it should not be forgotten that assimilation, as well as diet, could contribute to differences in Po-210.

THE NATURAL RADIATION DOSE

It is well known that Po-210 accounts for the bulk of the natural radiation dose equivalent received by most marine organisms. The wide range in marine natural radiation dose domains implied by the wide range in measured Po-210 concentrations has been commented on previously [13,18,19,22]. It is worth noting an obvious consequence of the allometric relationships we have discussed here: small organisms will on average be subjected to a larger dose from Po-210 than larger ones, by a factor of about 10 over a mass range of four orders of magnitude. A further factor of up to 30 or 40 can result from the intertaxon differences in Po-210 concentration.

CONCLUSION

When sufficient data covering a sufficient range of organism mass are available, both Po-210 and Pb-210 concentrations in marine organisms follow allometric relationships with negative slopes. For widely-based data, most of these slopes are consistent with the theoretically-based slope of -0.25. For narrowly-based data, indications are, for Po-210 at least, that the slopes are much more negative. Variations in concentration between different taxa are also significant, particularly for Po-210. These intertaxon variations are interpreted in terms of differences in diet and/or assimilation and can be a useful tool in food-web and dietary studies. The mass and intertaxon dependence of the Po-210 concentrations has obvious consequences for the variation in the natural radiation dose received by marine organisms.

More data are needed in many taxa, and many questions remain: Are the "true" regression slopes for widely-based data indeed -0.25 or is the alternative theoretical prediction of a slope of -0.33 a possibility? Why do narrowly-based data often give slopes which are significantly more negative than either theoretical predictions? Why is the overall range in Po-210 concentrations so much larger than for Pb-210? What are the relative contributions of diet and assimilation to the variations in concentration of both nuclides? How do the Po-210 and Pb-210 intertaxon concentration ranges compare with those for other trace elements in marine organisms?

ACKNOWLEDGEMENTS

We thank D. Stein for providing us with additional meso- and bathypelagic fish samples, J.A. Day for the polychaete samples and the Foundation for Research and Development and the University of Cape Town for financial support.

REFERENCES

1. Schmidt-Nielsen, K., Scaling: Why Is Animal Size So Important ?, Cambridge University Press, Cambridge, 1984.

2. Kleiber, M., Body size and metabolism. Hilgardia, 1932, **6**, 315-353.

3. Peters, R.H., The Ecological Implications of Body Size, Cambridge University Press, Cambridge, 1983.

4. Calder, W.A., Size, Function and Life History, Harvard University Press, Cambridge, Massachusetts, 1984.

5. McMahon, T., Size and shape in biology. Science, 1973, **179**, 1201-4.

6. Heussner, A.A., Energy metabolism and body size. I. Is the 0.75 mass exponent of Kleiber's equation a statistical artifact? Respir. Physiol., 1982, **48**, 1-12.

7. Feldman, H.A. and McMahon, T.A., The 3/4 mass exponent for energy metabolism is not a statistical artifact. Resp. Physiol., 1983, **52**, 149-163.

8. Platt, T. and Silver, W., Ecology, physiology, allometry and dimensionality. J.theor. Biol., 1981, **93**, 855-860.

9. Moloney, C.L. and Field, J.G., General allometric equations for rates of nutrient uptake, ingestion and respiration in plankton organisms. Limnol. Oceanol., 1989, **34**, 1290-99.

10. Boyden, C.R., Trace element content and body size in molluscs. Nature, 1974, **251**, 311-4.

11. Philips, D.J.H., Quantitative Aquatic Biological Indicators, Elsevier Applied Science, London, 1980.

12. Heyraud, M., Fowler, S.W., Beasley, T.M. and Cherry,R.D., Polonium-210 in euphausiids: a detailed study. Mar. Biol., 1976, **34**, 127-36.

13. Cherry, M.I., Cherry, R.D. and Heyraud, M., Polonium-210 and lead-210 in Antarctic marine biota and sea water. Mar. Biol., 1987, **96**, 441-9.

14. Cherry, R.D. and Heyraud, M., Polonium-210 in selected categories of marine organisms: interpretation of the data on the basis of an unstructured marine foodweb model. In Radionuclides: A Tool for Oceanography, eds. J.C. Guary, P. Guegueniat and R.J. Pentreath, Elsevier Applied Science Publishers, London, 1988, pp. 362-72.

15. Cherry, R.D., Heyraud, M. and James, A.G., Diet prediction in common clupeoid fish using polonium-210 data. J. Environ. Radioactivity, 1989, **10**, 47-65.

16. Heyraud, M. and Cherry, R.D., Polonium-210 and lead-210 in marine food chains. Mar. Biol., 1979, **52**, 227-36.

17. Cherry, R.D. and Heyraud, M., Polonium-210 content of marine shrimp: variation with biological and environmental factors. Mar. Biol., 1981, **65**, 165-75.

18. Cherry, R.D. and Heyraud, M., Evidence of high natural radiation doses in certain mid-water oceanic organisms. Science, 1982, **218**, 54-6.

19. Heyraud, M., Domanski, P., Cherry, R.D. and Fasham, M.J.R., Natural tracers in marine dietary studies: data for Po-210 and Pb-210 in decapod shrimp and other pelagic organisms in the Northeast Atlantic Ocean. Mar. Biol., 1988, **97**, 507-519.

20. Carvalho, F.P., Contribution à l'étude du cycle du polonium-210 et du plomb-210 dans l'environnement. Thèse de Doctorat ès-sciences, Université de Nice, 1990.

21. Heyraud, M., Food ingestion and digestive transit time in the euphausiid *Meganyctiphanes norvegica* as a function of animal size. J. Plank. Res., 1979, **1**, 301-11.

22. Carvalho, F.P., Po-210 in marine organisms: a wide range of natural radiation dose domains. Radiat. Protect. Dosimetry, 1988, **24**, 113-17.

RADIOTRACERS IN THE STUDY OF MARINE FOOD CHAINS. THE USE OF COMPARTMENTAL ANALYSIS AND ANALOG MODELLING IN MEASURING UTILIZATION RATES OF PARTICULATE ORGANIC MATTER BY BENTHIC INVERTEBRATES

ANTOINE GREMARE, JEAN MICHEL AMOUROUX, FRANCOIS CHARLES
Laboratoire Arago, UA CNRS n°117
F66650 Banyuls sur mer, France

ABSTRACT

The present study assesses the problem of recycling when using radiotracers to quantify ingestion and assimilation rates of particulate organic matter by benthic invertebrates. The rapid production of dissolved organic matter and its subsequent utilization by benthic invertebrates constitutes a major bias in this kind of study. However recycling processes may also concern POM through the production and reingestion of faeces. The present paper shows that compartmental analysis of the diffusion kinetics of the radiotracer between the different compartments of the system studied and the analog modelling of the exchanges of radioactivity beween compartments may be used in order to determine ingestion and assimilation rates. This method is illustrated by the study of a system composed of the bacteria *Lactobacillus sp.* and the filter-feeding bivalve *Venerupis decussata*. The advantages and drawbacks of this approach relative to other existing methods are briefly discussed.

INTRODUCTION

Once sedimented, particulate organic matter (POM) may enter different pathways (i.e., resuspension, burial, mineralisation, incorporation in the benthic food web). The relative importance of these different pathways is highly dependant on the activity of micro- and macrorganisms (mostly invertebrates) living at the water-sediment interface. The simplest of the interactions between POM and invertebrates corresponds to the utilization (i.e., the ingestion and assimilation) of POM by benthic invertebrates.

Quantification of ingestion and assimilation rates usually involves the incubation of the test organisms in the presence of labelled food sources [I]. As it has already been pointed out by several authors [2, 3], the use of radioisotopes implicitly assumes several hypotheses including the stability of the label and the absence of recycling. These conditions are very rarely met in closed experimental systems especially when dealing with live food sources (e.g., bacteria or diatoms) and/or large size macrobenthos.

This difficulty can be overcome either by using a labelling procedure which results in a greater stability of the label [1,4, 5, 6] or by using an experimental design (pulse-chase design) which minimizes recycling by reducing the duration of the contact between the labelled food and the test organism [1, 7, 8]. However special labelling procedures are often not applicable to live food sources [1, 4], whereas the use of a pulse chase design increases experimental artefacts due to stress [2] and is essentially restricted to small size macrobenthos (because of the limitation of incubation duration) [10].

Conover and Francis [2] and Smith and Horner [3] have suggested combining compartmental analysis (i.e., an experimental study of the changes of the repartition of the radioactivity within the study system) and analog modelling in order to overcome the problem of recycling within closed experimental systems. In this paper we will use some of our published data [11, 12] concerning the system *Venerupis decussata - Lactobacillus sp.* to show that this approach is indeed applicable to the quantification of ingestion and assimilation rates of benthic invertebrates feeding on POM. The advantages and drawbacks of this approach will then be compared with those of other existing methods.

THE IMPORTANCE OF RECYCLING

Computations of ingestion rates which are derived from radioisotopic studies are usually based on the following equation:

$$I = (X.M)/(R.T.B) \qquad (1)$$

where:
I (mgDW/mgDW/h) is the specific ingestion rate of POM
X (dpm) is the amount of radioactivity in the invertebrates at the end of the experiments
M (mgDW) is the mass of POM which is initially introduced
R (dpm) is the radioactivity of POM which is initially introduced
T (h) is the incubation duration
B (mgDW) is the biomass of invertebrates present in the microcosm

The use of this equation assumes that ingestion is the only pathway by which POM and Invertebrates may exchange radioactivity (Figure 1a). However, there is some experimental evidence showing that in many cases, because of the lack of stability in the labelling of POM, there may be other pathways for the exchange of radioactivity between POM and invertebrates.

Regarding DOM, previous experimental studies have shown that:

• monospecific suspensions of ^{14}C-labelled marine bacteria and diatoms rapidly produce radioactive DOM (35 % of the total radioactivity at t=4 h in the case of *Lactobacillus sp.*) [13], and
• benthic invertebrates (including both filter and deposit feeders) are able to utilize DOM originating from marine bacteria and diatoms [10, 13].

This suggests that recycling of DOM may constitute another pathway accounting for the exchange of radioactivity occuring between POM and benthic invertebrates (Figure 1b). Moreover, recycling of DOM is not the only process interfering with consumption of POM; both bacteria and invertebrates may produce radioactive CO_2 [14], whereas invertebrates may produce faeces, and consume DOM.

Therefore, the quantification of the amount of POM which has been ingested or assimilated by the invertebrates requires knowledge of the cumulative amounts of radioactivity which have transited through the pathways linking the different compartments of the system studied. This can be achieved through the coupling of compartmental analysis and analog modelling [2, 3].

(a)

(b)

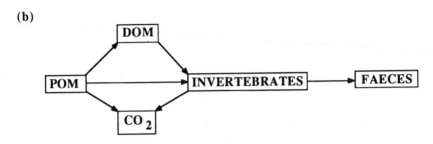

Figure 1. Different pathways of exchanges between Particulate organic matter (POM) and benthic invertebrates. (a) represents the unique pathway which is implicitly assumed when using equation (1). (b) represents a more realistic configuration of the exchanges accounting for the interaction with DOM, CO_2, and faeces.

COMPARTMENTAL ANALYSIS AND ANALOG MODELLING

Compartmental analysis
Compartmental analysis is the experimental analysis of the diffusion kinetics of a product between the different compartments of a system. In the present case, it consists in measuring temporal changes of radioactivity in the different compartments of the experimental chamber. The nature and the number of the compartments are dependent on the nature of the system studied. POM, DOM, CO_2, Invertebrates, and Sediment are usually considered during the study of transfers in benthic food chains [11, 15, 16]. Carrying out the compartmental analysis requires several sets of replicated experiments corresponding to different incubation durations to be run [11, 15, 16]. Analytical possibilities often limit the number of compartments that can be analyzed separately; however it is essential that there is no loss of radioactivity during the course of the experiments. Compartmental analysis generates a data set corresponding to the temporal changes of the distribution of radioactivity in the different compartments of the system studied (see Figure 3 for an example).

Analog modelling
Modelling consists in the simulation of the exchanges between the compartments of the system studied. The first step is the conceptualisation of the system in terms of both compartments and exchanges (see Figure 2 for an example). The second step consists in the formulation of the mathematical equations describing transfers and initial conditions. In the case of analog modelling, this requires the introduction of kinetic coefficients of mass transfer (K_i). The third step consists in the adjustment of the model to the experimental data (i.e., to the results of the compartmental analysis). This step leads to the determination of the values of the kinetic coefficients (see Figure 3 for an example).

Computation of ingestion and assimilation rates

Once modelling is achieved, it is then possible to compute ingestion and assimilation rates. Ingestion rate can be computed as follows:

$$I = (K.M)/B \tag{2}$$

where: I (mgDW/h/mgDW) is the ingestion rate of POM
 K (h^{-1}) is the kinetic constant corresponding to the ingestion of POM by the invertebrates
 M (mgDW) is the mass of POM which is initially introduced
 B (mgDW) is the biomass of invertebrates present in the microcosm

The amount of assimilated organic matter can be estimated in two different ways. It can be set either as the difference between the cumulative amounts of consumed organic matter (i.e., organic matter which has been consumed as DOM, POM, and faeces) and biodeposits produced, or as the cumulative amounts of the DOM and CO_2 produced by the bivalves. The first of these two procedures leads to an overestimation of assimilation since all the radioactivity contained in the bivalves at the end of the experiments is considered to be assimilated. On the contrary, the second of these two procedures leads to an underestimation of assimilation since all the radioactivity corresponding to the bivalves at the end of the experiments is considered not to be assimilated. In both cases, assimilation rate is set as the ratio between the amount of organic matter assimilated and consumed. This procedure can be summarized by the following set of equations:

$$\text{ASS+}(t) = (\int_{0t} K_1[BACT] + K_5[DOM] + K_{10}[SED] - \int_{0t} K_8[BIV]) \cdot 100 / (\int_{0t} K_1[BACT] + K_5[DOM] + K_{10}[SED]) \tag{3}$$

$$\text{ASS-}(t) = (\int_{0t} (K_6+K_7+K_9)[BIV]) \cdot 100 / (\int_{0t} K_1[BACT] + K_5[DOM] + K_{10}[SED])$$

where: Ass+ (t) and Ass- (t) (%) are the two estimations of assimilation rates at time t
 K_i (h^{-1}) are the kinetic coefficients of mass transfer
 $\int_{0t}$ is a sum from time 0 to time t
 [X] (dpm) is the activity of the compartment X at time t

APPLICATION TO THE STUDY OF THE SYSTEM *LACTOBACILLUS SP -VENERUPIS DECUSSATA*

We will use the data and the model presented by Amouroux et al. [11, 12] to illustrate this approach. The system studied is composed of a suspension of the bacteria *Lactobacillus sp* and of the filter-feeding bivalve *Venerupis decussata*. It is important to point out that the aim of the study by Amouroux et al. was not to compute ingestion and assimilation rates of *V. decussata* but to assess the impact of an organic pollutant on the functioning of this system [12].

Compartmental analysis

The compartments considered in the compartmental analysis are: Bacteria, Bivalves, DOM, CO_2, and Sediment. Details on the labelling, experimental, and analytical procedures may be found in [11].

The changes in the partitioning of radioactivity between the different compartments of the experimental system are shown in Figure 3. The radioactivity in the bacteria declined from 100 (t=0) to 9.3% (t=10 h) and then remained fairly constant (9.8% at t=40 h). Concurrently, radioactivity of Bivalves rose from 0 (t=0) to 31.0% (t=4 h) and then declined to 10.1% at t=40 h. Radioactivity in DOM increased from O at t=O to 14.0% at t=4 h and then remained fairly constant (10.4% at t= 40h). Radioactivity in CO_2 increased from 0 at t=0 to 26.2% at t=40 h. Radioactivity in the sediment increased rapidly during the first hours of the incubation (0 at t=0 vs 29.39% at t=4 h) before reaching 42.5% at t=40 h.

Analog modelling
The analog model used for the system *Lactobacillus* sp - *Venerupis decussata* is shown in Figure 2 [12]. The model is composed of 6 compartments (i.e., DOM, Non recyclable DOM, Bacteria, Bivalves, CO_2, Sediment).

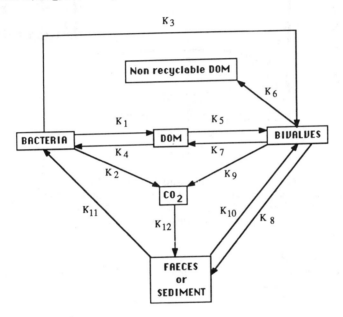

Figure 2. Six-compartment model of *'Lactobacillus sp - Venerupis decussata'* system showing the different exchanges and the corresponding kinetic constants of mass transfer. Redrawn after [12]

Exchanges of radioactivity between these compartments are simulated using the following set of differential equations:

$d[BACT]/dt = K_4[DOM] + K_{11}[SED] - (K_1+K_2+K_3)[BACT]$
$d[DOM]/dt = K_7[BIV] + K_1[BACT] - (K_4+K_5)[DOM]$
$d[BIV]/dt = K_5[DOM] + K_3[BACT] + K_{10}[SED] - (K_6+K_7+K_8+K_9)[BIV]$
$d[SED]/dt = K_8[BIV] + K_{12}[CO_2] (K_6+K_{11})[SED]$
$d[CO_2]/dt = K_2[BACT] + K_9[BIV] - K_{12}[CO_2]$
$d[NR\ DOM] = K_6[BIV]$

where: K_i are the kinetic coefficients of mass transfer
              [X] is the amount of radioacivity corresponding to the compartment X at time t

The fitting of the model to the experimental data collected during the compartmental analysis is shown in Figure 3 together with the corresponding values of kinetic coefficients.

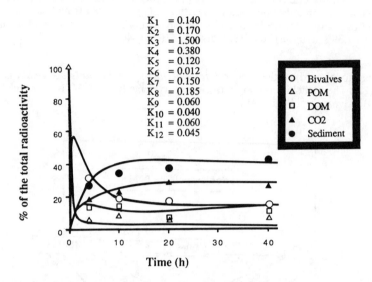

Figure 3. Time dependent variation in radioactivity of the different compartments of a *'Lactobacillus sp - Venerupis decussata '* system. Single points correspond to the results of the compartmental analysis. Continuous lines correspond to the results of analog modelling.
Redrawn after [11, 12]

Computation of ingestion and assimilation rates

The use of equation (2) leads to an ingestion rate of $2.8 \cdot 10^{-2}$ mgDW/mg DW/h for *Venerupis decussata* fed on *Lactobacillus sp*.

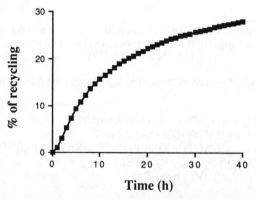

Figure 4. Proportion of recycled organic matter consumed by *Venerupis decussata*. Organic matter may be recycled through DOM and biodeposits. The percentage of recycling is set at the ratio (in %) of the cumulative amounts of radioactivity corresponding to the consumption of DOM and Faeces by bivalves, by the cumulative amounts of DOM, Faeces, and Bacteria consumed by bivalves.

The computation of the cumulated amounts of organic matter corresponding to the consumption of DOM, bacteria, and sediment by *Venerupis decussata* allow for the determination of the impact of recycling on the quantification of ingestion rates (Figure 4). At t=5 h, recycling of DOM and biodeposits account for 9.3% of the total consumption of organic matter by *Venerupis decussata*. This proportion reaches 22.2 and 28% at t=20 and 40h, respectively. Changes over time of the two estimations of assimilation rates are shown in Figure 5. The two estimations of the assimilation rate converge with increasing experiment duration. At t=40 h, the two estimations of assimilation rate are of 52.2 and 56.5%, respectively. Therefore, our conclusion is that the assimilation rate of *Venerupis decussata* fed on *Lactobacillus sp* is about 55%.

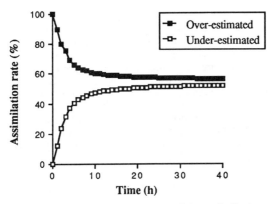

Figure 5. Changes over time of the two estimations of the assimilation rate of *Venerupis decussata* fed on *Lactobacillus* sp.

DISCUSSION

Measurements of ingestion rates

The bias resulting from the recycling of radioactive label has already been considered by several authors.

Three kind of solutions have been proposed depending on the experimental procedure of incubation which is used (i.e., closed or open systems). In a closed system, it is possible either to use more stable labelling procedures [4, 5, 6] or to correct the specific activity of the invertebrates by a coefficient accounting for the absorption of DOM. Such labelling procedures involve either different chemical compounds [4, 5, 6] or a preliminary hydrolysis of the labelled food [19] which limits their utilization to dead organic matter. The determination of the correction coefficient is based on the comparison of the specific activity in invertebrates feeding on POM and DOM, with the specific activity of either invertebrates feeding only on DOM [17] or dead invertebrates [18]. However this approach cannot account for the production of radioactive CO_2 both by the food source and by the invertebrates [10].

Open systems have often be used instead of closed systems to avoid such bias. Pulse-chase design experiments are designed to minimize recycling by using short incubation periods [11, 8] but [17], then either measuring invertebrate specific activities or transfering invertebrates to similar but unlabelled food conditions and measuring the specific activities of invertebrates and faeces. However there are three major drawbacks: (1) reducing the incubation period results in an increase of stress artefact, (2) this method is not applicable to all invertebrates but more or less restricted to small size stress tolerant organisms [10], and (3) this method requires *a priori* assumptions on the intensity of recycling (results of the present study suggests that recycling may account for as much as 10% of the total consumption of organic matter by *Venerupis decussata* after only 4 h of incubation).

The present study confirms results of previous works [20, 21] in showing that compartmental analysis and analog modelling can be used in closed experimental systems to determine ingestion rates of benthic invertebrates. The major advantage of this approach is that it takes into account all the possibilities of recycling without preliminar hypothesis. Its main drawback is that ingestion rates are not directly measured but deduced from experimental data through modelling.

Measurements of assimilation rates

There are two major kinds of approachs leading to the determination of assimilation rates in benthic invertebrates: Gravimetric methods, and indicator methods [1].

Gravimetric methods consist in the comparison of the amounts of a component which are ingested and rejected as faeces during a given meal (if ingestion is discontinuous) or a given period of time (if ingestion is continuous). These methods are based on the critical assumption that all the material present in the faeces is derived from food. Therefore, the secretion of mucus in faeces precludes the use of gravimetric approachs in most invertebrates [1]. In order to overcome this problem, gravimetric methods have been used in conjunction with radiotracers [1, 9]. The major drawbacks of gravimetric methods are: (1) that they are dependent on a good estimation of ingestion, (2) that they require a total recovery of the faeces produced (which is time consuming and is not possible for all invertebrates, (3) that assimilation rates are derived from a large number of experimental measurements (i.e., ^{14}C contents of the invertebrates and of the faeces).

Indicator methods consist in the assessment of gut passage on the ratio of an assimilable and a non assimilable compound. They are based on the use of the following equation:

$$Ass = [1 - (R_{food}/R_{faeces})] \cdot 100 \qquad (4)$$

where: Ass (%) is the assimilation rate
R_{food} is the ratio of the non assimilable by the assimilable compound in the food
R_{faeces} is the ratio of the non assimilable by the assimilable compound in the faeces

This type of approach has been first used with the ratio of organic to inorganic matter [22]. More recently it was associated to the use of radiotracers (i.e., radiotracer indicator methods) by the use of dual labelling, ^{14}C being the assimilable tracer, and ^{51}Cr being the non assimilable tracer [7]. The main advantage of this elegant method is that it is very straightforward and based on a very low number of experimental measurements. Its main drawbacks are due: (1) to the hard ß emission of ^{51}Cr which restricts its use to well equiped laboratories, (2) to the fact that $^{14}C/^{51}Cr$ ratios in the ingested and in the egested fractions cannot be carried out on the same animals [9], and (3) to the fact that it does not account for recycling during the incubation in the presence of the labelled food.

The present study confirms results of previous works [20, 21] in showing that compartmental analyis and analog modelling can be used in closed experimental systems to determine assimilation rates of benthic invertebrates. Using a similar approach, Amouroux et al [20] reported assimilation rates ranging from 51 to 75% for the deposit-feeding bivalve *Abra alba* fed on live diatoms, whereas Grémare et al [21] found assimilation rates ranging from 80 to 87% for the deposit-feeding polychaete *Eupolymnia nebulosa* fed on the same diatoms. These values are rather similar to those (60-75%) reported for the deposit-feeding gastropod *Hydrobia ventrosa* (gravimetric approach) [8], and those (38.4%) reported for *Hydrobia totteni* fed on diatoms (radiotracer indicator method) [9]. In the present study we found an assimilation rate of 55% for *Venerupis decussata* fed on the bacteria *Lactobacillus sp.* This figure can be compared with the 75% reported for *Hydrobia ventrosa* (gravimetric method) [8], the 72.1% reported for *Nucula annulata,* and the 42.1% reported for *Hydrobia totteni* fed on bacteria (radiotracer indicator method) [9]. Therefore, our overall conclusion is that the coupling of compartmental analysis and analog modelling is indeed applicable to the quantification of assimilation rates of benthic invertebrates. Its major advantages are that: (l) it makes no a *priori* assumptions on the stability of the label and on recycling, and (2) that ingestion and assimilation rates are determined simultaneously. Its main drawbacks are: (l) that assimilation rates are not measured directly but derived from experimental measurements through modelling, (2) that the assimilation rate is global and accounts for all the possible inputs of organic matter (and not only for the ingestion of POM).

REFERENCES

1. Lopez, G.R., Tanchichodok, P. and Cheng, I.J., Radiotracer methods for determining utilization of sedimentary organic matter by deposit-feeders. IN Ecology of marine deposit-feeders (Lecture notes on coastal estuarine studies; 31), eds. G. Lopez, G., Taghon, J. Levinton, Springer Verlag, Berlin., 1989, pp.149-170.

2. Conover, R.J. and Francis, R.V., The use of radioactive isotopes to measure the transfer of materials in aquatic food chains. Mar.Biol., 1973, 18, 272-283.

3. Smith, D.F. and Horner, M.J., Tracer kinetics applied to problems in marine biology. Can.Bull.Fish.Aquat.Sci., 1981, 210, 11-129.

4. Crosby, M.P., The use of a rapid radiolabelling method for measuring ingestion rates of detritivores. J.exp.mar.Biol.Ecol., 1985, 97, 273-283.

5. Wolfinbarger, L. and Crosby, M.P., A convenient procedure for radiolabeling detritus with ^{14}C-dimethylsulfate. J.exp.mar Biol.Ecol., 1983, 67, 185-198.

6. Lopez, G.R. and Crenshaw, M.A., Radiolabelling of sedimentary organic matter with ^{14}C formaldehyde: preliminary evaluation of a new technique for use in deposit-feeding studies. Mar.Ecol.Prog.Ser., 1982, 8, 283-289.

7. Calow, P. and Fletcher, C.R., A new radiotracer technique involving ^{14}C and ^{51}Cr for estimating assimilation efficiencies of aquatic primary consumers. Oecologia, 1972, 9, 155-170.

8. Kofoed, L.H., The feeding biology of *Hydrobia ventrosa* (Montagu). 1. The assimilation of different components of the food. J.exp.mar.Biol.Ecol., 1975, 19, 233-241.

9. Lopez, G.R. and Cheng, I.J., Synoptic measurements of ingestion rate, ingestion selectivity, and absorption efficiency of natural foods in the deposit-feeding molluscs *Nucula annulata* (Bivalvia) and *Hydrobia totteni* (Gastropoda). Mar.Ecol.prog.Ser., 1983, 11, 55-62.

10. Grémare, A., Consumption of diatoms and diatom filtrates by the tentaculate deposit-feeder *Eupolymnia nebulosa* (Annelida:Polychaeta). Mar.Biol., 1990, 106, 139-143.

11. Amouroux, J.M., Amouroux, J., Bastide, J. and Cahet, G., Interrelations in a microcosm with a suspension feeder and a deposit feeder. 1. Experimental study. Oceanol.Acta, 1990, 13, 61-68.

12. Amouroux, J.M., Amouroux, J., Bastide, J., Cahet, G. and Grémare, A., Interrelations in a microcosm with a suspension feeder and a deposit feeder. II. Modelling. Oceanol.Acta, 1990, 13, 69-78.

13. Amouroux, J.M., Preliminary study on the consumption of dissolved organic matter (exudates) of bacteria and phytoplankton by the marine bivalve *Venus verrucosa*. Mar.Biol., 1984, 82, 109-112.

14. Amouroux, J.M. and Amouroux, J., Comparative study of the carbon cycle in *Venus verrucosa* fed on bacteria and phytoplankton. III. Comparison of models. Mar.Biol., 1988, 97, 339-347.

15. Amouroux, J.M., Comparative study of the carbon cycle in *Venus verrucosa* fed on bacteria and phytoplankton. I. Consumption of bacteria *(Lactobacillus sp)*. Mar.Biol., 1986, 90, 237-241.

16. Amouroux, J.M., Comparative study of the carbon cycle in *Venus verrucosa* fed on bacteria and phytoplankton. II. Consumption of phytoplankton *(Pavlova lutheri)*. Mar.Biol., 1986, 92, 349-354.

17. Adams, S.M. and Angelovic, J.W., Assimilation of detritus and its associated bacteria by three species of estuarine animals. Chesapeake Sci., 1970, 12, 249254.

18. Guidi, L.D., The feeding response of the epibenthic amphipod *Siphonocoetes dellavallei* Stebbing to varying particle sizes and concentrations. J.exp.mar.Biol.Ecol., 1986, 98, 51-63.

19. Kemp, P.F., Direct uptake of detrital carbon by the deposit-feeding polychaete *Euzonus mucronata* (Treadwell). J.exp.mar.Biol.Ecol., 1986, 99, 49-61.

20. Amouroux, J.M., Grémare, A. and Amouroux, J., Modelling of consumption and assimilation in *Abra alba* (Mollusca:Bivalvia). Mar.Ecol.Prog.Ser., 1989, 51, 8797.

21. Grémare, A., Amouroux, J.M. and Amouroux, J., Modelling of consumption and assimilation in the deposit-feeding polychaete *Eupolymnia nebulosa*. Mar.Ecol.Prog.Ser., 1989, 54, 239-248.

22. Conover, R.J., Assimilation of organic matter by zooplankton. Limnol.Oceanogr., 1966, 11, 338-345.

NATURAL AND ARTIFICIAL RADIOACTIVITY IN COASTAL REGIONS OF UK

P McDonald*, G T Cook* and M S Baxter**

* Scottish Universities Research and Reactor Centre
East Kilbride, Glasgow, G75 0QU, UK.

** IAEA International Laboratory of Marine Radioactivity,
19 Avenue des Castellans, MC 98000, Monaco.

ABSTRACT

The transport and bioaccumulation of natural and artificial radionuclides in the coastal regions of UK have been investigated. The magnitude of radionuclide concentration variation throughout the UK coastline has been quantified and where appropriate source terms have been identified. Sites bordering the Irish Sea show the highest levels of artificial radioactivity which are derived from past Sellafield discharges. Enhanced concentrations of natural radioactivity were found in all Whitehaven samples (due to discharges from a phosphate ore processing plant at Whitehaven), in sediment at Blackhall Colliery near Hartlepool (from disposal of coal mining spoil directly into the North Sea) and in mussels at Aberdeen (due to discharges from the de-scaling of pipes and valves by the oil industry). Highest radionuclide concentrations were present in Whitehaven mussels -3124Bqkg^{-1} dry weight of ^{210}Po. Excluding sites bordering the Irish Sea, levels of natural radionuclides (^{210}Pb, ^{210}Po, ^{232}Th, ^{238}U) were greatly in excess of the artificial radionuclide concentrations in all sample types. The biological preference for accumulation of ^{210}Po relative to its grandparent ^{210}Pb is evident. ^{210}Po/^{210}Pb activity ratios range between 2 and 40 in mussels, winkles and seaweed but only between 0.24 and 1.3 in sediments and sea water. Radiologically, the highest radiation exposure to the public evident from the results would be from the ingestion of Whitehaven mussels (3.2mSv y^{-1}), the greatest single contribution being from technologically enhanced ^{210}Po.

INTRODUCTION

Since 1978, annual discharge rates from the BNFL nuclear fuel reprocessing plant at Sellafield, Cumbria, into the Irish Sea have been greatly reduced, the most recent discharges (1) tending towards insignificance relative to those in the mid-1970's. This discharge trend is complemented by a growing awareness since the mid-1980's of the radiological significance of natural radionuclides, particularly of ^{210}Po, in marine foodstuffs (2), (3), (4). High ^{210}Po accumulation in certain marine organisms has been known for many years (5), (6) but it is only of late, in estimating doses to man, that concentrations in mussels, winkles and crabs etc. have become of major interest (2), (7). Pentreath (4) showed that, on a daily diet of 600g fish and 100g each of crustacean, mollusc and seaweed (all consumed fresh), an annual dose of 2 mSv would be received from naturally occurring radionuclides (^{210}Po, ^{210}Pb, ^{40}K, ^{87}Rb, ^{226}Ra, U, ^{14}C), of which 75% would be attributed to ^{210}Po. The most recent calculated annual dose received by the critical group in the vicinity of Sellafield (anthropogenic nuclides only) was 0.19 mSv (8). Because of this growing but still preliminary awareness of the dosimetry of natural radionuclides, further investigations are required, particularly to understand spatial and temporal variability, determine areas of enhancement and identify potential sources. Not only are such studies needed for radiological assessment but also they may provide useful analogue information on the environmental behaviour of transuranic nuclides.

To maintain a perspective, we have undertaken a study to quantify both artificial and natural radionuclides in a consistent suite of selected materials (sea water, sediment, seaweed, mussels and/or winkles) from a variety of sites (polluted and natural) around the mainland UK coast. The natural radionuclides studied here are ^{40}K, ^{210}Pb, ^{210}Po, ^{232}Th and ^{238}U, while the artificial nuclides of interest are fission and activation products (in particular ^{137}Cs) and the transuranic nuclides (^{238}Pu, $^{239+240}$Pu and ^{241}Am).

SAMPLING AND ANALYSIS

The location of sampling sites are shown in Figure 1. For a balanced perspective sites had to be distributed around the coastline of mainland Britain in regions of a) natural uranium mineralisation, b) nuclear waste discharges, c) industrial discharges and d) no known potential radionuclide sources (control sites). Samples were collected at low tide at each site over a two week period in September 1989 with the exception of Aberdeen samples (April 1990). Sampling over a short time period effectively removes the potential for temporal variations when interpreting radionuclide concentrations in biological samples. On return to SURRC, seaweeds (Fucus vesiculosus) (~3kg wet weight) were washed and cleaned thoroughly, weighed, dried (45-50°C) and reweighed before being ground and homogenised. Sediment samples (~1kg wet weight) were dried at 45-50°C before being ground in a Tema grinder and homogenised using a motorised roller system. Mussels (Mytilus edulis) typically 35 to 55mm in length and winkles (Littorina littorea) of 15-30mm length, were removed from their shells and total soft tissues freeze-dried and then ground. Radioanalytical techniques used in this study have been described previously by McDonald et al., (9), (10). Briefly, gamma-emitting radionuclides (^{40}K, ^{137}Cs) are analysed using a high resolution reverse electrode germanium detector. (^{226}Ra concentrations in the majority of samples were below detection limits; 30-80Bqkg^{-1}). Plutonium isotopes and ^{210}Po are analysed using alpha-spectrometry. Pu is electrodeposited onto stainless steel discs while ^{210}Po is spontaneously deposited on silver discs). Replating ^{210}Po after a 5-6 month interlude determines ^{210}Pb concentrations. ^{232}Th and ^{238}U activity concentrations in all samples are determined by Inductively Coupled Plasma-Mass Spectrometry (ICP-MS) (VG Plasmaquad PQ-1).

RESULTS AND DISCUSSION

Sediments

From Table 1, it can be seen that artificial radionuclide activities are greatest in sediments bordering the Irish Sea (Sandyhills, Whitehaven, Ravenglass, Colwyn Bay) with the highest being found in Ravenglass sediment, the sample site closest to the Sellafield discharge point. Artificial radionuclides measured in Sandyhills sediment are present at higher activities than those in Whitehaven sediment despite being more distant from the Sellafield outfall. This apparent discrepancy is caused by the very sandy nature of the Whitehaven sample (cf Sandyhills), radionuclide concentrations being enriched in fine-grained sediments from the same area (11), (12). Beyond the Irish Sea sites, there is a clear indication that the more northerly sites are characterised by higher ^{137}Cs concentrations than those in the south of England. This is a direct consequence of net travel of the radiocaesium discharged from Sellafield northwards from the Irish Sea, along the west coast of Scotland and then into the North Sea. Plutonium activities in sediments beyond the Irish Sea area again show higher concentrations at the more northern sites. The $^{238}Pu/^{239+240}Pu$ activity ratios (0.13-0.23) are generally consistent with a nuclear reprocessing influence, except at Plymouth where the activity ratio indicates nuclear weapons test fallout (~0.04) (13). The very high $^{137}Cs/^{239+240}Pu$ activity ratio observed at Blackhall Colliery merits further confirmatory studies. ^{40}K is the most abundant radioactive species in all of the sediments, concentrations ranging between 25 and 760 Bqkg^{-1}. Concentration ranges of the other four natural radionuclides studied here are, interestingly, all of similar magnitude. It is also noteworthy

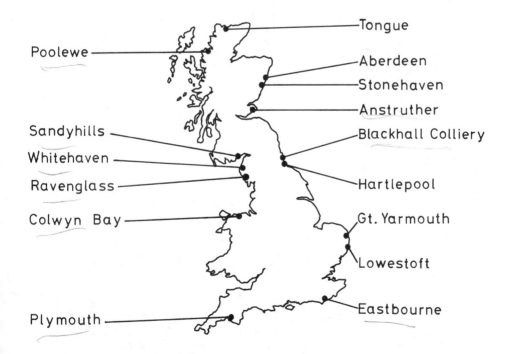

Figure 1. UK coastal study sampling locations.

TABLE 1
Radionuclide Concentrations in UK Coastal Sediments (Bqkg^{-1} dry weight ±2σ error).

Site	^{40}K	^{137}Cs	^{210}Pb	^{210}Po	^{232}Th	^{238}U	^{238}Pu	$^{239+240}$Pu
Tongue	597±12	26±1	13.9±1.0	20.1±0.4	15.6±0.6	13.4±0.4	0.16±0.02	1.03±0.08
Poolewe	495±14	7.6±0.5	7.81±0.80	10.9±0.2	9.3±1.5	13.1±0.4	0.056±0.007	0.44±0.04
Aberdeen	760±12	3.2±0.4	5.65±0.86	6.44±0.12	10.6±0.3	8.9±0.2	0.073±0.018	0.34±0.05
Stonehaven	668±15	8.2±0.7	13.1±1.4	12.5±0.4	14.3±0.6	14.0±0.4	0.097±0.011	0.54±0.04
Anstruther	193±8	7.8±0.4	8.58±0.86	9.31±0.30	9.9±0.4	12.1±0.4	0.052±0.007	0.31±0.02
Sandyhills	476±17	130±2	10.4±1.2	23.1±1.0	15.5±0.4	14.0±0.4	7.56±0.62	37.8±1.8
Whitehaven	103±8	29±1	12.8±1.0	56.6±0.6	5.3±1.0	22.1±0.4	0.67±0.16	3.13±0.32
Ravenglass	413±16	408±4	19.5±1.6	30.2±0.6	15.5±2.8	16.2±0.6	84.3±5.2	373±20
Blackhall	489±15	6.6±0.8	33.7±1.4	16.7±0.6	24.4±0.8	32.3±0.6	BDL	0.019±0.004
Hartlepool	175±10	1.5±0.2	5.34±0.66	5.63±0.26	3.4±0.2	5.2±0.1	BDL	0.135±0.028
Colwyn Bay	185±10	17±1	5.36±0.86	7.23±0.30	6.9±0.4	6.6±0.2	0.47±0.06	2.72±0.20
Great Yarmouth	77±14	<0.7	1.86±0.28	2.03±0.12	2.6±0.1	3.6±0.1	BDL	0.044±0.009
Eastbourne	25±6	<0.2	3.56±0.58	1.37±0.16	3.0±0.1	4.6±0.2	BDL	0.063±0.009
Plymouth	201±12	0.9±0.4	5.96±0.36	8.71±0.15	10.9±0.8	9.8±0.6	0.013±0.004	0.25±0.02

TABLE 2
Radionuclide Concentrations in *Fucus vesiculosus* Collected from UK Coastal Sites (Bqkg^{-1} dry weight ±2σ error).

Site	^{40}K	^{137}Cs	^{210}Pb	^{210}Po	^{232}Th	^{238}U	^{238}Pu	$^{239+240}$Pu
Tongue	955±42	4.5±1.7	1.33±0.09	6.21±0.16	0.12±0.01	12.2±0.2	0.19±0.05	0.85±0.01
Poolewe	700±40	4.2±1.5	2.18±0.14	10.9±0.2	0.03±0.01	9.5±0.6	0.42±0.16	2.22±0.46
Aberdeen	1285±46	3.3±1.4	4.16±0.42	9.54±0.42	0.18±0.01	4.5±0.1	0.15±0.08	0.79±0.19
Stonehaven	1198±36	2.4±1.2	1.75±0.44	8.82±0.22	0.16±0.02	9.9±0.5	BDL	0.28±0.05
Anstruther	1003±40	4.6±1.6	1.31±0.06	2.72±0.07	0.04±0.01	12.0±0.4	0.18±0.05	0.53±0.10
Sandyhills	1175±41	133±3	0.71±0.08	8.98±0.12	0.15±0.02	3.8±0.1	8.49±1.86	38.7±7.4
Whitehaven	1363±42	131±3	27.4±1.8	51.0±0.8	0.38±0.12	18.6±0.6	6.20±0.94	27.2±3.8
Ravenglass	1317±49	274±5	3.79±0.34	16.0±0.3	0.82±0.02	4.8±0.1	50±2	214±8
Hartlepool	1439±53	6.4±1.6	0.85±0.08	2.92±0.10	0.38±0.12	7.1±0.3	BDL	0.29±0.13
Colwyn Bay	1127±54	50±4	1.57±0.22	6.53±0.10	0.22±0.01	12.4±0.4	1.15±0.20	6.15±0.82
Great Yarmouth	374±29	4.4±0.8	1.68±0.13	6.50±0.14	0.29±0.01	5.1±0.1	BDL	0.51±0.16
Eastbourne	886±35	<1.5	1.25±0.18	8.66±0.14	0.08±0.01	8.9±0.2	BDL	0.17±0.04
Plymouth	985±47	<2.1	1.52±0.22	4.56±0.10	0.19±0.07	14.7±0.6	BDL	0.35±0.12

BDL - Below Detectable Limit   Table 1 ^{238}Pu 0.004 Bq kg^{-1}, Table 2 ^{238}Pu 0.05 Bq kg^{-1}.

that, the naturally occurring nuclides, ^{210}Pb, ^{210}Po, ^{232}Th and ^{238}U, show a general distribution pattern which coincides with the artificial nuclides in that sediments from northern locations contain, on average, higher activities than those from southern sites. For example Table 1 shows that ^{238}U and ^{232}Th activities are around three times higher in coastal sediments from northern UK. This coincides with the partitioning of catchment geology between igneous and metamorphic regimes.

Typical ^{232}Th and ^{238}U concentrations in marine sediments have been quoted at 18.5 and 74 Bqkg^{-1} respectively by Baxter (14) and at 20 and 11 Bqkg^{-1} respectively by Holm and Fukai (15). Clearly concentrations are variable as a function of geology and even the enhanced levels at Blackhall Colliery and Whitehaven fall within the range of natural variability.

Literature data for ^{210}Pb and ^{210}Po in sediments are particularly variable but, referring to Table 1, the ^{210}Po levels at Sandyhills, Whitehaven and Ravenglass are undoubtedly enhanced relative to the other UK coastal sites. Those at Ravenglass and Sandyhills may be a consequence of the direct transport of high ^{210}Po concentrations from Whitehaven.

Seaweeds

As with sediments, activities of artificial radionuclides in seaweeds were greatest in Ravenglass followed by Whitehaven, Sandyhills and Colwyn Bay (Table 2). Between Tongue and Stonehaven, ^{137}Cs and $^{239+240}$Pu activities (2.4-4.5 and 0.28-2.22 Bqkg^{-1} respectively) are in agreement with those reported by McDonald et al., (9) in Fucus collected in north of Scotland waters (^{137}Cs 3.7-5.3 Bqkg^{-1}, $^{239+240}$Pu 0.47-1.71 Bqkg^{-1}). Activities in Ravenglass seaweed, however, are much higher in this study (~3.5 times), than those reported by Hunt (8), suggesting that some sediment grains may have been occluded in the seaweed during analysis.

Typical ^{40}K, ^{210}Pb, ^{210}Po, ^{232}Th and ^{238}U concentrations in seaweed have been reported as 900-1100, 0.6-9.5, 0.6-14, 0.02-0.15 and 0.1-11 Bqkg^{-1} respectively (9), (16). From Table 2, the majority of seaweeds have activities within these ranges. Some elevation of natural radioactivity in seaweeds is, however, present at Whitehaven (^{210}Pb, ^{210}Po, ^{232}Th, ^{238}U), Ravenglass (^{232}Th) and Hartlepool (^{232}Th). ^{210}Po/^{210}Pb activity ratios range from 1.8 to 6.9 exhibiting an excess of ^{210}Po in seaweeds relative to its grandparent ^{210}Pb.

Mussels and Winkles

Table 3 shows that mussels and winkles collected from the Irish Sea shoreline exhibit the highest activities of artificial radionuclides with Ravenglass samples containing the highest levels. Further afield, ^{137}Cs levels are generally less than 10 Bqkg^{-1} and approaching the lower limit of detection for this sample matrix. $^{239+240}$Pu activities distant from the Irish Sea area are between 0.1 and 1 Bqkg^{-1}. The plutonium concentrations in mussels and winkles from the same site are similar.

Although $^{239+240}$Pu concentrations in mussels and winkles are similar, there is a quite definite bias for ^{210}Po, the mussel containing the higher activities. ^{210}Po enhancement is particularly evident in Aberdeen and Whitehaven mussels, typical natural ^{210}Po levels being between 100 and 200 Bqkg^{-1} in mussels and between 13 and 80 Bqkg^{-1} in winkles (2), (17). An enhanced ^{210}Po activity in winkles is observed at Whitehaven, as reported by Pentreath et al., (7). Elevated activities of ^{210}Pb coincide exactly with those of ^{210}Po in Aberdeen mussels and of ^{210}Po in Whitehaven mussels and winkles. Beyond the regions of ^{210}Pb and ^{210}Po enhancement, ^{210}Po levels are higher in mussels than in winkles, a trend which is not evident for ^{210}Pb suggesting that there are two different uptake mechanisms for these radionuclides. ^{210}Po/^{210}Pb activity ratios range between 9 and 40 in mussels

TABLE 3

Radionuclide Concentrations in Mussels and Winkles Collected from UK Coastal Sites (Bqkg^{-1} dry weight ±2σ error).

Site		^{40}K	^{137}Cs	^{210}Pb	^{210}Po	^{232}Th	^{238}U	^{238}Pu	$^{239+240}$Pu
Tongue	Mussel	211±154	7.9±5.6	8.0±0.8	197±2	0.161±0.01	2.08±0.02	BDL	0.49±0.09
	Winkle	440±100	<5.1	16.1±2.4	79.6±1.2	0.54±0.01	3.49±0.05	0.18±0.06	0.69±0.13
Poolewe	Mussel	281±177	<6.1	13.6±2.0	168±2	0.07±0.01	2.92±0.06	0.15±0.06	0.79±0.16
Aberdeen	Mussel	606±151	<6.9	75.5±5.4	625±12	0.69±0.02	2.90±0.06	0.09±0.04	0.76±0.13
	Winkle	526±104	<5.2	4.0±0.7	67±4	2.50±0.05	2.59±0.06	0.10±0.04	0.71±0.09
Stonehaven	Mussel	268±203	<6.4	17.4±1.8	158±2	0.19±0.01	3.50±0.04	0.10±0.04	0.36±0.08
	Winkle	493±123	<5.3	3.31±0.64	53.4±0.6	0.22±0.01	2.46±0.03	BDL	0.23±0.08
Anstruther	Winkle	136±44	<1.6	1.11±0.24	13.1±0.2	0.007±0.02	1.36±0.02	BDL	0.29±0.12
Sandyhills	Mussel	416±132	43±5	5.0±0.5	143±2	0.23±0.01	1.11±0.03	1.71±0.16	7.98±0.38
	Winkle	525±108	63±6	4.32±0.90	28.1±0.8	0.59±0.02	2.72±0.01	2.79±0.22	13.5±0.5
Whitehaven	Mussel	445±206	77±13	284±14	3124±18	0.72±0.01	37.1±0.6	5.35±0.38	21.2±0.9
	Winkle	456±105	58±5	68±4	399±2	0.61±0.01	18.9±0.2	3.67±0.30	17.3±0.7
Ravenglass	Mussel	396±125	75±6	6.7±1.3	160±2	0.30±0.01	2.72±0.04	18.3±1.6	80.7±4.6
	Winkle	481±110	203±9	5.3±0.9	60±1	0.81±0.001	2.18±0.04	20.0±1.4	84.4±4.2
Hartlepool	Mussel	372±180	<6.9	7.5±0.6	104±2	0.35±0.01	1.62±0.03	BDL	0.15±0.04
Colwyn Bay	Mussel	315±109	18±3	2.8±0.5	105±2	0.16±0.01	1.01±0.01	0.18±0.08	0.87±0.18
	Winkle	324±93	17±2	2.7±0.5	58.7±0.3	0.23±0.02	2.07±0.06	0.19±0.04	1.09±0.04
Lowestoft	Mussel	511±216	<11.6	4.7±0.9	186±2	0.95±0.02	3.45±0.05	BDL	0.47±0.11
Plymouth	Winkle	379±124	<6.2	2.1±0.5	51±1	0.11±0.01	2.69±0.08	BDL	1.04±0.40

BDL - Below Detectable Limit ^{238}Pu   0.04 Bqkg^{-1}

and between 5 and 24 in winkles, demonstrating that ^{210}Po is largely unsupported by ^{210}Pb in both species.

The mussel feeds on phytoplankton and other suspended material and, because of its ventilation rate, particle retention ability and ingestion efficiency, the mussel's filter-feeding mode is highly effective in collecting ambient food, especially where concentrations of food in the water column are normally high. Hence the sources of ^{210}Po to the mussel are threefold - sea water, phytoplankton and suspended sedimentary material, resulting in an abundance of ^{210}Po available for bioaccumulation.

^{40}K activities range between 100 and 600 Bqkg^{-1}, although because of small sample weights and high background considerations, large errors are associated with these data. This range, however, is much smaller than the ^{40}K range found in sediments. Relatively elevated activites of ^{232}Th were observed at Aberdeen, Whitehaven, Ravenglass and Lowestoft. ^{238}U, on the other hand, clearly demonstrates high concentrations in Whitehaven mussels and winkles (37 and 19 Bqkg^{-1} respectively) and lower concentrations elsewhere (1-3.5 Bqkg^{-1}.

Sea Water
Radionuclide concentrations determined in sea water are given in Table 4. For comparative purposes, the concentrations of the conservative nuclides, ^{137}Cs and ^{238}U, have been normalised to a typical coastal sea water salinity of 33°/oo. ^{137}Cs levels are of the same magnitude reported by Hunt (8) and Nies (18), ie North of Scotland waters ~50 mBql^{-1}, east coast of Britain ~25 mBql^{-1}, Colwyn Bay area 250-500 mBql^{-1}, Whitehaven and Sandyhills (north-east Irish Sea area) ~500 mBql^{-1}, Straits of Dover 7-15 mBql^{-1}. ^{137}Cs concentrations in Ravenglass sea water are consistent with the value (<1800 mBql^{-1}) given for this region by BNFL (1).

$^{239+240}$Pu concentrations are in agreement with those recently reported by Nies (19) for the east coast of Britain (0.032-0.094 mBql^{-1}) and by BNFL for the vicinity of Sellafield (10-15 mBql^{-1} for Puα). Other sites at which plutonium was detected (the Irish Sea area) exhibit concentrations intermediate between those of Ravenglass and the east coast. ^{238}Pu/$^{239+240}$Pu activity ratios in all samples exhibit a reprocessing influence.

Under natural conditions, ^{210}Pb, ^{210}Po and ^{238}U concentrations in surface sea water are 0.4-5, 0.2-4 and 37-44 mBql^{-1} respectively (4), (14), (15), (16), (19). The sea water samples exhibit excellent agreement with these values, enhancements, however, being at Whitehaven (^{210}Pb, ^{210}Po, ^{238}U). ^{210}Po/^{210}Pb ratios are typically less than unity indicating that ^{210}Po in sea water is, in general, totally supported by ^{210}Pb. Less commonly, however, local effects may produce ratios greater than 1.

The higher concentrations of artificial radioactivity can be accounted for by the vicinity of sampling sites to Sellafield. Similarly, the enhanced levels of natural radioactivity reflect nearby industrial inputs, the industries in this case being non-nuclear. At Whitehaven, the Albright and Wilson Marchon plant produces phosphoric acid and phosphates, processes which result in the discharges to sea of waste which is relatively high in natural uranium and its decay products. Such industrial practices result in contamination of the surrounding environment with enhanced activities of natural radioactivity (7), (20), (21). General activities of the natural radionuclides ^{226}Ra, ^{232}Th and ^{238}U in raw phosphate rock are 1200-2300 Bqkg^{-1}, 20-110 Bqkg^{-1} and 1300-2300 Bqkg^{-1} respectively (22). At Blackhall Colliery, 3 miles north of Hartlepool, spoil from coal mining is discharged directly into the North Sea. Because of the associated concentrations of natural radionuclides in coal, reviewed by Tadmor (23) (^{210}Pb 10-30 Bqkg^{-1}, ^{210}Po 10-30 Bqkg^{-1}, ^{226}Ra 10-2600 Bqkg^{-1}, ^{232}Th 2-100 Bqkg^{-1}, ^{238}U 10-160 Bqkg^{-1}), technologically enhanced levels are observed in the surrounding area. In the

TABLE 4

Radionuclide Concentrations in Filtered UK Coastal Waters (mBql^{-1} ±2σ error).

Site	Salinity(‰)	^{137}Cs	^{137}Cs corrected to 33‰ Salinity	^{210}Pb	^{210}Po	^{238}U	^{238}U corrected to 33‰ Salinity	^{238}Pu	$^{239+240}$Pu
Tongue	32.01	51±1	52±1	0.8±0.1	0.9±0.1	36±2	37±2	0.028±0.012	0.101±0.026
Poolewe	15.35	23±2	49±3	3.4±0.5	4.7±0.2	15±1	32±1	0.023±0.008	0.099±0.018
Aberdeen*	33.99	32±2	33±2	1.9	1.9	40±1	39±1	0.018±0.006	0.092±0.012
Stonehaven	34.33	14±2	14±2	1.3±0.2	1.1±0.1	41±1	40±1	BDL	0.038±0.012
Anstruther	33.40	28±2	28±2	1.6±0.2	0.9±0.1	40±1	40±1	BDL	0.032±0.010
Sandyhills	30.41	575±20	624±22	2.3±0.5	2.0±0.2	37±3	40±3	0.82±0.13	3.84±0.48
Whitehaven	32.32	594±10	606±10	32±1	7.8±0.6	83±3	85±3	0.43±0.06	1.75±0.18
Ravenglass	32.36	982±17	1000±17	4.7±0.6	1.5±0.2	43±3	43±3	1.95±0.50	8.6±1.0
Hartlepool	29.99	30±1	33±1	18±1	8.7±0.3	45±2	50±3	0.008±0.004	0.044±0.010
Colwyn Bay	31.10	323±6	343±6	15±1	8.7±0.4	42±2	45±2	0.063±0.022	0.427±0.068
Great Y'mth	33.44	32±2	32±2	4.7±0.6	4.0±0.2	40±2	39±2	BDL	BDL
Eastbourne	34.41	8.5±1.4	8.2±1.3	1.0±0.2	0.7±0.1	39±2	37±2	BDL	BDL
Plymouth	33.72	BDL	-	1.0±0.1	1.0±0.1	41±2	40±2	BDL	BDL

* - Aberdeen ^{210}Pb and ^{210}Po sea water concentrations based on 1st ^{210}Po - plating data. ^{210}Pb and ^{210}Po activity ratio assumed to be unity. These measures have been taken as a result of the lack of time available for reliable ^{210}Po replating data (5 months at least).
BDL - Below Detection Limit ^{137}Cs 3mBql^{-1}, Pu 0.006mBql^{-1}

vicinity of Aberdeen, the removal of barium sulphate scales from pipes and valves used in the offshore oil industry in the North Sea is undertaken. This scale material is high in natural radionuclides, particularly ^{226}Ra, ^{228}Ra and their daughters (typical activities are ~37 kBqkg^{-1} but can be as high as 3.7 MBqkg^{-1})(24). Therefore discharge of such scale into Aberdeen water would explain the high ^{210}Po concentrations found in mussels from the area. However, elevated levels of natural radionuclides were not found in other sample types from this site. This inconsistency requires further investigation.

Radiological Assessment
The radiological assessment performed here is based on the ingestion of mussels and winkles (10 kg y^{-1} fresh weight). Excluding Irish Sea sites, doses, received from artificial radionuclides, would be typically less than 1 μSvy^{-1} (present recommended NRPB principal limit on UK public exposure is 1 mSvy^{-1} (25)). For the Irish Sea sites artificially derived exposures would vary between 1 and 180 μSvy^{-1}.
The calculated doses received from natural radionuclides are dominated by contributions from ^{210}Pb and ^{210}Po. ^{232}Th and ^{238}U exert a minor influence. At Aberdeen and Whitehaven (regions of ^{210}Pb and ^{210}Po enhancement), natural dose rates of 690 μSvy^{-1} and 3230 μSvy^{-1} would be received from the consumption of mussels. Elsewhere typical natural dose rates lie between 25 and 200 μSvy^{-1}. It is interesting to note that in areas where no enhanced levels of radionuclides are present, calculated doses received from natural ^{210}Po through mussel and winkle consumption (20-158 μSvy^{-1}) are similar to the doses received by the members of the Sellafield critical group from plutonium and americium (155 μSvy^{-1}; (8)).

CONCLUSIONS

This preliminary research has provided a broad data set on radionuclides in a range of coastal samples throughout the UK. It indicates areas of artificial radionuclide enhancement - Sandyhills, Whitehaven, Ravenglass and Colwyn Bay - associated with Sellafield discharges, the highest being at Ravenglass. Areas of natural radionuclide enhancement are found at Aberdeen (due to discharges from the oil industry-descaling of pipes etc), Blackhall Colliery (from coal spoil disposal directly into the North Sea) and Whitehaven (from Albright and Wilson phosphate processing plant discharges). Excluding sites on the Irish Sea coast, natural radionuclide concentrations render artificial radionuclide concentrations insignificant. Subsequent radiological assessments for these sites are dominated by contributions from natural radionuclides, in particular ^{210}Pb and ^{210}Po. It is interesting to note generally that activities of natural radioactivity in coastal sediments are high in northern UK and low in southern UK, reflecting the rough geological division of the country, an observation specific for UK geology but not implicit for the rock types concerned. It is evident here that mussels have a greater ability than winkles or seaweed to accumulate ^{210}Po, this being a direct consequence of their respective feeding mechanisms. ^{210}Po/^{210}Pb activity ratios in potential sources - sea water and sediment - range between 0.24 and 1.3. In the organisms, such quotients are 2 to 40. This selective uptake of ^{210}Po suggests that it is more biologically compatible than ^{210}Pb, probably reflecting an affinity for sulphur-containing compounds in such systems (ie amino acids, proteins) (26), (27).
Overall, this research reinforces the increasing environmental awareness of technologically enhanced inputs of natural radionuclides by conventional industries and sets in context the relative contributions to enhanced marine

radioactivity in UK by the nuclear and non-nuclear industries.

ACKNOWLEDGEMENTS

The authors wish to thank Mr K McKay for assistance in processing samples for analysis using the ICP-MS technique. This research was funded by BNFL, Risley (contract H54588B). The views expressed in this paper are those of the authors and do not necessarily reflect those of BNFL, Risley.

REFERENCES

1. BNFL, Annual Report on radioactive discharges and monitoring of the environment, 1989. British Nuclear Fuels plc, Health and Safety Directorate, Risley, 1990.
2. McDonald, P., Fowler, S.W., Heyraud, M. and Baxter, M.S., Polonium-210 in mussels and its implications for environmental alpha-autoradiography. J. Environ. Radioactivity, 1986, 3, 293-303.
3. Carvalho, F.P., ^{210}Po in marine organisms: a wide range of natural radiation dose domains. Radiation Protection Dosimetry, 1988, 24, 113-7.
4. Pentreath, R.J., Radionuclides in the aquatic environments. In Radionuclides in the Food Chain, ed. M.W. Carter, Springer-Verlag, 1988, pp. 99-119.
5. Hoffman, F.L., Hodge, V.F. and Folsom, R.R., ^{210}Po radioactivity in organs of selected tunas and other marine fish. J. Radiation Res., 1974, 15 103-6.
6. Heyraud, M. and Cherry, R.D., Polonium-210 and lead-210 in marine food chains. Mar. Biol., 1979, 52, 227-36.
7. Pentreath, R.J., Camplin, W.C. and Allington, D.J., Individual and collective dose rates from naturally occurring radionuclides in seafood. In Proceedings of 4th International Symposium on Radiation Protection -Theory and Practice, 1989, pp. 297-300.
8. Hunt, G.J., Radioactivity in surface and coastal waters of the British Isles, 1989. Aquatic Environment Monitoring Report No 23. Ministry of Agriculture Fisheries and Food, Directorate of Fisheries Research, Lowestoft, 1990.
9. McDonald, P., Cook, G.T. and Baxter, M.S., A radiological assessment of Scottish edible seaweed consumption. Environmental Management and Health, 1990, 1, 18-27.
10. McDonald, P., Cook, G.T., Baxter, M.S. and Thomson, J.T., Radionuclide transfer from Sellafield to south-west Scotland. J. Environ. Radioactivity, 1990, 12, 285-98.
11. Jefferies, D.F., Fission-product radionuclides in sediments from the north-east Irish Sea. Helgolander wiss. Meeresunters, 1968, 17, 280-90.
12. Duursma, E.K. and Eisma, D., Theoretical, experimental and field studies concerning reactions of radioisotopes with sediments and suspended particles of the sea. Part C, Applications to field studies. Neth. J. Sea Res., 1973, 6, 265-324.
13. Hardy, E.P., Krey, P.W. and Volchok, H.L., Global inventory and distribution of fallout plutonium. Nature, 1973, 241, 444-5.
14. Baxter, M.S., The disposal of high-activity nuclear waste in the oceans. Mar. Poll. Bull., 1983, 14, 126-32.
15. Holm, E. and Fukai, R., Actinide isotopes in the marine environment. J. Less-Common Metals, 1986, 52, 227-36.

16. Cherry, R.D. and Shannon, L.V., The alpha radioactivity of marine organisms. Atom. Energy Rev., 1974, 12, 1-45.
17. McDonald, P., Baxter, M.S. and Fowler, S.W., Distribution of alpha- and gamma- emitting radionuclides in mussels, winkles and prawns. J. Environ. Radioactivity, in press.
18. Nies, H., The contamination of the North Sea by artificial radionuclides during the year 1987. J. Environ. Radioactivity, 1990, 11, 55-70.
19. Woodhead, D.S., Contamination due to radioactive materials. Mar. Ecol., 1984, 5, 1111-1287.
20. Koster, H.W., Radiation doses caused by natural radionuclides in the north-east Atlantic. In The Radiological Exposure of the Population of the European Community from Radioactivity in North European Marine Waters - Project 'MARINA'. Commission of the European Communities, 1989, pp. 269-84.
21. McCartney, M., Kershaw, P.J. and Allington, D.J., The behaviour of ^{210}Pb and ^{226}Ra in the eastern Irish Sea. J. Environ. Radioactivity, 1990, 12, 243-65.
22. UNSCEAR, Ionising radiation and biological effects. Annex D, 1982, pp. 113-7.
23. Tadmor, J., Radioactivity from coal-fired power plants: a review. J. Environ. Radioactivity, 1986, 4, 177-204.
24. Heaton, B., Radioactive scale in offshore oil installations. In Radiation Protection Practice, 7th International Congress of the International Radiation Protection Association, IRPA7, Sydney, Australia, 1988, Vol 2, pp. 872-5.
25. NRPB, Committed doses to selected organs and committed effective doses from intakes of radionuclides. NRPB-GS7. National Radiological Protection Board, Oxford, 1987.
26. Heyraud, M., Cherry, R.D. and Dowdle, E.B., The subcellular localisation of natural ^{210}Po in the hepatopancreas of the Rock Lobster (Jasus Calandii). J. Environ. Radioactivity, 1987, 5, 249-60.
27. Harada, K., Burnett, W.C., LaRock, P.A. and Cowart, J.B., Polonium in Florida groundwater and its possible relationship to the sulfur cycle and bacteria. Geochim. Cosmochim. Acta, 1989, 56, 143-50.

THE ECOLOGICAL HALF-LIFE OF Cs-137 IN JAPANESE COASTAL MARINE BIOTA

YUTAKA TATEDA and JUN MISONOU
Central Research Institute of Electric Power Industry
270-11, 1646 Abiko, Abiko city, Chiba, JAPAN

ABSTRACT

The origin of Cs-137 in the Japanese marine ecosystem is largely attributed to fallout from nuclear tests in the 1960's. The data concerning fallout in Japan indicate that radiocesium introduced into the Japanese environment has recently leveled off. The change of the atom ratio between Cs-137 and stable Cs in the marine organisms of different trophic level and seawater was studied to estimate the ecological half-life of radiocesium in the Japanese coastal marine ecosystem from 1984 to 1990.
In the analysis of the Cs-137/stable Cs atom ratio in the different trophic level, it was found that the ratio for each trophic level had nearly the same values. The Cs-137/stable Cs atom ratio in marine biota and seawater decreased exponentially from 5.6E-9 to 3.6E-9 during the six years period. The ecological half-life of Cs-137 in the Japanese marine ecosystem was estimated to be approximately 14 years.

INTRODUCTION

Cs-137 radioactivity in the Japanese environment has decreased over the last 30 years. The distinctive feature of Cs-137 in Japanese environment is that the origin of this radionuclide has been fallout from atomic bomb tests of which the 1981 tests in China were the most recent, and also from the Chernobyl nuclear reactor accident in 1986 (1,2). The emission of radionuclides from Japanese nuclear facilities is highly regulated and the Cs-137 emission has been negligible. Given this situation, Cs-137 in the marine environment was anticipated to be in a state of equilibrium with stable Cs, because the Cs-137 activity level in recent fallout has leveled off in Japan (3), and Cs-137 introduced into the marine coastal environment has been well-mixed by marine processes. Also as the main chemical form of Cs is the dissolved ionic state (4), Cs-137 is more likely to reach the state of equilibrium with stable Cs. If we are going to use Cs-137 as a tracer of stable Cs behaviour in the marine ecosystem, it is necessary to check the change in the ratio of Cs-137 to stable Cs in environmental samples. Since there are few studies on Cs-137/stable Cs ratios in Japanese marine environmental samples (5,6), we analyzed Cs-137 and stable Cs in

Japanese coastal marine organisms and seawater from 1984 to 1990 to estimate the ecological half-life of Cs-137 in marine ecosystem in Japan.

MATERIALS AND METHODS

From 1984 to 1990, marine biological samples and seawater were collected from 8 locations in coastal areas of Japan (Fig. 1). All sites contained typical Japanese coastal marine biota. 65 samples of algae, including 18 species, 16 samples of invertebrates, including 10 species, and 37 samples of fish, including 18 species, were collected by means of SCUBA diving, traps, and set net fishing. 12 samples of filtered seawater were collected at the same sites.

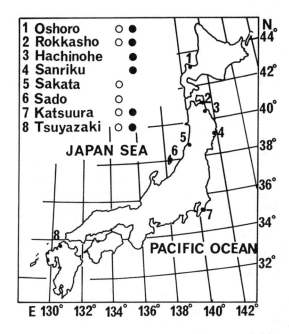

Figure 1. Location of sampling sites. The open circles are seawater stations, while the filled circles are biota stations.

Biological samples were cleaned, dissected and dried for 24 hours at 110°C weighed, and ashed at 250°C for 24 hours, then 450°C for 48 hours. The Cs-137 in the whole-body samples of algae, the soft parts of invertebrates, the muscles of fish and seawater was separated radiochemically by the ammonium-molybdophosphate method. The beta-ray of Cs-137 in samples was measured with a low-background, gas-flow counter (ALOKA LBC-451). The stable Cs concentration in biological samples was determined by non-destructive neutron activation analysis with irradiation at JRR-4 in the Japan Nuclear Institute, and gamma-ray counting with a Ge(Li) detector (ORTEC GEM-10195) and a multi-channel analyzer (TRACOR NORTHERN TN-1750). Seawater samples were frozen dried and packed in quartz glass tube, and the stable Cs content was analyzed using the same method. The credibility of analytical methods were assured by determinations of Cs-137 in the reference standard materials of IAEA/AG-B-1, IAEA/MA-B-3/1, and of Cs in the NIES No.1, No.6 and NBS SRM 1572.

To find the relation between the Cs-137/stable Cs ratio and the trophic level in the food chain, the trophic level of sample organisms in the food chain was assigned by taxnomical sorting of their stomach contents, and the reported food items of the organisms.

RESULTS

Cs-137/stable Cs Atom Ratio in Japanese Coastal Seawater

The Cs-137 concentration, the stable Cs concentration, and the Cs-137/stable Cs atom ratio in seawater at all collection sites are shown in Table 1. The Cs-137 concentration in Japanese coastal seawater was approximately 5.1 mBq/l in 1984, and decreased to 3.5 mBq/l in 1990. The stable Cs concentration in filtered seawater ranged from 0.25 to 0.34 µg/l. The Cs-137/stable Cs atom ratios were in the order of magnitude of 1E-09.

TABLE 1. The Cs-137/stable Cs atom ratio in seawater (0.45 µm filtered) at collection sites

Date	Location	Cs-137 mBq/l	Stable Cs µg/l	Cs-137/Cs atom ratio
1984/05/16	Katsuura	5.0 ± 0.3(a)	0.25 ± 0.01(b)	6.0E-09
05/22	Sakata	5.1 ± 0.2	0.29 ± 0.01	5.3E-09
1985/06/27	Tsuyazaki	4.7 ± 0.2	0.34 ± 0.02	4.2E-09
05/21	Sado	4.8 ± 0.2	0.25 ± 0.02	5.8E-09
1987/07/15	Oshoro	4.5 ± 0.3	0.3(c)	4.6E-09(d)
06/26	Tsuyazaki	3.7 ± 0.2	0.3	3.7E-09
05/20	Rokkasho	5.0 ± 0.2	0.3	5.0E-09
1989/07/12	Oshoro	4.2 ± 0.2	0.3	4.2E-09
07/05	Tsuyazaki	3.6 ± 0.3	0.3	3.6E-09
08/03	Rokkasho	3.6 ± 0.2	0.3	3.6E-09
1990/07/18	Oshoro	3.5 ± 0.2	0.3	3.5E-09
07/03	Rokkasho	3.7 ± 0.2	0.3	3.8E-09

(a) One sigma counting uncertainty
(b) 5% counting error
(c) Not analyzed. Value was cited from Bowen (7).
(d) Calculated value based on average stable Cs concentration in seawater: 0.3µg/l.

Cs-137 Concentration, Stable Cs Concentration and Cs-137/stable Cs Atom Ratio in Japanese Coastal Marine Organisms

The Cs-137 concentration, the stable Cs concentration and the Cs-137/stable Cs atom ratio in marine organisms collected are shown in Table 2. The Cs-137 and stable Cs concentraions in marine organisms are known to be high in fish muscle, low in invertebrate soft parts, and to vary widely in algae (4). Our results showed the same trend. The Cs-137/stable Cs atom ratio were mostly in a order of magnitude of 1E-09. All organisms were classified in trophic levels as follows; (1): producers, (2): herbivores, (3): detritivores and plankton feeders, (3): primary carnivores, (4): secondary carnivores, (5): tertiary carnivores.

TABLE 2. Cs-137 concentration, stable Cs concentration and Cs-137/stable Cs atom ratio in Japanese coastal marine organisms.

Year Date	Location	Species	Trophic level	Cs-137 Bq/kg (wet weight)	Stable Cs ng/g	Cs-137/ stable Cs atom ratio
		1984 Algae				
7/10	Oshoro	Ulva pertusa	1	0.21 ± 0.1(a)	7.7	6.3E-09
7/10	Oshoro	Sargassum yezoense	1	0.13 ± 0.02	15	2.6E-09
6/23	Tsuyazaki	Sargassum horneri	1	0.13 ± 0.04	10	3.7E-09
7/10	Oshoro	Undaria pinnatifida	1	0.33 ± 0.06	19	5.2E-09
7/10	Oshoro	Laminaria religiosa	1	0.13 ± 0.01	6.9	5.5E-09
6/19	Katsuura	Sargassum horneri	1	0.1 ± 0.02	5.2	5.6E-09
7/10	Oshoro	Sargassum horneri	1	0.61 ± 0.2	33	5.6E-09
6/23	Tsuyazaki	Eisenia bicyclis	1	0.25 ± 0.04	13	6.1E-09
6/19	Katsuura	Sargassum ringgoldianum	1	0.17 ± 0.02	7.7	6.2E-09
6/19	Katsuura	Hizikia fusiformis	1	0.11 ± 0.02	5.2	6.2E-09
7/10	Oshoro	Sargassum thunbergii	1	0.45 ± 0.1	22	6.3E-09
6/19	Katsuura	Sargassum patens	1	0.16 ± 0.02	6.4	7.6E-09
6/23	Tsuyazaki	Hizikia fusiformis	1	0.22 ± 0.03	7.2	9.1E-09
6/23	Tsuyazaki	Sargassum ringgoldianum	1	0.26 ± 0.05	6.6	1.2E-08
7/10	Oshoro	Laurencia nipponica	1	0.14 ± 0.03	13	3.3E-09
7/10	Oshoro	Neodilsea yendoana	1	0.37 ± 0.05	27	4.1E-09
		1985 Algae				
7/16	Oshoro	Enteromorpha intestinalis	1	0.25 ± 0.06	22	3.5E-09
5/22	Katsuura	Hizikia fusiformis	1	0.07 ± 0.01	6.8	2.9E-09
6/25	Tsuyazaki	Sargassum patens	1	0.09 ± 0.02	8.4	3.1E-09
6/25	Tsuyazaki	Eisenia bicyclis	1	0.11 ± 0.01	9.3	3.5E-09
7/16	Oshoro	Undaria pinnatifida	1	0.14 ± 0.03	12	3.5E-09
7/16	Oshoro	Sargassum yezoense	1	0.13 ± 0.03	11	3.8E-09
6/25	Tsuyazaki	Hizikia fusiformis	1	0.12 ± 0.01	9.3	3.8E-09
7/16	Oshoro	Laminaria religiosa	1	0.17 ± 0.01	12	4.2E-09
5/22	Katsuura	Sargassum ringgoldianum	1	0.16 ± 0.02	11	4.4E-09
6/25	Tsuyazaki	Sargassum ringgoldianum	1	0.09 ± 0.01	36	7.6E-09
7/16	Oshoro	Neodilsea yendoana	1	0.24 ± 0.03	16	4.6E-09
		1985 Fish				
8/19	Katsuura	Sebastes inermis	3	0.31 ± 0.01	17	5.5E-09
8/19	Katsuura	Parapristipoma trilineatum	3	0.19 ± 0.01	10	5.6E-09
8/19	Katsuura	Hexagrammos otakii	4	0.2 ± 0.01	11	5.7E-09
8/19	Katsuura	Sebastes pachycephalus	4	0.19 ± 0.01	8.7	6.4E-09
8/19	Katsuura	Ditrema temmincki	3	0.25 ± 0.01	11	6.8E-09
8/19	Katsuura	Sebastiscus marmoratus	4	0.4 ± 0.02	17	6.9E-09
		1986 Algae				
7/3	Oshoro	Enteromorpha intestinalis	1	0.08 ± 0.02	5.7	4.2E-09
6/23	Tsuyazaki	Codium fragile	1	0.03 ± 0.01	1.3	6.4E-09

TABLE 2. (continued)

Year Date	Location	Species	Trophic level	Cs-137 Bq/kg (wet weight)	Stable Cs ng/g	Cs-137/ stable Cs atom ratio
		1986 Algae				
7/ 3	Oshoro	Ulva pertusa	1	0.03 ± 0.01	1.0	8.0E-09
6/25	Sanriku	Ulva pertusa	1	0.14 ± 0.02	2.9	1.5E-08
6/25	Sanriku	Desmaleia ligulata	1	0.03 ± 0.00	2.3	4.0E-09
6/23	Tsuyazaki	Sargassum patens	1	0.14 ± 0.03	10	4.4E-09
7/ 3	Oshoro	Sargassum horneri	1	0.11 ± 0.02	6.6	4.9E-09
6/25	Sanriku	Sargassum thunbergii	1	0.16 ± 0.02	10	5.0E-09
5/26	Katsuura	Sargassum horneri	1	0.32 ± 0.02	17	5.6E-09
6/25	Sanriku	Hizikia fusiformis	1	0.12 ± 0.01	6.4	5.7E-09
6/25	Sanriku	Laminaria japonica	1	0.05 ± 0.00	2.7	6.0E-09
7/ 3	Oshoro	Sargassum yezoense	1	0.13 ± 0.02	6.3	6.2E-09
7/ 3	Oshoro	Sargassum thunbergii	1	0.16 ± 0.04	7.3	6.7E-09
6/23	Tsuyazaki	Eisenia bicyclis	1	0.22 ± 0.02	9.3	7.0E-09
5/26	Katsuura	Sargassum thunbergii	1	2.05 ± 0.05	88	7.0E-09
6/25	Sanriku	Sargassum horneri	1	0.1 ± 0.01	4.2	7.1E-09
6/25	Sanriku	Costaria costata	1	0.08 ± 0.01	3.3	7.3E-09
7/ 3	Oshoro	Laminaria religiosa	1	0.28 ± 0.04	12	7.3E-09
6/23	Tsuyazaki	Sargassum ringgoldianum	1	0.12 ± 0.03	4.8	7.8E-09
6/23	Tsuyazaki	Hizikia fusiformis	1	0.16 ± 0.01	6.0	8.0E-09
6/23	Tsuyazaki	Sargassum horneri	1	0.14 ± 0.02	5.1	8.1E-09
7/ 3	Oshoro	Undaria pinnatifida	1	0.17 ± 0.02	6.1	8.2E-09
5/26	Katsuura	Sargassum ringgoldianum	1	0.3 ± 0.01	9.6	9.4E-09
5/26	Katsuura	Sargassum patens	1	0.18 ± 0.02	4.4	1.3E-08
7/ 3	Oshoro	Laurencia nipponica	1	0.11 ± 0.01	7.9	4.2E-09
6/25	Sanriku	Neodilsea yendoana	1	0.1 ± 0.01	4.1	7.6E-09
7/ 3	Oshoro	Neodilsea yendoana	1	0.27 ± 0.02	9.7	8.6E-09
		1986 Invertebrates				
5/26	Katsuura	Haliotis discus	2	0.05 ± 0.01	2.7	5.2E-09
6/ 3	Hachinohe	Patinopecten yessoensis	2-3	0.04 ± 0.01	2.1	5.3E-09
6/23	Tsuyazaki	Loligo bleekeri	3	0.08 ± 0.01	4.1	5.8E-09
6/ 3	Hachinohe	Paroctopus dolfeni	3-4	0.06 ± 0.01	3.0	6.5E-09
7/ 3	Oshoro	Stichopus japonicus	2-3	0.03 ± 0.01	1.1	7.2E-09
6/ 3	Hachinohe	Halocynthia roretzi	2-3	0.12 ± 0.01	5.2	7.3E-09
5/26	Katsuura	Loligo bleekeri	3	0.08 ± 0.01	3.2	7.7E-09
5/26	Katsuura	Turbo cornutus	2	0.06 ± 0.01	2.1	8.7E-09
6/ 3	Hachinohe	Todarodes pacificus	3	0.17 ± 0.01	4.6	1.1E-08
		1986 Fish				
6/23	Tsuyazaki	Siganus fusescens	2	0.19 ± 0.01	18	3.2E-09
6/ 3	Hachinohe	Sebastes baramenuke	5	0.24 ± 0.01	16	4.6E-09
5/26	Katsuura	Kareius bicoloratus	3	0.13 ± 0.01	7.9	4.9E-09
6/23	Tsuyazaki	Hexagrammos agurammus	3	0.23 ± 0.02	14	5.1E-09
6/ 3	Hachinohe	Limanda schlencki	3	0.32 ± 0.01	17	5.6E-09
6/ 3	Hachinohe	Paralichthys olivaceus	5	0.22 ± 0.01	11	5.8E-09
7/ 9	Katsuura	Ditrema temmincki	3	0.27 ± 0.01	14	5.8E-09
5/28	Katsuura	Paralichthys olivaceus	5	0.29 ± 0.01	15	5.9E-09
6/ 3	Hachinohe	Microstomus achne	3	0.22 ± 0.01	11	6.1E-09
6/ 3	Hachinohe	Ditrema temmincki	3	0.27 ± 0.01	13	6.5E-09
5/26	Katsuura	Hexagrammos otakii	3	0.29 ± 0.01	13	6.5E-09

TABLE 2. (continued)

Year Date	Location	Species	Trophic level	Cs-137 Bq/kg (wet weight)	Stable Cs ng/g	Cs-137/ stable Cs atom ratio
		1986 Fish				
7/ 3	Oshoro	Pleurogrammus azonus	4	0.31 ± 0.01	14	6.9E-09
5/26	Katsuura	Scombrops boops	5	0.59 ± 0.01	23	7.6E-09
6/ 3	Hachinohe	Pleurogrammus azonus	4	0.4 ± 0.01	14	8.5E-09
6/ 3	Hachinohe	Hexagrammos otakii	3	0.38 ± 0.07	14	8.5E-09
		1987 Invertabrates				
6/11	Hachinohe	Halocynthia roretzi	2-3	0.03 ± 0.01	3.9	2.1E-09
10/2	Hachinohe	Anthocidaris crassipina	2	0.1 ± 0.01	10	3.1E-09
7/ 9	Hachinohe	Todarodes pacificus	3	0.06 ± 0.01	4.1	4.0E-09
10/2	Hachinohe	Patinopecten yessoensis	2-3	0.09 ± 0.01	6.6	4.3E-09
5/ 8	Hachinohe	Erimacrus isenbeckii	3	0.04 ± 0.01	2.5	4.4E-09
10/7	Hachinohe	Haliotis discus	2	0.04 ± 0.01	2.6	4.5E-09
		1987 Fish				
10/2	Hachinohe	Hippoglossoides dubius	3	0.1 ± 0.01	11	3.1E-09
7/ 2	Hachinohe	Ditrema temmincki	3	0.06 ± 0.00	4.4	4.1E-09
5/19	Hachinohe	Paralichthys olivaceus	5	0.28 ± 0.01	16	5.3E-09
6/22	Hachinohe	Limanda yokohamae	3	0.19 ± 0.01	11	5.3E-09
5/19	Hachinohe	Pleurogrammus azonus	4	0.37 ± 0.01	21	5.4E-09
5/18	Hachinohe	Hexagrammos otakii	3	0.44 ± 0.01	23	5.9E-09
		1988 Invertabrates				
11/9	Hachinohe	Todarodes pacificus	3	0.04 ± 0.01	2.6	5.0E-09
		1988 Fish				
11/9	Hachinohe	Oncorhyncus keta	5	0.08 ± 0.01	10	2.3E-09
11/9	Hachinohe	Paralichthys olivaceus	5	0.23 ± 0.01	12	5.8E-09
		1989 Algae				
7/ 2	Oshoro	Ulva pertusa	1	0.03 ± 0.02	3.0	2.9E-09
7/ 2	Oshoro	Laminaria religiosa	1	0.09 ± 0.02	7.2	3.8E-09
7/ 2	Oshoro	Undaria pinnatifida	1	0.06 ± 0.02	3.0	3.4E-09
7/ 2	Oshoro	Neodilsea yendoana	1	0.13 ± 0.02	7.8	4.8E-09
7/ 2	Oshoro	Laurencia nipponica	1	0.04 ± 0.01	3.8	3.3E-09
		1989 Fish				
7/ 2	Oshoro	Hexagrammos otakii	3	0.11 ± 0.01	7.1	4.8E-09
		1990 Algae				
7/18	Oshoro	Ulva pertusa	1	0.12 ± 0.02	9.6	3.8E-09
7/18	Oshoro	Undaria pinnnatifida	1	0.05 ± 0.02	6.1	2.4E-09
7/18	Oshoro	Sargassum yessonese	1	0.07 ± 0.02	6.8	3.2E-09
7/18	Oshoro	Sargassum miyabei	1	0.10 ± 0.02	7.1	4.3E-09
7/18	Oshoro	Neodilsea yendoana	1	0.12 ± 0.02	9.6	3.9E-09
		1990 Fish				
7/18	Oshoro	Hexagrammos otakii	3	0.19 ± 0.01	11	5.1E-09
7/18	Hachinohe	Hexagrammos otakii	3	0.26 ± 0.01	13	6.0E-09

(a) One sigma counting uncertainty

DISCUSSIONS

Cs-137/stable Cs Atom Ratio in the Different Trophic Level

57 biological samples were collected in 1986, including all trophic levels, from producers, as organisms of the lowest trophic level in the food chain, to carnivores, the highest trophic level. The concentration factors of Cs-137, stable Cs and the Cs-137/stable Cs atom ratio throughout the all trophic levels in 1986 are shown in Fig. 2.

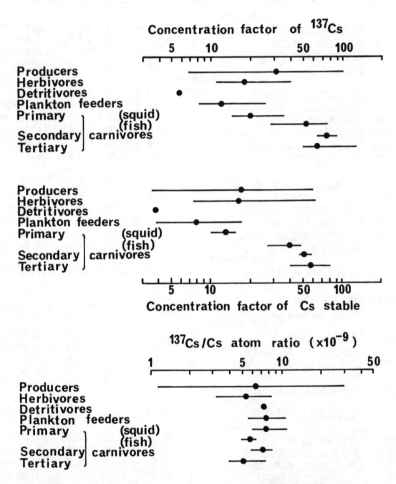

FIGURE 2. Cs-137 and stable Cs concentration factors and the Cs-137/stable Cs atom ratio in the different trophic level.

Both of the concentration factors of Cs-137 and stable Cs were high in the higher trophic level animals and low in the lower trophic level species. However the concentration factors of Cs-137 and stable Cs were not necessarily lowest in the producers. Osterberg et al. (8) reported that Cs-137 was detected in higher trophic level pelagic species but not detected in lower trophic level organisms in the Pacific Ocean. Our results showed the same tendency in higher trophic level organisms, but not the same in producers(algae), compared to Osterberg's pelagic phyto- and zooplankton.

Though the concentration factors varied one order of magnitude, the Cs-137/stable Cs atom ratios at all trophic levels had almost equal values. This indicates that Cs-137 reached a state of equilibrium with stable Cs at every trophic level, despite the fact that these samples were collected within only 90 days after the Chernobyl accident, while the fallout increased during these months (1). This suggests the rather short residence time of Cs-137 at each step in the food chain. Because, it means that Cs-137 which was introduced from the Chernobyl fallout to marine environment transfered along the food chain during rather short time range, in which the uptake from food pathway is considerably large (9,10).

Annual Change in the Cs-137/stable Cs Atom Ratios in Marine Biota and Seawater

The Cs-137/stable Cs atom ratio in marine biota showed a nearly log-normal distribution for each year from 1984 to 1990. An example for 1986 is shown in Fig. 3. This indicates that the Cs-137/stable Cs atom ratio can be used as an index of the Cs-137 level in biota of ecosystems, whether the data are for algae, invertebrates, fish.

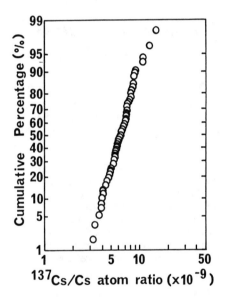

FIGURE 3. The Cs-137/stable Cs atom ratio in biota at 1986 by cumulative percentage.

The geometric means of Cs-137/stable Cs atom ratios in seawater and biota for each year shown in Fig. 4. The results indicate that the Cs-137/stable Cs atom ratios in seawater and biota decreased every year, and the Cs-137/stable Cs atom ratios for both seawater and biota had nearly the same values in a given year from 1984 to 1990, except for 1986, the year the Chernobyl accident occurred. Our results corroborate the report of Robertson et al. (11) for the close agreement of Cs-137 specific activity in seawater and fish in the N.E. Pacific Ocean, and of Suzuki et al. (5)

or same results in the Japanese fish and seawater from 1964 to 1971. However Suzuki's results were believed to be affected by the fallout from a series of Chinese nuclear tests started in 1968, which caused monthly and annual fluctuations of Cs-137 fallout (3).

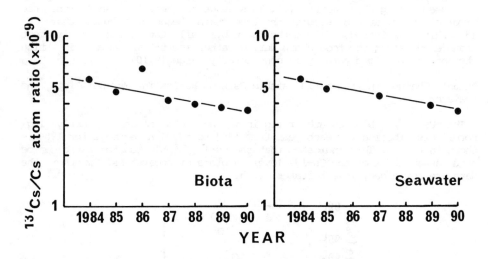

FIGURE 4. Regression curves of the Cs-137/stable Cs atom ratio in biota and seawater. Seawater samples were not analyzed in 1986 and 1988. Data from biota samples in 1986 were not used to calculate the effective half-life.

The regression curve of Cs-137/stable Cs atom ratio in seawater and biota showed exponential decrease. Judging from the facts that the concentrations of Cs-137 in fallout levelled off after the last nuclear bomb test in China in 1982, our results indicate that a constant diffusion of Cs-137 to stable Cs in the marine environment. With a correction of the half-life of Cs-137 (30.17 years), the ecological half-lives of Cs-137 in seawater and biota are calculated by the effective half-lives of Cs-137 obtained from regression curves of Cs-137/stable Cs atom ratios in biota and seawater.

TABLE 3. The ecological half-life of Cs-137 in the Japanese marine ecosystem.

		Biota	Seawater
Loss coefficient (year^{-1})	b	0.063	0.069
Effective half-life (year)	Tef	9.87	9.67
Ecological half-life (year)	Tec	14.7	14.2

$R_t = R_0\, e^{-bt}$ where
R_0: the Cs-137/stable Cs atom ratio at the initial time.
R_t: the Cs-137/stable Cs atom ratio at the time t (year).
$1/\text{Tef} = 1/\text{Tec} + 1/30.17$

The ecological half-lives of Cs-137 in biota and seawater were estimated to be 14.7 years and 14.2 years, respectively. Therefore, the ecological half-life of Cs-137 in marine ecosystem is estimated to be approximately 14 years.

ACKNOWLEDGEMENTS

For collecting of samples, we thank the staff of Kyushu University, Hokkaido University, Kitazato University, Tohoku Regional Fisheries Research Laboratory, and Chiba University. For the technical assistance in analysis, we thank the staff of Japan Chemical Analysis Center and the Japan Nuclear Institute. For the assistance in pretreatment of samples, we thank Miss. N. Sugita.

This research was granted by the Central Research Institute of Electric Power Industry.

REFERENCES

1. Aoyama, M., Hirose, K. and Sugimura, Y., Deposition of gamma-emitted nuclides in Japan after the reactor IV accident at Chernobyl USSR., J. Radioanal. Nucl. Chem., 1987, 116, 291-306
2. Higuchi, H., Fukatsu, H., Hashimoto, T., Nonaka, N., Yoshimizu, K., Omine, M., Takano, N. and Abe, T., Radioactivity in surface air precipitation in Japan after the Chernobyl accident., J. Environ. Radioactivity, 1988, 6, 131-144
3. Katsuragi, Y. and Aoyama, M., Seasonal variation of Sr-90 fallout in Japan through the end of 1983., Meteorology and Geophysics, 1986, 37, 15-36
4. Coughtrey, P.J. and Thorne, M.C., Radionuclide distribution and transport in terrestrial and aquatic ecosystems. A critical review of data. vol. 1., A. A. Balkema, Rotterdam, 1983
5. Suzuki, Y., Nakamura, R. and Ueda, T., Cesium-137 contamination of marine fishes from the coasts of Japan., J. Rad. Res., 1973, 14, 382-391
6. Umezu, T., Minamisato, Y. and Tabata, K., Dissimilarity of Co-60/Co, Cs-137/Cs and Sr-90/Sr ratios in Beryx splendens from Pacific and Atlantic Ocean., Bull. Jap. Soc. Sci. Fish., 1986, 52, 1985-1993
7. Bowen, H.J.M., Environmental chemistry of elements., Academic Press, London, 1979
8. Osterberg, C., Pearcy, W.G. and Curl, H. Jr., Radioactivity and its relationship to oceanic food chains., J. Mar. Res., 1964, 22, 2-12
9. Suzuki, Y., Nakahara, M., Nakamura, R. and Ueda T., Roles of food and seawater in the accumulation of radionuclides by marine fish., Bull. Jap. Soc. Sci. Fish., 1979, 45, 1409-1416
10. Hewett, C. J. and Jefferies, D. F., The accumulation of radioactive caesium from water by brown trout (Salmo trutta) and its comparison with plaice and rays., J. Fish. Biol., 1976, 9, 479-489
11. Robertson, D. E., Rancitelli, L. A. and Perkins, R. W., Multielement analysis of seawater, marine organisms and sediments by neutron activation without chemical separations. In Proc. of the Int. symp. on the application of neutron activation analysis in oceanography., Institute Royal des Sciences Naturelles de Belgique, Brussels, 1968

THE USE OF Pb-210/Ra-226 AND Th-228/Ra-228 DIS-EQUILIBRIA IN THE AGEING OF OTOLITHS OF MARINE FISH

J.N. SMITH and R. NELSON
Marine Chemistry Division
Bedford Institute of Oceanography
Dartmouth, Nova Scotia, Canada B2Y 4A2

S.E. CAMPANA
Marine Fish Division
Bedford Institute of Oceanography
Dartmouth, Nova Scotia, Canada B2Y 4A2

ABSTRACT

Naturally-occurring isotopes of radium are ideally suited as tracers for chemical uptake in the calcified tissues of marine organisms since radium is a water soluble, bio-geochemical analogue for calcium. Assays designed to exploit this uptake mechanism can be used to determine the longevity of certain species of fish. Measurements of Pb-210/Ra-226 dis-equilibria in the otoliths of redfish have revealed that this species of fish can live to ages in excess of 75 years in coastal waters off Nova Scotia, Canada. Measurements of the Th-228/Ra-228 disequilibria in the otoliths of the much shorter-lived silver hake and flying fish may provide estimates of longevity on time scales of 0-10 years, which could then be used to evaluate the accuracy of currently-used ageing models based on otolith annulus counts. The radioisotopic ageing technique relies on the extraction of a rectangular block centred around the core of the otolith whose analysis for radionuclide ratios permits an unambiguous estimate of the age of the fish. Age determinations of fish based on natural radioisotopes can result in significant improvements in the assessment and management of certain fisheries resources.

INTRODUCTION

Fundamental to the study of marine biology is the age and growth rate of specific organisms and yet, for many fish this is a poorly known parameter. Most ageing techniques rely on the identification of annual increments associated with the growth of teeth, otoliths (ear bone), scales, vertebrae or other bony tissues of the organism. The difficulties in these techniques generally lie with uncertainties in the regularity with which the growth increments occur and in the identification of the chronological markers themselves. For example, several north temperate fish species are thought to have lifespans over 60 y [1], but the reported maximum ages have varied widely depending upon the choice of ageing methodologies and the interpretation of the investigator [2]. Errors in estimates of fish ages can confound any attempt at optimal management of a fish stock and interfere with the most basic understanding of the life cycles of the species.

Natural radionuclide tracer measurements offer a new and comparatively accurate methodology for determining the growth rates of marine organisms. The techniques discussed in this report rely on the uptake by fish otoliths of either Ra-226 or Ra-228 from seawater and the subsequent decay of these isotopes to their daughter products (Pb-210 and Th-228, respectively) in a closed chemical system. Measurements of the parent and daughter product ratios then permit a determination of the fish age since the time of radium uptake. Pb-210/Ra-226 disequilibria provide a sensitive dating range of from 0-100 y while the Th-228/Ra-228 disequilibria is useful for fish ages of 0-20 y.

Radionuclide ageing techniques rely on two basic geochemical features of the naturally-occurring, U-238 decay series, (1) the high solubility of radium in seawater compared to its parent and daughter radio-isotopes, and (2) the relatively efficient uptake of radium from seawater by marine tissues as a proxy for calcium. These conditions combine to insure that an initial state of radioactive dis-equilibrium is established by the incorporation of radium isotopes into the tissues of organisms, unaccompanied by parent or daughter radio-isotopes. Successful applications of the method then rely on a sufficiently high signal to noise ratio in the daughter/parent activities as they decay to a state of radioactive equilibrium.

The measurement of radionuclide ratios in whole otoliths or other accreting structures does not generally supply conclusive evidence of the age of the fish. The basic problem is that as the otolith accretes, it continually forms a new outer sheath of material having a parent (eg. Ra-226) concentration equivalent to it's initial, or t (time) = 0 value. However, the material formed in previous years will have already begun to generate Pb-210 by radioactive decay. As a result, a gradient in the Pb-210 activity will be formed along any axis from the centre of the otolith to an external surface. The Pb-210 activity at the otolith centre will have a maximum value while newly formed material at the external surface will have a minimum or zero value. Hence, the analysis of a whole otolith will give an intermediate Pb-210 activity having some value between zero and the maximum value representative of the core of the otolith. Upon introduction into the decay equation, this number will always give an under-estimate of the true age.

The first Pb-210/Ra-226 disequilibria measurements on otoliths were undertaken by Bennett et al. [3] on the whole otoliths of redfish (<u>Sebastes diploproa</u>). They attempted to overcome the above difficulties by formulating an otolith mass-growth model and then calculating the degree of Pb-210 dilution by the continuous addition of mass during the life time of the fish. The difficulty with this approach was that it involved a certain degree of circular reasoning. The accuracy of the age determination depended on the validity of the mass-growth model, but the formulation of a mass-growth model was based on an initial estimate of the fish age. Campana et al. [4] circumvented this entire issue by shaving away the outer layers of material leaving only the central core of the otolith formed near the time of hatch. Under these conditions, a simple, first order radioactive decay law can be used to determine fish age, since the rate of mass accumulation at the otolith surface, and, indeed, variations in the rate, have no impact on the properties of the core.

The underlying assumptions of the methodology employed by Campana et al. [4] are reviewed in the present paper. Comparisons are made between the whole otolith and core techniques. In addition, more recent applications of this aging technique to shorter-lived fish using a different isotope pair (Ra-228/Th-228) are discussed.

METHODOLOGY

Otolith Analyses

Redfish were collected from depths of 200-900 m along the edge of the Scotian Shelf in October, 1985 and 1986. Sagittal otolith pairs (n=2775) were extracted from fresh specimens and stored dry in paper envelopes. One otolith from each pair was cracked, charred and coated with oil for the determination of fish age via annulus counts. Annuli were counted under 50 x magnification at least twice by each of two independent readers, using criteria established by Beamish et al. [5]. The remaining otolith of each pair from fish of age categories; 6, 10-15, 20-25, 30-35, 40-45, 50-55 and 60+ y were selected for core extraction and radiochemical analyses.

Otolith cores were extracted as rectangular blocks centred around the otolith nucleus. The location of the nucleus was assessed in relation to intact otolith morphology and confirmed through sectioning of test samples. While a block is a crude approximation of otolith shape the relatively large annuli in young otoliths imply that slight errors in measurement or extraction would have little effect on the mean age of the core. Cores were isolated with a low-speed, diamond blade saw, thinned with a metallurgical polishing machine and cleansed of surface residue. Cores were subsequently pooled within age categories to form samples weighing approximately 1.0 g (21-28 cores).

Flying fish (Hirundichthys affinis) and silver hake (Merluccius bilinearis) specimens were obtained from commercial fisheries in Barbados and Nova Scotia, respectively in February, 1990. Otolith pairs were collected and examined by the methods noted above.

Radionuclide Measurements

For Pb-210/Ra-226 analyses, one gram otolith samples from redfish were pre-treated to remove organic or inorganic deposits by first soaking them in 30% hydrogen peroxide solution, followed by successive washings with 1 N HCl solution and distilled water, using high purity reagents. Po-208 tracer solution plus 10 ml of 50% conc. HCl + HNO_3 were added to the samples and then heated to dryness. Ten ml of conc. HCl were added, the solution again heated to dryness and then re-dissolved in successive additions of 1 ml of H_2O_2 and 30 ml of 0.5 N HCl. The polonium isotopes were plated at 85-90 deg. C. for approximately 4-6 hours onto a silver disc and the Po-208 and Po-210 activities (equal to the Pb-210 activity for samples several years old) subsequently measured by alpha particle spectrometry. The Pb-210 blank was 0.0005 dpm and background counts were less than 1/day.

Following plating of the polonium isotopes, the solution was transferred to an extraction cell connected to a radon gas extraction vacuum line. The radon was subsequently transferred (with a 95 % transfer efficiency) to a scintillation counting vial and then the radon (and Ra-226) activity determined using a photomultiplier assembly. The counting efficiency of the entire system for Ra-226, determined using NBS Ra-226 standard solutions, was 80 ± 5%. Ra-226 analyses of both the core and whole otolith samples revealed no significant differences. As a result, the whole-otolith analyses of 3 samples (13 g each) were used to determine a mean Ra-226 concentration (0.033 ± 0.002 dpm/g) for the otolith material.

Following pre-treatment procedures, the flying fish and silver hake otolith samples for Th-228/Ra-228 analyses were digested in aqua regia for four hours, together with Ra-226 and Th-229 yield tracers, cooled and 20 ml of hydrogen peroxide were added drop wise. An additional 30 ml of aqua regia were then added and the samples evaporated to dryness. The residue was redissolved in 100 ml of conc. HNO_3 and 200 mg of $Pb(NO_3)_2$ carrier added. The solution was reduced in volume to 75 ml, 25 ml of fuming nitric acid were added, and the radium separated by co-precipitation with $Pb(NO_3)_2$ [6]. The precipitate was centrifuged (saving the supernatant for Th analyses), dried in a counting vial and the Ra-226 and Ra-228 activities determined by gamma counting using the 339 kev and 911 kev gamma energies of Ac-228 for the Ra-228 measurement and the 186 kev (Ra-226) and 295 kev (Pb-214) gamma energies for the Ra-226 measurement. The gamma analyses were conducted using a Canberra HPGE well detector having a relative efficiency of 20%, a 103.5 cc active volume and a well 20 mm in diameter and 35 mm deep.

The supernatant was boiled to dryness and the residue dissolved in 8 N HNO_3. The thorium isotopes were separated by ion exchange using BIORAD AG 1x8 resin, 100-200 mesh and eluted from the column with 12 N HCl. The sample was then redissolved in H_2SO and plated onto a stainless steel disk at 1.1 amp and 50 volts for two hours. The alpha activities of Th-228 and the yield tracer, Th-229, were measured by alpha particle spectrometry. Blanks determined using the same procedures, reagents, carriers and tracers as in the otolith analyses gave values of 0.010 ± 0.005 dpm for Th-228 and 0.050 ± 0.020 for Ra-228.

AGEING MODELS

Model Assumptions
Radium disequilibria aging models for whole otoliths are generally based on the following assumptions: (1) There is negligible, post-formational, internal migration of radionuclides across internal otolith annuli, ie. the otolith core constitutes a closed chemical system; (2) The uptake of the daughter radionuclide from seawater or other external sources is small compared to the parent (radium) uptake; (3) The uptake of parent and daughter radionuclides is in constant proportion to the mass accumulation rate of the otolith.

Condition (1) ensures that the chronological, radioactive disequilibria features of the otolith are preserved and is supported by numerous experiments indicating that internal chemical migration within the acellular, otolith structure is negligible [7]. However, it should be noted that this condition is apparently invalid in some cellular systems such as the vertebrae of sharks where post-formational transport of Pb-210 has been observed [8]. Condition (2) ensures that the signal to noise ratio of the daughter/parent disequilibria signal is sufficiently great to ensure the practical viability of the method. This assumption can be validated by measurements on the otoliths of young fish and small discrepancies can be accommodated within a given model. Condition (3) is the most problematic to whole-otolith applications of radium aging because it generally presupposes a knowledge of the otolith mass-growth rate as determined by unvalidated annulus counts. As noted above, the coring technique obviates the necessity of relying on condition (3), except for the initial period of core formation. Indeed, Fenton et al. [9] detected a major violation of Condition (3) in their measurements of Pb-210/Ra-226 disequilibria in the otoliths of blue grenadier from Southeastern Australian waters. They measured an order of magnitude decrease in the Ra-226 concentration of whole otoliths of fish having ages (determined by annulus counts) of 1 to 10 years old. They ascribed this variability to the life habits of the fish, with the younger, juveniles spending their time in radium-enriched, nearshore waters and the adults occupying the deeper, offshore waters, diminished in Ra-226.

Core Model

The processes governing radium uptake and decay in otoliths are;

$$N_1 \xrightarrow{\lambda_1} N_2 \tag{1}$$

$$N_2 \xrightarrow{\lambda_2} \text{product} \tag{2}$$

where N_1 represents the parent radionuclide having a decay constant, λ_1, and N_2 represents the daughter radionuclide having a decay constant, λ_2. The relationship between the radionuclide activity in the otolith and it's age can be determined by first considering the formation of a small nucleus or core of the otolith. The parent and daughter specific activities, A_1 and A_2, and their ratio in this core at some future time or "age", t, are given by the standard equations for this general decay scheme;

$$A_1 = A_1° e^{-\lambda_1 t} \tag{3}$$

$$A_2 = A_1° \left(\frac{\lambda_2}{\lambda_2 - \lambda_1}\right) \{e^{-\lambda_1 t} - e^{-\lambda_2 t}\} + A_2° e^{-\lambda_2 t} \tag{4}$$

As the otolith grows, new material is accreted to the outer surface having an initial activity ratio, $R° (= A°_2/A°_1)$, while the inner nucleus of the otolith has an activity ratio given by A_2/A_1. In the general case, the total activity, S_1, of N_1 (or S_2, of N_2) in the otolith is calculated by integrating the product of the mass-growth rate, G(t) and equation 3 (or equation 4) with respect to time (equation 5). In the study of Campana et al. [4], it was assumed that during the period of core formation (up to an age of T or approximately 10% of its full growth), condition 3 above was valid and the otolith underwent a linear increase of mass with time at a constant rate of Ra-226 incorporation. This constant mass-growth (CMG) model can be applied to the general case by substituting a constant mass-growth rate (equal to the total mass, M, of the core divided by it's age, T) for G(t) in equation 5, ie.

$$S_1(T) = \int_0^T G(t) A_1 dt \quad \overset{(\text{CMG model})}{=} \quad M/T \int_0^T A_1° e^{-\lambda_1 t} dt \tag{5}$$

Integration of equation 5, and the equivalent equation for S_2, followed by dividing S_2 by S_1 gives for the activity ratio, A_2/A_1, of the otolith core at t = T;

$$\frac{A_2(T)}{A_1(T)} = \frac{S_2}{S_1} = \frac{\lambda_2}{(\lambda_2 - \lambda_1)} + \left\{R° - \frac{\lambda_2}{(\lambda_2 - \lambda_1)}\right\} \frac{\lambda_1(1 - e^{-\lambda_2 T})}{\lambda_2(1 - e^{-\lambda_1 T})} \tag{6}$$

Following the formation of the core at t = T, it's future activity ratio remains unaffected by the changes in the otolith growth and is governed solely by the rate law. Since equation 6 gives the activity ratio, A_2/A_1 at t = T, then the activity ratio of the otolith core at any future time is given by substitution into equations 3 and 4;

$$\frac{A_2(t)}{A_1(t)} = \frac{\lambda_2}{(\lambda_2 - \lambda_1)} \{1 - e^{-(\lambda_2 - \lambda_1)(t - T)}\} +$$

$$\left[\frac{\lambda_2}{(\lambda_2 - \lambda_1)} + \{R° - \frac{\lambda_2}{(\lambda_2 - \lambda_1)}\} \frac{\lambda_1(1 - e^{-\lambda_2 T})}{\lambda_2(1 - e^{-\lambda_1 T})}\right] e^{-(\lambda_2 - \lambda_1)(t - T)} \quad (7)$$

for t ≥ T. By cutting the otolith back to its core, future variability in either the otolith accretion rate or in the activity ratio, $A°_2/A°_1$, of newly accreted material are eliminated as necessary considerations in the model. Errors associated with the use of a linear mass-growth model for the core can be minimized (to the extent possible consistent with maintaining a reasonable signal to noise ratio) by simply reducing the size and age of the core. In fact, recent evidence from flying fish and silver hake studies (S. Campana, unpubl. man.) indicates that throughout part of their life history, otolith mass increases exponentially with time, a functional relationship which would produce large errors in whole otolith studies based on linear mass-growth models, but would have a minor effect on core model results.

Pb-210/Ra-226 and Th-228/Ra-228

For the Pb-210 ($t_{1/2}$ = 22.3 y)/Ra-226 ($t_{1/2}$ = 1620 y) pair, the half-life of Ra-226 is sufficiently greater than that of Pb-210 that the parent does not appreciably decay during the period of Pb-210 ingrowth and equations 6 and 7 simplify to;

$$\frac{A_{Pb-210}}{A_{Ra-226}} = 1 - (1 - R°)\frac{(1 - e^{-\lambda t})}{\lambda t} \quad (8)$$

$$\frac{A_{Pb-210}}{A_{Ra-226}} = (1 - e^{-\lambda(t-T)}) + \{1 - (1 - R°)\frac{(1 - e^{-\lambda T})}{\lambda T}\}e^{-\lambda(t-T)} \quad (9)$$

for the cases (0 < t ≤ T) and (T < t), respectively and where λ is the decay constant for Pb-210. (Note that the last exponential term in equation 9 was inadvertantly omitted from the equivalent equation (2) in [4]).

At large values of t, a condition of secular equilibrium is attained with the parent and daughter activities becoming equal. The Pb-210/Ra-226 ratio predicted using a linear mass-growth model for the formation of the otolith core at T = 5 y (eqn. 8), and the core model (eqn. 9), thereafter are illustrated in Figure 1b, for values of T = 0-100 y and values of R°

= 0, 0.1 and 0.2. For comparison, the results in Figure 1a were determined using the linear mass-growth model (eqn. 8) for the entire age range. The activity ratio for the core model is more sensitive to fish age over the 0-100 y range compared to the linear growth model, although both models have a limiting activity ratio of 1 at large ages.

For the Th-228 ($t_{1/2}$ = 1.91 y)/Ra-228 ($t_{1/2}$ = 5.76 y) pair, equations 6 and 7 are applicable and the Th-228/Ra-228 activity ratio exponentially approaches a constant value of 1.5 as a function of t. The decay curves for T = 1.5 y and R° = 0, 0.1 and 0.2 are illustrated in Figure 1c. In contrast to the Pb-210/Ra-226 pair, the Th-228/Ra-228 pair are most useful over the fish age range of 0-10 y.

RESULTS AND DISCUSSION

Redfish
From their radiochemical assay of whole otoliths and the use of a two stage, linear otolith growth model, Bennett et al. [3] concluded that redfish lived to considerably greater ages than was evident from the annuli visible on the otolith surface. Attempts in the present study to reproduce the measurements of Bennett et al. [3] met with mixed success. Whole otoliths of the Atlantic redfish, Sebastes mentella from different age ranges were analyzed for their Pb-210/Ra-226 ratio. The results are compared to the predictions of a single stage, linear growth model (equation 8) in Figure 1a. The measurements from the younger fish are in good agreement with model predictions. However, the results for fish in excess of 20 y old are not consistent with model predictions, although the 2 sigma uncertainties of all the data generally overlap the R°= 0 curve. Discrepancies between the experimental results and the linear-growth model predictions may be due to non-linear growth, changes in the Ra-226 uptake rate or to some contamination of the otolith surface which was not completely removed in this initial set of experiments.

These above-mentioned sources of error were avoided in a subsequent study of Atlantic redfish by restricting the Pb-210/Ra-226 analyses to the otolith core [4]. The measured Pb-210/Ra-226 activity ratios for the cores are plotted against the age of the fish determined by annular counts. The agreement between the experimental results and the annulus counts using a core age of 5 y is very good. The core age of 5 y was chosen based on the relative benefits of having a young core age for modelling purposes, an adequate sample size and good confidence in the core age based on annulus counts. From their results with juvenile Sebastes, Bennett et al. [3] concluded that R°<0.2 while the results of this study for the two youngest otolith cores (Figure 1b) are consistent with R°<0.1.

The agreement between these results and the curve predicted from annulus counts indicates that the criteria outlined by Beamish et al. [5] for the interpretation of otolith annuli are correct. These criteria have been widely accepted by those ageing Pacific fishes [1], but have been rejected by those ageing some Atlantic species as inducing age overestimation. The core method results also indicate that previous suggestions of longevity in Atlantic redfish in excess of 70 y are correct [10] as the present examinations identified individuals that were 75 y old [4].

Silver Hake and Flying Fish
Flying fish (Hirundichthys affinis) is the main commercial species harvested from the Caribbean Sea. Although daily growth rings are evident in young flying fish, they become

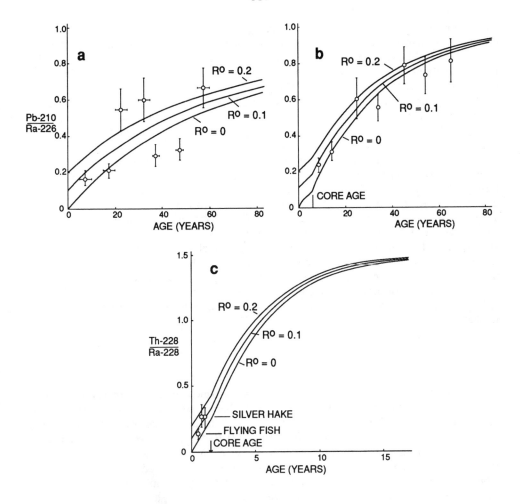

Figure 1. (a) Decay curves for the Pb-210/Ra-226 activity ratio as a function of fish age for the linear growth model (T = 0-80) and for given initial ratios, $R°$. Experimental results for whole otoliths from different age classes or redfish are also shown (error bars reflect 1 sigma uncertainties). (b) Decay curves for the Pb-210/Ra-226 activity ratio as a function of fish age for given initial ratios, $R°$. A linear growth model was employed during the period of T = 5 y for the formation of the otolith core. Experimental results for sets of otolith cores taken from redfish in given age classes are also shown. (c) Decay curves for the Th-228/Ra-228 activity ratio as a function of fish age for given initial ratios, $R°$. A linear growth model was employed during the period of T = 1.5 y for the formation of the otolith core. Initial results for whole otoliths of silver hake and flying fish are also shown.

too narrow in adults to be useful as an ageing tool. As a result, no technique has been previously available to determine growth rates which are the most significant parameter for fishery models. The single result of a Th-228/Ra-228 activity measurement on otolith samples from adult flying fish is illustrated in Figure 1c. Adult flying fish are thought to be short-lived, having ages of less than 2 years. The present results are consistent with activity ratios predicted by the linear growth model for the Th-228/Ra-228 pair (equation 6) for an age of 0.5 y. These initial whole otolith results provide the first evidence of a short lifetime (1-2 y) for flying fish. Measurements are presently being conducted on otolith cores of older flying fish to confirm their relatively short life spans.

Silver hake (Merluccius bilinearis) is also a commercially important fish, harvested mainly by Soviet and Cuban fishermen, and constitutes one of the largest fish biomasses off Eastern Canada. Although age estimates have been made on annuli, the annuli are unvalidated. Silver hake are thought to live longer than flying fish, but have ages of less than 10 y old. Whole otolith samples from juvenile silver hake collected off the Scotian Shelf, having ages of 1-2 y based on length frequencies, were analyzed for the Th-228/Ra-228 activity ratio. The two results of these analyses are generally consistent with a linear growth model over the initial stage of otolith growth (Figure 1c). However, an exponential growth model would provide better agreement between the experimental and predicted results, and this functional relationship will be incorporated into future models. The results for both flying fish and silver hake tend to confirm the applicability of the Th-228/Ra-228 ageing technique to short-lived fish.

CONCLUSIONS

(1) Otoliths constitute a closed chemical system owing to their acellular composition, and thus are suitable for radioisotopic dating. However, their masses and specific radionuclide activities are low and it is necessary to have low and stable detector backgrounds and long counting times in order to obtain useful results. (2) Radioisotopic dating techniques have important applications to the confirmation or rejection of fish ages determined using traditional techniques based on annulus counts. The otolith core model approach has major advantages over the multi-stage, linear growth model previously utilized to age redfish. (3) Pb-210/Ra-226 activity ratios have proven effective in determining the longevity of long-lived fish species such as redfish. Initial applications of the otolith core ageing method to juvenile silver hake and adult flying fish indicate that the Th-228/Ra-228 activity ratios are consistent with model predictions and thereby support the potential utility of this isotopic pair in ageing short lived fish.

REFERENCES

1. Chilton, D.E. and Beamish, R.J., Age determination methods for fishes studied by the Groundfish Program at the Pacific Biological Station, Can. Spec. Publ. Fish. Aquat. Sci., 1982, **60**, 1-102.

2. Campana, S.E., Comparison of age determination methods for the starry flounder, Trans Am. Fish. Soc., 1984, **113**, 365-369.

3. Bennett, J.T., Boehlert, G.W. and Turekian, K.K., Confirmation of longevity in Sebastes dipoproa from Pb-210/Ra-226 measurements in otoliths, Mar. Biol., 1982, **71**, 209-215.

4. Campana, S., Zwanenburg, K.C.T. and Smith, J.N., Pb-210/Ra-226 determination of longevity in redfish, Can. J. Fish. Aquat. Sci., 1990, **47**, No. 1, 163-165.

5. Beamish, R. J., New information on the longevity of Pacific ocean perch (Sebastes alutus), J. Fish. Res. Board Can., 1979, **36**, 1395-1400.

6. Koide, M. and Bruland, K.W., The electrodeposition and determination of radium by isotopic dilution in seawater and in sediments simultaneously with other natural radionuclides, Analytica Chim. Acta, 1975, **75**, 1-19.

7. Campana, S.E. and Neilson, J.D., Microstructure of fish otoliths, Can. J. Fish. Aquat. Sci., 1985, **42**, 1014-1032.

8. Welden, B.A., Cailliet, G.M., and Flegal, A.R., Comparison of radiometric with vertebral age estimates in four California elasmobranchs. In The Age and Growth of Fish, eds. R.C. Summerfelt and G.E. Hall, Iowa State Univ. Press, Ames, 1987, pp. 301-315.

9. Fenton, G.E., Ritz, D.A. and Short, S.A., Pb-210/Ra-226 disequilibria in otoliths of Blue Grenadier, Macruronus novaezelandiae; Problems associated with radiometric ageing, Aust. J. Mar. Freshwater Res., 1990, **41**, 467-473.

10. Sandeman, E.J., A contribution to the problem of the age determination and growth-rate in Sebastes, ICNAF Spec. Publ., 1961, **3**, 276-284.

SOME PRIORITIES IN MARINE RADIOACTIVITY RESEARCH

MURDOCH S BAXTER

International Atomic Energy Agency
International Laboratory of Marine Radioactivity
Principality of Monaco

ABSTRACT

The presentation will combine the author's personal views on current research priorities with those of the IAEA Monaco laboratory. Current results and programmes will be outlined to illustrate the main points.

Posters

CHARACTERIZING THE MIXING OF WATER FLOWING INTO THE NORTH SEA USING ARTIFICIAL GAMMA EMITTERS
(Tramanor cruise, July 1988)

P. Bailly du Bois, P. Guéguéniat, R. Gandon and R. Léon.

Laboratoire de Radioécologie Marine, IPSN-DPEI-SERE
B.P. 508, 50105 CHERBOURG (France)

Work carried out by the CEA Radioecology Laboratory (La Hague) during a campaign on board the IFREMER oceanographic research vessel "Cryos" in July 1988 has enabled the measurement of conservative chemical and physical parameters over the whole North Sea. Continuous salinity profiles were obtained and the radionuclides ^{125}Sb, ^{137}Cs and ^{134}Cs were extracted from samples collected at 256 stations; the extraction procedure involved using manganese dioxide at pH 3.5 and Co-K ferrocyanide.

The values at each sampling station were interpolated over the entire area of the North Sea to yield the contoured distribution maps presented here (5 in all).

We can distinguish successively:

a) A water mass with weak gradients (35-34.5 g/l), corresponding to the entry of Atlantic waters which extends over about half the North Sea from the Shetlands to the Dogger Bank. The levels of ^{137}Cs express the gradual mixing with the waters labelled by the nuclear reprocessing plant of Sellafield; the amount of ^{125}Sb is under one Bq/m^3.

b) A water mass with a slight drop in salinities (34-34.5 g/l), clearly marked by ^{137}Cs (20-40 Bq/m^3), wich enter in the North Sea between Fair Isle and Scotland. It correspond to Atlantic waters labelled by the Sellafield discharges in the Irish Sea. Thoses waters goes along the East English coast, mixing with the first ones.

c) The penetration of waters from the Channel. It could not be identify by salinity because of the various fluvial and marine inputs crossing the Dover Strait. Otherwise, it could be clearly visualized by ^{125}Sb (20-40 Bq/m^3) coming from the reprocessing plant of la Hague. The Channel's plume extend from the Dover Strait to the Skagerrak, with a slight mixing with the central waters near the coasts up to the Denmark (less than 50% of central waters); the transitional area ($^{125}Sb = 1-15$ Bq/m^3) extend 60Km off Rotterdam and as far as 200 Km to the west of Denmark. The values of ^{137}Cs in the Channel plume are low, and due essentially to the fluvial inputs labelled by the Tchernobyl accident ($^{137}Cs = 4-25$ Bq/m^3, $^{134}Cs = 0-5$ Bq/m^3).

d) The entrance to the Baltic. It is characterized by a strong brackish influence (minimum 22 g/l), and a clear mark with the Tchernobyl spike ($^{137}Cs = 30-71$ Bq/m^3, $^{134}Cs = 3-13$ Bq/m^3) in the northern part of the Skagerrak and along the Norwegian coast, associated with elevated water temperatures (more than 18° C). A vertical profile performed in the middle of the Skagerrak shows that these features only concern the topmost 30 m. Beneath this layer, we have: T=5-6° C, S=34.3-35 g/l, $^{137}Cs = 30$ Bq/m^3, $^{134}Cs = 2$ Bq/m^3; these values suggests a mixing with waters labelled by older releases from Sellafield.

e) The North Sea way out along the Norvegian trench. It is a mixing of the four inputs with the following levels: S = 27-34.5 g/l, $^{137}Cs = 10-45$ Bq/m^3, $^{134}Cs = .7-7$ Bq/m^3, $^{125}Sb = 1-10$ Bq/m^3. The antimony show the influence of the Channel until 150 Km from the Norvegian coast, and the ^{137}Cs display the limit of waters leaving the North Sea at 200 Km from the Norvegian coast.

TRANSPORT D'ELEMENTS A L'ETAT DE TRACES DANS LES EAUX COTIERES DE LA MANCHE : ETUDE DE LA DISTRIBUTION SPATIALE D'UN TRACEUR RADIOACTIF (^{106}Ru-Rh) DANS LES MOULES ET LES FUCUS.

P. GERMAIN*, J.C. SALOMON**

* Laboratoire de Radioécologie Marine, IPSN-DPEI-SERE
  B.P. 508, 50105 CHERBOURG (France)
** IFREMER Centre de Brest, B.P. 70 - 29280 PLOUZANE (France)

Afin de mieux connaître les mouvements de métaux dans les eaux côtières de la Manche, nous avons étudié, dans le cadre d'études radioécologiques, la distribution d'un traceur radioactif artificiel (^{106}Ru-Rh) rejeté par l'usine de La Hague dans des espèces indicatrices Fucus serratus et Mytilus edulis.

Malgré les nombreux paramètres agissant sur les variations des niveaux de radioactivité chez les espèces, certaines observations relatives à la distribution du ^{106}Ru-Rh se sont répétées systématiquement, quelles que soient l'époque, les quantités et la qualité des produits rejetés par l'industrie nucléaire. Ces résultats sont dus aux caractéristiques du transport et de la dilution, comme l'indique la comparaison entre, d'une part, la distribution du ^{106}Ru-Rh chez les espèces, et d'autre part, les champs de courants résiduels lagrangiens et les trajectoires à long terme obtenus à l'aide d'un modèle mathématique.

Ainsi, à l'ouest du point de rejet de l'usine de La Hague, ^{106}Ru-Rh n'est plus décelé à Roscoff, à 180 km ; ceci est lié à la veine d'eau atlantique qui longe la côte bretonne jusqu'à Lannion. A l'est, ^{106}Ru-Rh est détecté dans le Pas de Calais et au delà. Les trajectoires du modèle montrent effectivement une structure en bandes parallèles se dirigeant vers le Pas de Calais à partir du cap de La Hague, ce qui explique aussi l'absence de ^{106}Ru-Rh le long des côtes anglaises. Au sud du point de rejet (Carteret), il apparaît une rupture brutale des niveaux de ^{106}Ru-Rh, correspondant à la frontière entre les tourbillons des Minquiers et de Flamanville ; puis les niveaux évoluent peu le long du continent et des côtes des îles (Jersey, Guernesey...) du fait d'une série de tourbillons. Les gradients observés à l'est du Cotentin témoignent du rôle du tourbillon de Barfleur et de sa confrontation avec l'eau venant de La Hague, ainsi que de l'influence des rivières.

STUDIES ON THE TRANSPORT OF COASTAL WATER FROM THE ENGLISH CHANNEL TO THE BALTIC SEA USING RADIOACTIVE TRACERS (MAST PROJECT)

HENNING DAHLGAARD
Risø National Laboratory, DK - 4000 Roskilde, Denmark

HARTMUT NIES,
Bundesamt für Seeschiffahrt und Hydrographie, Hamburg, Germany,

ALBERT W. VAN WEERS,
Netherlands Energy Research Foundation ECN, Petten, Netherlands

PIERRE GUEGUENIAT,
CEA, Laboratoire de Radioecologie Marine, La Hague, France

P. KERSHAW,
Ministry of Agriculture Fisheries and Food, Fisheries Laboratory, Lowestoft, UK

ABSTRACT

Radionuclides discharged from La Hague is used to trace transport and dispersal of coastal water masses. A time-series of radionuclide measurements in water samples taken in the English Channel, at the Netherlands coast, in the German North Sea sector and in Danish waters is being performed. Rates of water transport, dilution of the coastal current with other water masses and transit times will be elucidated. Radionuclide ratios facilitate tracing of specific discharges in spite of dilution. This will contribute significantly to the knowledge of transit routes, transit times and mixing. The data will be used to improve models capable of describing the transport quantitatively. The results from the sampling programme will furthermore provide a unique data set for validating other models.

The programme runs under the EEC Marine Science and Technology programme (MAST) during a 3 year period commencing October 1990. The work will mainly be based on the analysis of technetium-99 and antimony-125 in seawater, but other radionuclides will also be considered. The programme will be outlined and preliminary data presented.

STUDIES OF CHERNOBYL ^{90}Sr AND Cs ISOTOPES IN THE NORTHWEST BLACK SEA
(Poster)

Ken O. Buesseler, Hugh D. Livingston and Susan A. Casso
Woods Hole Oceanographic Institution, Woods Hole, MA 02543 (U.S.A.)

William R. Curtis
U.S. E.P.A. Office of Radiation Programs, Washington, DC 20460 (U.S.A.)

Jon A. Broadway
U.S. E.P.A. Office of Radiation Programs, NAREL, Montgomery, AL (U.S.A.)

Gennady G. Polikarpov, Ludmilla G. Kulibakina and Andrey Karachintsev
Institute of Biology of Southern Seas, Sevastopol, Crimea (U.S.S.R)

During 1990, an oceanographic cruise was conducted to study the fate of Chernobyl fallout radionuclides in the Black Sea. The focus of this program was a study of the waters and sediments in the Danube fan region. Work was conducted aboard the R/V *Professor Vodjanitsky*, operated by the Institute of Southern Seas Biology, Sevastopol (U.S.S.R.), with participation by scientists from the Woods Hole Oceanographic Institution and the U.S. Environmental Protection Agency. As part of this program, samples of marine sediments, waters and biota were analyzed by the three labs in an intercomparison of radioanalytical techniques.

The program included the collection of samples on a transect away from the mouths of both the Danube and Dnepr Rivers. These samples will be used to study the transport of fresh waters from the Dnepr, now labeled with Chernobyl ^{90}Sr, into the Black Sea and surrounding shelf waters. ^{134}Cs/^{137}Cs watercolumn data will also be used to examine the continued mixing of surface waters to depth across the sharp pycnocline which separates the oxic and anoxic waters in this basin. Finally, sediment cores were collected from the shelf, and the activities of Chernobyl and other bomb fallout and naturally occurring radionuclides analyzed for deposition information and for studies of biological uptake of radionuclides.

THE SCOPE FOR STUDIES OF ARTIFICIAL RADIONUCLIDES IN THE SEAS AROUND THE USSR

G. G. Polikarpov

Institute of Biology of Southern Seas
Sevastopol, Crimea, 335000, USSR

Soviet studies of anthropogenic radionuclides in the seas around the USSR have mostly been concerned with the Black Sea and the Baltic Sea. There have been reports of a few studies of artificial radioactivity in the Sea of Azov and the Barents Sea. Participants in the working group meeting of the Advisory Committee on the Protection of the Sea (ACOPS) held in Sevastopol in November 1990, discussed, amongst other items, the results of these studies.

At the present time data are available as follows:

^{3}H, ^{90}Sr, ^{137}Cs, Pu, and many fission and activation product radionuclides in the Black Sea, the Sea of Azov and the Baltic Sea;

^{137}Cs in the Barents Sea, the White Sea and the Kara Sea;

^{90}Sr, ^{137}Cs in the Bering Sea, the Sea of Okhotsk and the Sea of Japan; and

There are no data available for the Laptev Sea and the East Siberian Sea.

^{90}Sr is available as a tracer of Dnepr River water masses in the northwestern region of the Black Sea.

The ^{137}Cs concentrations in the surface waters of the Black Sea are 15 times greater than those in the Sea of Japan.

International efforts are needed to undertake large-scale investigations of anthropogenic radionuclide inputs into the Arctic Ocean from the Urals and Siberia via the Rivers Ob and Yenisei.

TENEURS EN $^{239+249}$Pu, ^{137}Cs, ^{90}Sr DES EAUX DE MER AU VOISINAGE DE L'ATOLL DE MURUROA ET EN POLYNESIE FRANCAISE

Y. BOURLAT et J. RANCHER

Service Mixte de Sécurité Radiologique
CEA-DIRCEN BP 16, 91311 Montlhéry Cedex FRANCE

POSTER

Le poster présente l'étude de l'évolution depuis 1987 des concentrations en $^{239+240}$Pu, ^{137}Cs, ^{90}Sr dans les eaux du lagon, le long des flancs de l'atoll (0-800 m), ainsi que dans l'océan proche et lointain (0-2500 m).

Pour le ^{137}Cs et le ^{90}Sr océaniques, les valeurs les plus fortes apparaissent dans la couche de surface, respectivement 2.6 Bq/m^3 et 1.8 Bq/m^3. Ces concentrations décroissent rapidement avec la profondeur pour atteindre les limites de détection au dela de 800 m.

L'allure de ces différents profils s'explique tout a fait a partir des données hydrologiques de l'océan tropical caractérisé à ces latitudes par une couche de mélange isotherme d'une centaine de métres d'épaisseur et par une thermocline permanente isolant les equx superficielles des systèmes intermédiaires au-dela de 500 m.

Le plutonium à un comportement différent dans la colonne d'eau et présente un maximum qui se situe dans notre cas à la base de la thermocline aussi bien aux abords de MUROROA qu'au large de TAHITI. A proximité immédiate de l'atoll et en surface seulement, on note des valeurs jusqu'a 2 a 3 fois le bruit de fond océanique, traduisant l'influence des eaux du lagon dont l'activité moyenne est de 0.4 Bq/m^3.

L'ensemble de ces données réparties dans le flux océanique de part et d'autre de l'atoll, indique clairement que la distribution de ces radionucléides dans l'environment dépend quasi exclusivement des facteurs extérieurs que l'on retrouve a l'échelle planétaire : retombées mondiales dues aux expérimentations aériennes des années 1950-60, circulations océaniques et cycles géochimiques.

URANIUM-SERIES NUCLIDES IN SEDIMENT, WATER, AND
BIOTA OF SAQVAQJUAC INLET,
A SUBARCTIC ESTUARY, N.W. COAST HUDSON BAY

Gregg J. Brunskill, Ray H. Hesslein, Harold Welch
Fisheries and Oceans Canada
Freshwater Institute
501 University Crescent
Winnipeg, Manitoba R3T 2N6
Canada

Saqvaqjuac Inlet (63°39'N., 90°39'W.) is a fjord-like estuary of the Saqvaqjuac River on the northwest coast of Hudson Bay. Surface water salinities vary from 5-30‰, and deeper water is 33-34‰ and sometimes anaerobic. Isolated and shallow-silled small bays of the inlet have anaerobic, hypersaline (up to 100‰) deep waters. River discharge and atmospheric deposition are major sources of Pb-210 and Po-210, and these nuclides are inversely related to salinity. Sea water inputs from the tides, and sediment diagenesis appear to be sources of U-238, Ra-226, and Th-230, and they are positively related to salinity. Hypersaline anaerobic waters are usually depleted in U and Ra, and enriched in Pb-210 and Po-210. Bioconcentration factors for these nuclides in estuarine biota will be given, and the highest factors (10^5 to 10^6) were found for Pb-210 and Po-210 in hepatopancreas of crabs and pyloric caecae of Greenland cod.

LES MESURES DIRECTES ININTERROMPUES GAMMA-SPECTROMÉTRIQUES DANS LE MILIEU DE MER-UN DES MÉTHODES PERSPECTIVES DE L'ETUDE ET DU CONTRÔLE DE LA RADIOACTIVITÉ DE L'EAU MARITIME

O.V.Rumiantcev

L'Institut de la géochimie et de la chimie
analytique de V.I.Vernadski, l'Académie des
Sciences de l'URSS, Moscou

La présence dans l'eau maritime des gamma-irradiateurs naturels et artificiels orée dans ce milieu le gamma-champ.

L'étude des caracteristiques de ce champ sur l'échelle spatiale et provisoire différente par le méthode "in situ" permet de recevoir une information précieuse sur la contamination radio-active du milieu maritime et sur son changement. Cette information complétée par des résultats de l'analyse des échantillons est nécessaire aussi pendant la recherche d'autres processus hidrophisiques.

Dans le rapport on examine les particularités du processus des mesures directes et de l'interprétation des résultats aussi que les questions de la métrologie.

On discute les tâches fondamentales qu'on peut résoudre a l'aide du méthode "in situ" pendant le mesurage " dans le point " et a'bord du navire.

On présente les exemples des moyens de la mesure directe de la gamma-activite y compris coincidants avec les échantilloneurs.

Dans le rapport on généralise l'expérience de l'utilisation (1986-1989) des méthodes directes pendant le moniteur radioactif des pièces d'eau douce (riviéres, réservoirs) polluées par les produits de l'accident de Tchernobyl.

IODINE AND ITS VALENCY FORMS AS THE TRACERS OF WATER MOVEMENT

Novikova S.K.

The Vernadsky Institute of Geochemistry and
Analytical Chemistry USSR Academy of Sciences.

The research of the chemical fields' structures of the ocean, in addition to methods allowing to characterize the marine environment with T-S diagrams, can serve as a characteristics on their chemical parameters for estimate of differences of sea water masses.

For classification of sea waters on their chemical characteristics, there is expediently to take as a base the intensity of chemical processes in the sea water layers and the distribution of chemical parameters which the chemical and physical processes are told on, including the stratification and circulation of sea waters.

There are red-ox processes where the index of their intensity is the valency state of the chemical elements in the ocean.

As a method of researches of the sea waters' chemical characteristics, the experimental data of iodine valency forms and the ratios of their absolute concentrations were used in the aquatoria of the Farere-Iceland threshold.

The attempt of stratificational estimate of water masses (nuclei) was made on the concentration ratios' values of the main iodine valency forms as additional characteristics of the chemical differences of sea waters in this region.

AN AUTOMATED RADIOCHEMICAL PURIFICATION PROCEDURE FOR THE MASS SPECTROMETRIC ANALYSES OF THORIUM AND PLUTONIUM IN ENVIRONMENTAL SAMPLES
(Poster)

Ken O. Buesseler

Woods Hole Oceanographic Institution, Woods Hole, MA 02543 (U.S.A.)

Current procedures for the determination of radionuclides at low activities require the ability to separate and purify the isotopes of interest prior to their analysis. Ion exchange procedures have proved quite useful in this regard. An automated analytical system has now been developed at Woods Hole which allows for the purification of radionuclides via ion exchange procedures. This system will be described as it has been used for the preparation of marine samples for thermal ionization mass spectrometry (TIMS). An all Teflon-based series of valves, pumps and tubing is controlled via a micro-processor in order to process up to five samples through anion exchange columns. The inherent advantages of this system include a high sample throughout at a considerable savings of labor. Reproducible and high analytical yields are achieved as well as extremely low and constant blanks for Th and Pu which are required for TIMS.

Using this combination of automated purification and TIMS, ^{232}Th, ^{230}Th, ^{239}Pu and ^{240}Pu can all be determined from single liter seawater samples. This is orders of magnitude more efficient than alpha-counting procedures for these long-lived radionuclides.

A PRELIMINARY STUDY TO ASSESS THE EFFECT OF SOME SEAWATER COMPONENTS ON THE SPECIATION OF PLUTONIUM

D. McCUBBIN and K. S. LEONARD
Ministry of Agriculture, Fisheries and Food,
Directorate of Fisheries Research,
Pakefield Road, Lowestoft, Suffolk NR33 0HT, UK

ABSTRACT

Plutonium is known to exist in two different oxidation states, Pu(IV) and Pu(V), in 0.45 μm filtered seawater. Since Pu(IV) and Pu(V) exhibit very different sediment sorption behaviour, the transport of Pu in the aquatic environment is dependent upon oxidation state and the rate of interconversion between the species. A number of laboratory experiments have been carried out to determine possible parameters which influence the rate of Pu redox reactions and the extent of sorption by suspended particulate in the marine environment. Results suggest that both the rate of oxidation of Pu(IV) and reduction of Pu(V) increase with increasing suspended particulate concentration. Also, although the sorption of Pu(IV) does not appear to be dependent upon the major cations present in seawater, the sorption of Pu(V) is decreased in the presence of Ca^{2+} and Mg^{2+} ions.

Particle cycling rate constants from nutrient particle and thorium isotope data in the NW Atlantic Ocean

R. J. Murnane[1] and J. K. Cochran[2]

[1] Atmospheric and Oceanic Sciences Program, Princeton University, Princeton, NJ 08544
[2] Marine Sciences Research Center, State University of New York, Stony Brook, New York 44794-5000

A model consisting of three phases - solution, small suspended particles (P_s) and large, sinking particles (P_L) has been used to determine particle cycling rate constants as a function of depth in the NW Atlantic (Nares Abyssal Plain). The rate constants were determined by inverting the model using available data on dissolved phosphate, dissolved (A_d) and particulate thorium isotopes (A_s and A_L), suspended particle concentrations, and mass and thorium activity fluxes. The model envisions small particle formation (at a rate β_1) in its upper two layers (see figure below). At all depths, small particles decompose (at a rate β_{-1}) and aggregate into large particles (at a rate β_2), and large particles disaggregate into small particles (at a rate β_{-2}) and sink with a velocity ω. Thorium adsorbed onto particles behaves in a similar manner. Thorium in solution is adsorbed onto small particles at a rate that is a function of the particle concentration ($k_1 P_s$) and desorbed (at a rate k_{-1}). Thorium is produced by a dissolved parent and decays (λ) in each phase. Particles and thorium can be advected, diffused, and change with time. Initial estimates for the rate constants were made based on prior work and the available data. Measured values and initial estimates were weighted for the inversion by the variance associated with each term. The sum of the time dependence and transport terms for each phase are included in the inversion, but assumed equal to zero with a large variance. The rate constants obtained with the inversion ($k_1 = 5.1 \cdot 10^4$ L g^{-1} y^{-1}; $k_{-1} = 2$ y^{-1}; and $0.1 < \beta_{-1} < 42$; $3.5 < \beta_2 < 4$; $156 < \beta_{-2} < 234$ y^{-1}) are dependent on the initial estimates, but well within the variance of the initial estimates. We predict profiles using rate constants produced with the inversion and a one-dimensional, steady state finite difference model, that roughly match observations. The departure of the forward results from observations highlights the importance of the time dependence and transport terms for this data set.

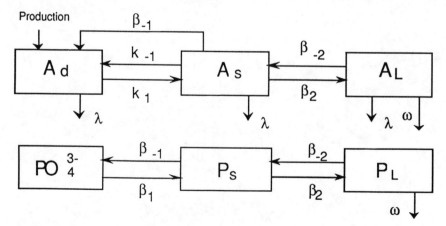

CHERNOBYL RADIONUCLIDES AS TRACERS OF SEDIMENTATION PROCESSES IN THE NORTHERN ADRIATIC SEA (ITALY)

R.Delfanti*, V.Fiore*, C.Papucci*, L.Moretti**, E.Tesini**,
S.Salvi***, S.Bortoluzzi****, M.Nocente***and P.Spezzano****

*ENEA, C.P.316, La Spezia, Italy;**ENEA, Via Mazzini 2, Bologna;
ENEA, C.P.1, Camugnano (BO);*ENEA, Saluggia (VC), Italy.

The tracer signal of Chernobyl radionuclides in the Northern Adriatic Sea was easily detected, due both to the high, direct fallout deposition over this area, and to the contribution of the major Italian rivers, whose drainage basins were also heavily contaminated by Chernobyl fallout.

The poster reports, for the period 1987-90, the concentrations and distributions of Cs-134 and Cs-137 in surface water, particulate matter and sediments from the Northern Adriatic Sea and from two of the major rivers of this region. Ratios between the concentrations of 137-Cs and 90-Sr in seawater before and after the Chernobyl accident are also reported and discussed.

One year after the Chernobyl accident most of the Cesium isotopes were dissolved in the water masses. However, high concentrations and inventories of Chernobyl radionuclides were measured in the sediments at the river-sea interface and in the coastal areas facing the river mouths: these zones can be considered as temporary reservoirs of sediments and associated pollutants transported by the rivers.

The radionuclide vertical profiles in the sediment cores and the inventories of the Cesium isotopes point out remarkable differences in sedimentation regimes in the study area, allowing the identification of zones characterized by a calm sedimentation regime and zones where erosion processes are prevailing over sedimentation.

A STUDY OF Pb-210 METHOD APPLICABILITY TO GUANABARA BAY
SEDIMENTATION RATES

Moreira, I.[1], Godoy, J.M.[1,2] and Mendes, L.B.[1].
1. Chemistry Department - Pontifícia Universidade Católica do Rio de Janeiro - RJ, Brasil. 2. IRD. Rio de Janeiro - RJ, Brasil.

The Guanabara Bay, with a total area of 400 km² (of which about 80% has depths of less than 10 m) is situated near Rio de Janeiro and serves as an estuary for more than 100 rivers and channels. There are about 5000 industries installed around this estuarine system. The industrial development of this area took place about 30 years ago, and since this time, the Guanabara Bay has become a sink for many untreated domestic and industrial sewages. The occupation of the region occurred simultaneously with anthropogenic events such as dumping, denudation of the land and correction of many river beds. The main sources of pollution are industrial activities. Many studies related to pollution in this estuary have been reported, including some concerning heavy metals in sediments.

The Pb-210 Constant Rate of Supply method (C.R.S.) was applied to two different sampling points from the Guanabara Bay, Rio de Janeiro. Although the sedimentation rate pattern was the same from both locals, with a change for actual higher rates, the time when this change has ocurred was different from one place to the other. The method has given coherent higher sedimentation rates for the point closer to a tributary river (P2) than those for the other (P1). The obtained P1 sediment layer ages and the mercury profile agree very well with the local industrial history. For the sampling point P2, the copper profile for the more recent years also agrees with environmental authorities register available.

A 20 kg gravity corer with plexiglass tubes of 100 cm lenght and 3.6 cm internal diameter was used for sediment sampling. The samples were dried at 105°C to determine the water content. About 3g of the dry material of each sample were digested with a mixture of HNO_3 and HF. The Pb-210 content was determined through its decay product Bi-210, and Ra-226 by the gross alpha counting method. In both cases a low-background counter, was used. The Pb-210 and Ra-226 detection limits were 12 mBq/g and 3 mBq/g, respectivelly.

The aim of this study was to verify the applicability of the Pb-210 method, used for sedimentation rate determinations, in particular the Constant Rate of Supply method (C.R.S.), and apply it as a tool on the interpretation of sediment's heavy metal profiles.

ARTIFICIAL RADIONUCLIDES IN THE SURFACE SEDIMENT OF THE IRISH SEA

M McCartney*, D C Denoon, P J Kershaw and D S Woodhead

Ministry of Agriculture, Fisheries and Food
Directorate of Fisheries Research
Fisheries Laboratory
Lowestoft
Suffolk
NR33 0HT   UK

ABSTRACT

Artificial radionuclides have been discharged into the eastern Irish Sea from the Sellafield nuclear fuel reprocessing plant, in Cumbria, for over 30 years. The temporal and spatial variations of many of these radionuclides in the surface sediment of the Irish Sea have been determined from a large number of samples collected over the period 1968-1988.

Changes in the activities of various radionuclides in the surface sediment, in response to the Sellafield signal, throw light on the behaviour of the radionuclides, the receiving environment and the discharge pathway. Generalised dispersion pathways are derivable from the concentration gradients of individual radionuclides. Differences in the distribution of conservative (^{137}Cs) and non-conservative (^{239,240}Pu) radionuclides are apparent, indicating the latter's greater dependence on particulate transport. More detailed information on the geochemical processes can be obtained from studying the activity quotients of specific nuclide pairs. For example, activities of ^{239,240}Pu in the surface sediment of the Irish Sea changed little over the period 1983-1988, whereas ^{137}Cs activities decreased markedly. This resulted in a widespread decrease in the ^{137}Cs/^{239,240}Pu activity quotients. Although the ^{137}Cs release rate over this period was reduced far more than that of ^{239,240}Pu, the decrease observed in the environment cannot be explained by physical decay alone. Some other process, resulting in the loss of caesium from the surface sediment of the Irish Sea, must be operating. The decrease cannot be attributed to either the removal of contaminated sediment from the area or the addition of clean sediment. Both of these processes would have resulted in a concomitant decrease on the ^{239,240}Pu activities in the surface sediment. The most likely explanation is that caesium is being preferentially desorbed from the contaminated sediment into the relatively clean seawater.

* Present address: Scottish Universities Research and Reactor Centre, East Kilbride, Glasgow G75 0QU, UK

RADIOCAESIUM IN NORTH- EAST IRISH SEA SEDIMENTS: INTERSTITIAL WATER CHEMISTRY

S J MALCOLM

DIRECTORATE OF FISHERIES RESEARCH
FISHERIES RESEARCH LABORATORY
LOWESTOFT,   NR33 0HT
UK

Although ^{137}Cs is not usually regarded as a particle reactive radionuclide significant quantities are present in the sediments of the north-east Irish Sea as a result of the authorized discharges from the Sellafield reprocessing plant. Over the last decade the amounts of ^{137}Cs discharged from the plant have decreased considerably following the introduction of new technology but it has been observed that the concentration of ^{137}Cs in the water column of the north-east Irish Sea has not decreased proportionately to the decrease in the discharges and modelling studies have suggested that the seabed may now be acting as a source for this nuclide to the water.

The post depositional behaviour of ^{137}Cs has not been examined in detail in th marine environment as the concentrations (resulting from fallout) present in the interstitial water of most sediments is very low. In the north-east Irish sea higher concentrations of ^{137}Cs are present permitting the study of the interstitial water chemistry.

Interstitial water samples were colected from Reineck box cores at a site near to the Sellafield discharge point using conventional anoxic handling and squeezing techniques. ^{137}Cs concentrations in both the interstitial water samples and associated sediments were determined by GeLi spectrometry. Additional data on the composition of the interstitial water was obtained by a variety of wet chemical techniques.

The concentration of ^{137}Cs in the interstitial water increases with depth in the seabed following the increasing concentration in the solid phase. There is, however, an excess of ^{137}Cs in the interstitial water which may result from ion-exchange involving ammonium produced during nitrate reduction in the sediments.

The implication of this initial study is that caesium may be diffusing from the seabed into the overlying water. Irrigation of the sediments which is known to occur may increase the rate of this transport.

Quantification of the upward ^{137}Cs flux from the sediments to the water column of the north-eastern Irish Sea will require further study.

CORRELATION BETWEEN PARTICLE SIZE AND RADIOACTIVITY IN INTER-TIDAL SEDIMENTS IN NORTHERN IRELAND
F K LEDGERWOOD AND R A LARMOUR

ENVIRONMENTAL PROTECTION DIVISION
DEPARTMENT OF THE ENVIRONMENT FOR NORTHERN IRELAND

In Northern Ireland the Department of the Environment monitors the impact of Sellafield liquid discharges on the coastal environment. Part of this programme involves looking at the activity associated with inter-tidal sediments.

In 1990 we determined the particle size distribution of the sediments sampled using a Malvern particle size analyser type 2600C.

Our poster presentation gives details of particle size distribution and associated transuranic, Caesium 134 and Caesium 137 activities for a number of sediment samples.

As expected there is a strong correlation between activity and particle size and this is the over-riding factor in varying inter-tidal sediment activity around the Northern Ireland coast.

A few results are reproduced in the following table.

The radiochemical analyses were carried out by Ministry of Agriculture Fisheries and Food, Fisheries Research Laboratory, Lowestoft.

	PERCENTAGE OF SAMPLE IN GIVEN SIZE RANGE			
PARTICLE SIZE MICRONS	SAMPLE 1	SAMPLE 2	SAMPLE 3	SAMPLE 4
1503.9 - 697.6	0.0	4.2	0.0	0.0
697.6 - 427.6	0.0	7.9	0.0	3.5
427.6 - 300.8	0.0	18.7	0.0	4.9
300.8 - 224.8	0.7	32.7	0.2	6.1
224.8 - 172.4	0.2	28.2	0.1	6.2
172.4 - 133.8	0.2	5.7	0.5	6.6
133.8 - 104.0	1.5	0.6	2.9	10.7
104.0 - 80.9	2.7	0.4	2.2	6.9
80.9 - 63.1	4.8	0.8	4.0	4.9
63.1 - 49.2	7.4	0.1	7.0	7.2
49.2 - 38.6	6.8	0.0	8.6	8.1
38.6 - 30.4	5.3	0.1	7.2	4.2
30.4 - 24.1	4.3	0.1	3.9	1.9
24.1 - 19.3	5.0	0.1	3.7	2.4
19.5 - 15.5	8.2	0.1	7.6	7.7
< 15.5	52.9	0.3	52.1	18.7
MEAN RADIOACTIVITY CONCENTRATION (DRY) $BqKg^{-1}$				
Caesium 134	7.31	0.3	5.01	1.95
Caesium 137	236	13.56	178	105
Americium 241	7.5	0.55	14.78	6.34
Plutonium 239 + 240	14.5	-	10.5	7.17

^{210}Pb CHRONOLOGIES OF RECENT SEDIMENTS IN TIDAL LAKES OF THE CHENIER MARSHES OF LOUISIANA.

J.R. Meriwether[1], X. Xu[1], R. Lee[1], W. Sheu[1], M. Broussard[2], S.F. Burns[3], R.H. Thompson[4], and J.N. Beck[2]

The chenier marshes of southwestern Louisiana which adjoin the northern Gulf of Mexico contain many embedded lakes and the higher ridges from which the region derives its name. This region, along with the Mississippi delta to the east are being lost to the sea at an alarming rate. It has been estimated that Louisiana is losing approximately fifty square miles of its rich wetlands each year. Depleting factors include subsidence, coastal erosion, the dredging of access canals to oil/gas wells and levee containment of the major rivers. Accretion occurs as a result of inorganic sedimentation and decomposition of marsh plants. The research reported here concerns the measurement of sedimentation rates in the chenier region and evaluation of the materials, including pollutants, incorporated in the sediment over the last hundred years.

Cores have been taken from four lakes in the chenier region. The 10-cm diameter cores, 60-80 cm long, were sliced into 2-cm sections. The gamma ray spectrum of each section, after at least a three week grow in period, was obtained. The 46.5 keV ^{210}Pb and the 609 keV ^{214}Bi (a measure of the equilibrium ^{210}Pb) gamma rays were analyzed. The transmission of 46.5 keV gammas through each sample was used to correct for sample self-absorption. Sedimentation rates vary from 0.2 to 1.0 cm yr^{-1}. The cumulative residual unsupported ^{210}Pb below a depth, used in the constant-rate-of-supply model, is shown in the figure on the left.

Portions of each core section were used to determine thirty different elements using both atomic absorption and neutron activation analysis. The elemental concentrations were correlated with the sediment deposition chronology. The barium deposition in White Lake is shown in figure on the right.

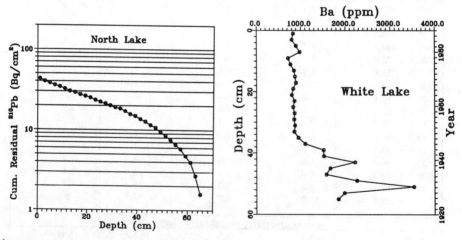

[1]The University of Southwestern Louisiana, Lafayette, LA, USA 70704-4210
[2]McNeese State University, Lake Charles, LA, USA 70609-2255
[3]Portland State University, Portland, OR, USA 97207-0751
[4]Louisiana Tech University, Ruston, LA, USA 71272-3015

The Chemical Associations and Behaviour of Long-Lived Actinide Elements in Coastal Soils and Sediments

M C Graham[1], M S Baxter[2], F R Livens[3] and R D Scott[1]

This postgraduate research programme concentrates on improving the understanding of the chemical associations and behaviour, in soils and sediments, of radionuclides discharged to the environment from industrial sources (eg from BNFL, Sellafield) and of their naturally occurring analogues. To a large extent, the chemical and physical forms of radionuclides in the environment control their behaviour.

Artificial radionuclides are largely held in surface horizons and environmental studies are often directed towards identification of biological pathways and chemical associations. Recent research has highlighted the importance of organic matter in influencing the behaviour of radionuclides in sediments. However, many aspects of the association of metals and organic components are poorly understood and the aim of this three-year study is to investigate these topics using both field and simplified model systems.

Analyses, by α-spectrometry, of environmental samples containing low concentrations and low specific activities of actinide elements often require rigorous chemical separations and lengthy activity measurements. Samples for ICP-MS do not require such extensive preparation or analysis time. Low detection limits and simultaneous multi actinide analysis make it an attractive method for the determination of long-lived actinide elements. It is proposed to combine the methods of ICP-MS and gel chromatographic techniques in order to investigate the speciation of natural and artificial actinides in soils and marine and intertidal sediments in the coastal areas of the Irish Sea.

1. SURRC, East Kilbride, Glasgow, G75 0QU, UK.
2. IAEA, International Laboratory of Marine Radioactivity, Monaco, MC 98000
3. ITE, Merlewood Research Station, Grange-over-Sands.

Saltmarshes: An interface for exposure to radionuclides from marine discharges at Sellafield

S B Bradley and S R Jones

Environmental Protection Group, BNFL, Sellafield,
Seascale, Cumbria CA20 1PG

Abstract

Radionuclides are discharged into the Irish Sea as low level radioactive effluent from nuclear reprocessing at Sellafield. A portion of the discharged effluent is transported to the coastal environment and some activity is known to accumulate with fine grained sediments supplied to saltmarshes. These saltmarshes provide an area where Man can be exposed to radionuclides discharged previously. In the future, occupancy of saltmarshes may become an important pathway for external exposure to radionuclides.

This paper outlines work already undertaken on saltmarshes in the Esk estuary, some 15 km from Sellafield. Field measurements of the activity in sediments within the estuary have shown the spatial variation in concentrations, while information from routine monitoring exercises shows the temporal variation over the past 12 years. The pattern for concentrations of some nuclides to be higher in the inner reaches of the estuary than at the entrance to the sea has been demonstrated in several studies. This shows the importance of physical processes for sorption of nuclides to sediments and the cycling of sediments within the estuary to affect the accumulation of radionuclides.

Initiatives currently being pursued by BNFL on the Esk estuary include:

1. Monitoring of dose rate above sediments on a routine basis to determine long-term trends.
2. Habit survey for the estuary.
3. Collection and analysis of sediments being deposited at present.
4. Chemical/physical studies on sediments from saltmarshes and sea-washed pasture to determine the effects of different storage periods on the removal of nuclides from sediments.
5. Frequent, detailed gamma-surveys over areas of saltmarsh to determine the variability over the year.

Additional work is underway to characterise the estuarine environments of the R.Duddon, R.Lune, R.Wyre, R.Ribble, R.Mersey and R.Dee. These are co-operative studies with researchers in a number of university departments, and they emphasise the multi-disciplinary approach which is important in studies of environmental processes.

EVALUATION OF METAL DYNAMICS IN SEDIMENTS OF A TROPICAL COASTAL LAGOON BY MEANS OF RADIOTRACERS AND SEQUENTIAL EXTRACTIONS

Horst Fernandes, Antonio Eduardo de Oliveira, Sambasiva Patchineelam*, Katia Cardoso and Lene Holanda

Comissão Nacional de Energia Nuclear/IRD; CxP. 37750 CEP 22642; Rio de Janeiro-Brasil

*Departamento de Geoquímica/Universidade Federal Fluminense; Morro do Valonguinho, s/no.; CEP 24210; Niterói-Brasil

Jacarepaguá tropical coastal lagoon, located in the south of Rio de Janeiro City, Brazil, undergoes a severe process of eutrofization and heavy metal pollution. Dredging of bottom sediment is an usual practice in the system which may account for metal remobilization, and consequently, increase of their concentration in the overlying water.
To access metal bioavailability in these sediments, sequential extractions were carried out, aiming at providing a comparison between metal distribution in sediments of the tributary rivers and of the lagoon itself. The distribution pattern of the lagoonal sediments indicated a preferential enrichment of Mn, Fe and Ni in the oxidizable fraction of 77%, 56% and 28% respectively, in comparison with fluvial sediments (6% for Mn, 8% for Fe and 7% for Ni). Cu showed a distinct pattern of distribution being equally enriched in the oxidizable fraction of the sediments of both media (about 77%). Zn and Pb were equally distributed in the reducible fraction of both media (46% and 75% respectively). The obtained data suggest the existence of a process of metal translocation in the geochemical phases of the sediments.
In order to confirm this hypothesis, a pool of radionuclides - ^{54}Mn, ^{65}Zn, ^{109}Cd and ^{210}Pb - was added to lagoonal water samples, which were put in contact with fresh oxidized lagoon sediment samples. Radionuclides uptake from the water was quite rapid e.g., 100% of bulk metal concentration after 15 minutes of contact, except for ^{54}Mn which showed 75% of retention in the sediment in the same period.
The reducible fraction proved to be the major support for the studied metals, except for ^{54}Mn, which was mainly associated to the exchangeable one (60%). The oxidizable fraction accounted for the retention of no more than 10% of the added radionuclides. This evidence supports the hypothesis of Fe and Mn oxides and associated metals dissolving in anoxic lagoon sediments. As the redox potential of Mn is higher than Fe, it is acceptable that Mn-oxides may be the main carriers for Ni. On the other hand, it may be probable that Zn and Pb would be associated to Fe-oxides that are only moderately reducible and by this reason, not so mobile as Mn-oxides. In a second step of the process, the metals would become captured by sulfide anions, in the oxidizable fraction of the sediment, instead of migrating to the water column.

^{210}Pb UPTAKE BY A TROPICAL BROWN SEAWEED (Padina gymnospora)

Magalhães, V.F.[1];   Karez, C.S.[1];   Pfeiffer, W.C.[1]   and Guimarães, J.R.D.[2].

[1] Laboratório de Radioisótopos, Instituto de Biofísica Carlos Chagas Filho, UFRJ, Ilha do Fundão, Rio de Janeiro-Brazil, cep:21949.
[2] IRD-CNEN - Av. das Américas Km 11.5, Barra da Tijuca, Rio de Janeiro, Brazil, cep:22793.

The Sepetiba Bay, a 519 Km^2 semi-enclosed water body, 60 Km from Rio de Janeiro is increasingly polluted by Zn and Cd discharges, and with the on-going construction of a petrochemical complex, Pb is expected to become a critical pollutant as well.

Laboratory radiotracer experiments were therefore performed to assess the role of P.gymnospora, a locally abundant brown seaweed, as an indicator of Pb pollution. ^{210}Pb uptake was fast: the seaweeds reached 56% of maximum activity in 30 minutes, probably by passive adsorption on cell walls, and uptake stabilized after 16 hours. The bioaccumulation factor BF, (dry weight basis) was 1.2×10^4, comparable to those obtained by other authors for in situ studies with stable Pb.

Upon transfer to non-spiked seawater, Pb release by the seaweed showed a bi-phasic exponential pattern, the fast component having a biological half-life (T_b) of 7.5 hours. The second and slower elimination component had a T_b of 8.2 days, and the combination of both resulted in a T_b of 7.2 hours.

These results indicate a strong affinity of P.gymnospora for Pb, adsorption playing an important role in Pb binding. This binding is however easily reversible, suggesting that P.gymnospora will be a useful indicator of short-terms variations in the Pb concentration of Sepetiba Bay waters.

Strombus pugilis (MOLLUSC: GASTROPODE) AS A POTENTIAL INDICATOR OF Co-60 IN A MARINE ECOSYSTEM

> Rosane B.C. Moraes & Leticia Maria Mayr
> Departamento de Biologia Marinha
> Universidade Federal do Rio de Janeiro
> 21941 - Rio de Janeiro - Brazil

During normal operation Angra dos Reis Nuclear Power Plant (Rio de Janeiro, Brazil) releases, into the sea, radioactive effluents which need to be monitored. Among the released radionuclides, the Co-60 is set off because of its long half-life also because is it an essential element for several metabolic processes in biota.

In order to estimate the viability and importance of the eatable gastropod Strombus pugilis for monitoring this radionuclide in the Angra dos Reis region, field investigations on the animal availability in two sampling stations were performed. In the laboratory the bioaccumulation factors and biological half-lifes of Co-60 in the animals were determined. The experiments of incorporation employing water were performed with radiocobalt in cationic form and also with the form named "metabolized". The releasing rate of the radionuclide was studied after two types of contamination: acute and chronic.

The transference of the Co-60 from sediment to animal was not significant (3,1), while the bioaccumulation factors from the Co-cationic and "Co-metabolized" water's were 756 and 337, respectively, considering the dry weigth of the soft parts.

The fast radiocobalt incorporation, the elevated bioaccumulation factors, the biological half-life of the species, and the human consumption of S. pugilis in the region indicate Strombus pugilis as an important organism to be included in the monitoring procedures around the Nuclear Power Plant.

THE UPTAKE AND DISTRIBUTION OF α-EMITTING RADIONUCLIDES IN MARINE ORGANISMS

P McDonald*, M S Baxter† and S W Fowler†
*Scottish Universities Research and Reactor Centre, East Kilbride
Glasgow, G75 0QU, UK
†IAEA, International Laboratory of Marine Radioactivity,
19 Avenue des Castellans, MC 98000 Monaco

ABSTRACT

Alpha-autoradiographic and radiochemical techniques were applied to mussels (Mytilus edulis), winkles (Littorina littorea) and Dublin Bay prawns (Nephrops norvegicus) radiolabelled under both laboratory and environmental conditions in order to assess the differences and similarities in radionuclide accumulation. In the laboratory, the marine organisms were exposed ^{237}Np, ^{239}Pu and ^{241}Am using food and water labelling media. In general, tissues and organs associated with digestion had greater radionuclide accumulative properties than other soft tissues. The most intense α-track distributions were found, however, in the winkle's chitinous operculum. Other markedly active tissues were the viscera of the mussel and winkle, the pallial complex of the winkle and the hepatopancreas of the prawn, each of which exhibited varying degrees of heterogeneity in their α-track distributions. Laboratory experiments show that the accumulation and distribution of transuranic radionuclides in marine invertebrates are highly influenced by the presence of scleroproteins, chitinous material and mucous secretions.

Environmentally labelled organisms, collected from the vicinity of the British Nuclear Fuels plc reprocessing plant at Sellafield, Cumbria, England, were monitored for their baseline α-activities of naturally occurring ^{210}Po. ^{210}Po was found to be in excess of Pu and Am activities in both the mussel and the prawn but less in the winkle. ^{210}Po/$^{239+240}$Pu ratios were ~3 in mussels, >7 in prawns and ~0.5 in winkles. ^{210}Po levels in remote British and French coastal sites showed that ^{210}Po in Cumbrian mussels were not locally enhanced by Sellafield discharges (ie are natural) and are largely unsupported by grandparent ^{210}Pb.

ABYSSAL MODULE INCUBATOR 6000 METERS (A.M.I. 6000), UN DISPOSITIF DE MARQUAGE PAR L'EAU D'ORGANISMES MARINS DE GRANDES PROFONDEURS

Gilles GONTIER, Dominique CALMET, Sabine CHARMASSON.

Service d'Etudes et de Recherches sur l'Environnement du CEA,

Station Marine de Toulon,

Base IFREMER, BP 330, 83507 La Seyne sur mer, France

RESUME

Ce module a été mis au point pour permettre la simulation "in situ" d'échanges de radioactivité entre la phase aqueuse et les organismes des grands fonds océaniques. En effet, devant la difficulté d'élever en laboratoire ces organismes, la conception d'un dispositif de marquage "in situ" s'est avérée nécessaire.
Sa mise en oeuvre peut s'effectuer à partir d'un câble treuillé ou bien, du fait que cet engin est autonome, par gréement sur une ligne de mouillage afin de maîtriser la hauteur par rapport au fond.
L'A.M.I.6000 est constitué de 3 nasses indépendantes et son principe de fonctionnement se décompose en 5 opérations:
- Après mouillage de l'A.M.I., les nasses s'orientent dans l'axe du courant.
- Capture des animaux benthiques, attirés par des appâts, par la fermeture électro-mécanique des ouvertures des nasses.
- Injection des marqueurs et homogénéisation de l'eau par mise en action de pompes hyperbares commandées par une centrale de programmation.
- Echantillonnage de l'eau dans les nasses par seringues hyperbares.
- Remontée de l'A.M.I. en surface par largage d'un lest (commande acoustique de la surface).
Capable d'intervenir jusqu'à une profondeur de 6000m, cet engin a été utilisé pour la première fois sur le site de stockage de déchets de faible activité dans l'Atlantique Nord-Est (46°04N, 16°42W), par une profondeur de 4700m, lors de la mission EPICEA 2 (navire océanographique JEAN CHARCOT, sept.-oct. 88). Il a permis de piéger une espèce, Eurythenes gryllus, crustacé amphipode qui effectue des migrations de grandes amplitudes dans la colonne d'eau (-5000 à -1000m) et qui peut donc déterminer des déplacements de matières du fond vers la surface. Lors de cette mission, 131 individus de 26 à 120 mm ont été piégés et marqués par un mélange de radionucléides (^{144}Ce, ^{137}Cs, ^{60}Co, ^{54}Mn et ^{65}Zn) sur des périodes d'incubation de 8 à 120 heures.
Ce dispositif peut être également mis en oeuvre pour l'étude de la fixation d'autres types d'éléments (ex: métaux), pour d'autres espèces vivant dans les grands fonds. L'emploi d'un tel dispositif en bactériologie pourrait s'avérer intéressant, compte tenu du volume important de chacune des 3 nasses d'incubation (60 litres).

RADIONUCLIDES IN THE STUDY OF MARINE PROCESSES: IS THERE A ROLE FOR REGULAR MONITORING?

G. J. Hunt
Ministry of Agriculture, Fisheries and Food
Fisheries Laboratory, Lowestoft, Suffolk NR33 0HT

In setting radioactivity monitoring programmes, full consideration is needed of the objectives which the programmes are required to fulfil. The main objectives are:
1. to enable estimation of public radiation exposure (individual and collective);
2. to demonstrate compliance with conditions of operators' disposal Authorisations;
3. to provide independent surveillance for inadvertent or unrecorded releases.

Additionally, the information provided by monitoring programmes can have two important benefits. One of these is the ability to provide reassurance to the public. The other is the potential for gaining scientific information which can be used to improve dose predictions and/or the understanding of environmental processes. This potential should always be borne in mind when setting monitoring programmes. One of the advantages of regular monitoring by comparison with some research programmes is the length of time series which can be covered.

Two examples of the information which can be gained from regular monitoring are presented in relation to the dispersion of radionuclides in the Irish Sea from the British Nuclear Fuels plc (BNFL) Sellafield plant. First, MAFF monitoring of actinides in marine materials dates from the early 1970s. A semi-empirical model has been used to investigate the effective time over which liquid discharges from Sellafield remain reflected in marine materials (Hunt, G.J., Sci. Tot. Env. 46 (1985) 261-278). This model was based on a simple exponential expression of actinide availability, and to a first order, reasonable fits to empirical data were obtained. With the benefit of more data, further investigations are being made of the adequacy of this simple approach. Table 1 compares availability times estimated from data available up to 1983 with those estimated using the same simple methodology using data now available (up to 1989). There are small but significant increases (on the basis of paired 't' tests, $p < 0.01$ for Pu, $p < 0.05$ for Am). This suggests a more complex process affecting actinide availability than described by a simple exponential, with the possibility of a contribution from terms with longer time constants than were first identifiable.

Table 1 Availability times for Pu and Am in marine materials near Sellafield

Material	Sampling area/ landing point	Availability times (years)			
		Plutonium		Americium	
		Data to 1983	to 1989	Data to 1983	to 1989
Winkles	St Bees	4	5	7	8.5
Mussels	St Bees	3	4	6	8
Crabs	Sellafield shore area	1.5	3.5	4	5
Plaice	Whitehaven	2	4.5	4	5.5
Porphyra	Braystones	2	2	6	5
Surface sediment	Newbiggin	4	5	6	6.5

The second example of the use of long-term regular monitoring relates to the apparent excess radiocaesium concentration in sea water near Sellafield compared with expectation on the basis of current discharge rates (Hunt G.J. and Kershaw P.J., J. Radiol. Prot. 10 (1990) 147-151). More data are now available which show that from 1983 (when an excess is first observable) to the end of 1990, about 700 TBq ^{137}Cs have been remobilised in the tidal vicinity near Sellafield, and from 1983-89 about 1600 TBq have been remobilised in the Irish Sea as a whole. The rate of remobilisation of ^{137}Cs in the Irish Sea is now about 5 to 10 times the current rate of discharges from Sellafield.

© Crown copyright

COMPARISON OF GENOTOXIC RESPONSES OF SOME BENTHIC MARINE INVERTEBRATES TO RADIATION

Florence L. Harrison, Roger E. Martinelli and John P. Knezovich

Environmental Sciences Division
Lawrence Livermore National Laboratory
Livermore, CA 94550

The impact of radiation on the genetic material of a group of benthic inverterbates is under investigation to identify sensitive methods to be used to quantify genotoxic responses both under laboratory conditions and field conditions. Responses that have been compared are the induction of chromosomal aberrations, sister chromatid exchanges, lethal mutations, and DNA strand breakage. Of these, DNA strand breakage appears to be a sensitive and cost-effective method applicable for rapidly assessing radiation-induced damage.

An index of DNA strand breakage and repair was obtained using an alkaline-unwinding assay. This assay, which was developed for vertebrate systems, was adapted for use with the polychaete worm *Neanthes arenaceodentata*, the intertidal bivalve *Mytilus edulis*, and the tube-dwelling amphipod *Ampelisca abdita*. For each assay, the test organisms were exposed to a range of doses of acute radiation using a sealed ^{137}Cs source at a dose rate of 4.2 Gy min^{-1}. Then, the amounts of double-stranded (dsDNA), optimum-unwound (ouDNA), and single-stranded (ssDNA) DNA were determined by quantifying the amounts of fluorescence of the DNA: Hoechst 33258 complex.

There was a significant increase in DNA strand breakage between control and radiation-exposed animals. The DNA of *Neanthes arenaceodentata* appeared to be very sensitive to radiation; significant DNA strand damage was found at exposures greater than 0.3 Gy. Currently under investigation is the mitigation of radiation-induced damage by enzymatic DNA repair. Information on repair is needed to understand the potential impact of radiation on field populations of representative marine organisms.

This work was performed under the auspices of the U.S. Dept. of Energy through Lawrence Livermore National Laboratory under Contract W-7405-Eng-48 and was funded by the U.S. Environmental Protection Agency Office of Radiation Programs.

A MODEL OF DIODROMOUS FISH SPAWNING

N.C.Luckyanov, Computer Center of USSR AS, Moscow

A model is developed which describes the diodromous fishes spawning process in rivers with regulated flow. In the model the following processes are taken into account: velocity of fishes upstream movement, velocity of water downstream movement, nonuniformity of spawning area distribution along the river, riverbed profile, etc.

The model was used to assess the influence of different water outflow through the dam strategies on spawning efficiency on example of Volga river sturgeon. But very soon it was found out that there were no reliable quantitive data describing behaviour of spawning fishes in river. Thus now we have a very difficult problem to develop a method possibly with radioactive tracers to assess physiological parameters of spawning in a very broad river (few km wide) on a rather long distance (up to 300 km) fishes the number of which is about 1 mln.

In our calculations we used quantitative data and results of our calculations reasonably described the behaviour of the studied population.

INDEX OF CONTRIBUTORS

Aarkrog, A., 3
Albrecht, H., 24
Amouroux, J.M., 319
Arnal, I., 197

Bacon, M.P., 164, 285
Badie, C., 197
Bailly du Bois, P., 363
Baumann, M., 299
Baxter, M.S., 52, 61, 177, 329, 360, 381, 386
Beck, J.N., 380
Beechey, J., 84
Begg, F.H., 52
Belastock, R.A., 164
Blanco, M., 23
Bojanowski, R., 3
Bortoluzzi, S., 375
Bothner, M.H., 164
Bourlat, Y., 368
Bourlés, D.L., 209
Bradley, P.E., 61
Bradley, S.B., 382
Brand, T., 222
Breton, M., 74
Broadway, J.A., 366
Broussard, M., 380
Brunskill, G.J., 369
Buat-Ménard, P., 116
Buesseler, K.O., 285, 366, 372
Burns, S.F., 380

Calmet, D., 387
Campana, S.E., 350
Cardoso, K., 383
Casso, S.A., 366
Caulfield, J.J., 265
Charles, F., 319
Charmasson, S., 387
Chen, Q.J., 12

Cherry, R.D., 309
Clifton, R.J., 255
Cochran, J.K., 285, 374
Cook, G.T., 52, 329
Cunningham, J.D., 265
Curtis, W.R., 366

Dahlgaard, H., 12, 365
Delfanti, R., 94, 375
de Meijer, R.J., 210
Denoon, D.C., 377
de Oliveira, A.E., 383
Dulac, F., 116

Edmond, J.M., 209
Ellett, D.J., 61

Fernandes, H., 383
Fernandez, J.M., 197
Fiore, V., 375
Fowler, S.W., 116, 286, 386

Gandon, R., 363
Gérino, M., 187
Germain, P., 364
Godoy, J.M., 376
Gontier, G., 187, 387
Graham, M.C., 381
Greenfield, M.B., 210
Grémare, A., 319
Guegueniat, P., 74, 363, 365
Guimarães, J.R.D., 384

Hamilton, E.I., 234
Hamilton-Taylor, J., 165
Harrison, F.L., 389
Herrmann, J., 24
Hesslein, R.H., 369
Heyraud, M., 309
Holanda, L., 383

Holm, E., 3, 245
Honeyman, B.D., 107
Hunt, G.J., 388
Hutton, R.C., 177

Jones, S.R., 382

Karachintsev, A., 366
Karez, C.S., 384
Kelly, M., 165
Kershaw, P.J., 234, 365, 377
Knezovich, J.P., 389
Knox, S., 165
Kulibakina, L.G., 366

Lambert, C.E., 116
Larmour, R.A., 265, 379
La Rosa, J., 116, 286
Ledgerwood, F.K., 265, 379
Lee, R., 380
Léon, R., 363
Leonard, K.S., 373
Lesscher, H.M.E., 210
Livens, F.R., 381
Livingston, H.D., 3, 285, 366
Lobo, A.M., 245
Loijens, M., 154
Long, S., 265
Lopez, J.-J., 286
Luckyanov, N.C., 390

Magalhães, V.F., 384
Malcolm, S.J., 378
Martinelli, R.E., 389
Mayr, L.M., 385
McCartney, M., 52, 377
McCubbin, D., 373
McDonald, P., 329, 386
McEnri, C., 265
McKee, B.A., 130
Melquiond, J.-P., 187
Mendes, L.B., 376

Meriwether, J.R., 380
Miquel, J.C., 116
Misonou, J., 340
Mitchell, P.I., 23, 37, 265
Molero, J., 23
Moore, W.S., 130
Moraes, R.B.C., 385
Morán, A., 23
Moreira, I., 376
Moretti, L., 375
Murnane, R.J., 374

Nelson, R., 350
Nguyen, H.V., 116
Nielsen, S.P., 3, 12
Nies, H., 24, 365
Nocente, M., 375
Novikova, S.K., 371

O'Colmain, M., 265

Papucci, C., 94, 375
Patchineelam, S., 383
Patel, B., 276
Patel, S., 276
Penedo, F., 165
Persson, R.B.R., 3
Pfeiffer, W.C., 384
Polikarpov, G.G., 366, 367
Prandle, D., 84
Put, L.W., 210

Queirraza, G., 94

Raisbeck, G.M., 209
Rancher, J., 368
Reyss, J.L., 116
Ritchie, G.D., 142
Romero, L., 245
Roos, P., 3
Roveri, M., 94
Rumiantcev, O.V., 370

Rutgers van der Loeff, M.M., 129
Ryan, T.P., 37, 265

Salomon, J.C., 74, 364
Salvi, S., 375
Sampson, K.E., 177
Sánchez-Cabeza, J.A., 23, 37, 245
Santschi, P.H. 107
Schell, W.R., 37
Schmidt, S., 116
Scott, E.M., 52, 61
Scott, R.D., 177, 381
Sheu, W., 380
Shimmield, G.B., 142, 222
Small, L.F., 286
Smith, J.M., 350
Solano, M., 23
Spezzano, P., 375
Stora, G., 187

Tateda, Y., 340
Tesini, E., 375

Teyssie, J.L., 286
Thompson, R.H., 380
Titley, J.G., 165
Todd, J.F., 130
Turner, D.R., 165

van Weers, A.W., 365
Vidal-Quadras, A., 23, 37
Vives Batlle, J., 37, 265

Wefer, G., 299
Welch, H., 369
Williams, G.L., 165
Williams, R., 234
Wollast, R., 154
Woodhead, D.S., 377

Xu, X., 380

Yiou, F., 209

Zhen, Z., 197